AF598050

INVERTEBRATE PALEONTOLOGY OF ISRAEL
AND ADJACENT COUNTRIES

WITH EMPHASIS ON THE TRIASSIC
AND JURASSIC BRACHIOPODA

INVERTEBRATE PALEONTOLOGY OF ISRAEL AND ADJACENT COUNTRIES

WITH EMPHASIS ON THE TRIASSIC AND JURASSIC BRACHIOPODA

Howard R. FELDMAN, Ph.D.

Biology Department
Lander College for Women
The Anna Ruth and Mark Hasten School
A Division of Touro College

Research Associate,
Division of Paleontology (Invertebrates),
American Museum of Natural History,
New York, NY

New York 2013

Library of Congress Cataloging-in-Publication Data:
A catalog record for this book is available from the Library of Congress.

ISBN 978-1-61811-305-4 (hardback)
ISBN 978-1-61811-306-1 (electronic)

©Touro College Press, 2013
Published by Touro College Press and Academic Studies Press.
Typeset, printed and distributed by Academic Studies Press.
Cover design by Ivan Grave

Touro College Press
Michael A. Shmidman and Simcha Fishbane, Editors

43 West 23rd Street
New York, NY 10010, USA
touropress.admin@touro.edu

Academic Studies Press
28 Montfern Avenue
Brighton, MA 02135, USA
press@academicstudiespress.com
www.academicstudiespress.com

For my wife Susan for her understanding, great patience, support and tolerance, despite my having taken over the house with thousands of fossil specimens, each one of which is unique and tells a story.

TABLE OF CONTENTS

FOREWORD

It has been a real pleasure to review Dr. Howard R. Feldman's published record! His contributions to basic geological-paleontological science are very meritorious. I first came in contact with Dr. Feldman when he was working on his graduate degree in paleontology, centered on the study of Devonian brachiopods from New York. At that time I was favorably impressed by his diligence, and his rapidly developing capabilities in basin paleontology and stratigraphy. The promise of these early years has been fulfilled now in his professionally mature years. After his initial baptism working with Devonian brachiopods, he took advantage of the opportunity to do field work, fossil collecting and stratigraphy in the Jurassic of the Middle East, the Sinai Peninsula and southern Israel in particular. Fruitful collaboration with the appropriate Israeli geologists and careful work on resulting brachiopod collections from Sinai and Israel, along with some material from Jordan, has resulted in a series of publications. These publications feature detailed, critical morphology (including ontogenetic information when available), and taxonomy of the brachiopods, plus mature consideration of their paleontology and biogeography. Feldman's work pays careful attention not only to material collected by him, but also critical attention to the earlier work of others in this area, updating the older work and leaving no stone unturned in the effort to place both his own material and those previously published by others in their proper context. Additionally, his paleoecological work on the community paleoecology of his materials is exemplary, as is his work on associated trace fossils in the faunas.

Feldman is now a very well rounded professional, adept not only in carefully describing Mesozoic brachiopods, but also evaluating their biogeographic, paleoecologic and paleogeographic implications. Feldman is one of the very small numbers of paleontologists today capable of effectively studying and publishing on Mesozoic brachiopods; I can think of no one else in North America thus occupied; he has earned an important position in our profession.

Arthur J. Boucot, Ph.D.
Department of Zoology
Oregon State University

INTRODUCTION

Much of the groundwork for paleontologic research in the Levant and Sinai was conducted by scientists of the Geological Survey of Israel. What follows is a brief summary of some important studies by those scientists and others that dealt mainly with faunas in Israel and Sinai and laid the foundation for future research, particularly on the invertebrate faunas.

The Geological Survey of Israel (GSI) was established in 1949 (Grader and Reiss, 1958). This event represented a major step toward building up the country's economy in that it helped develop its natural resources, such as oil, gas and minerals. Grader and Reiss reported that in 1958 the GSI was made up of seven divisions: Geochemistry, Hydrogeology, Mapping, Mineralogy, Oil, Paleontology and Seismology & Geomorphology. Today The GSI is organized into six divisions: Directors Office, including Administration and Logistics, Water and Mineral Resources, Geochemistry and Environmental Geology, Geological Mapping and the Subsurface Environment, Engineering Geology and Geological Hazards, and Earth Sciences Information Systems.

The Paleontology Division of the Geological Survey of Israel, headed by Professor Zeev Reiss in 1959, was divided into micropaleontology and megapaleontology sections. Its research program was important in contributing to various aspects of early geological exploration in Israel, including the mapping program, water, oil and mineral exploration (Grader and Reiss, 1958). For example, Reiss and Issar (1961) reported on subsurface Quaternary correlations in the Tel Aviv area and described six stratigraphic complexes each of which was characterized by a distinct assemblage of foraminiferans. In the 1950s there were relatively few publications by scientists of the Geological Survey of Israel on the megafaunas of the country (see for example Avnimelech, 1952; Avnimelech et al., 1954; Remy and Avnimelech, 1955; Parnes, 1958).

In the 1960s, Parnes (1961, 1963, 1964, 1965) described *Pseudopygurus* Lambert from southern Israel, Coniacian ammonites from the Negev and a Middle Jurassic fauna from Makhtesh Ramon, also in the Negev. Lerman (1960) described Triassic pelecypods from southern Israel and Sinai. Avnimelech (1961) reported on a pachydiscid ammonite from Campanian chert of Israel and an isocrinid fragment from the Cretaceous of the upper Galilee. Freund (1961) reported on the distribution of Lower Turonian ammonites from Israel and neighboring countries and Freund and Raab (1969) described Lower Turonian ammonites from Israel. Raab (1962) described Jurassic-Early Cretaceous ammonites from the southern coastal plain. Lewy wrote a series of papers on ammonites from southern Israel (1967, 1969a, 1969b). Mishnaevsky (1966, 1967) studied the ostreides in the Cenomanian of central and southern Israel and Egypt. Reiner (1968) wrote on the Callovian gastropods from Hamakhtesh Hagadol in southern Israel in which he described twenty nine species including one renamed and three new species.

In the 1970s the exploration of Gebel El-Maghara, northern Sinai, enabled the scientists of the Geological Survey of Israel to study this vast and geologically diverse area. Eighteen papers were presented at the 1972-73 seminar of the Geological Survey of Israel edited by Gill (1974). The topics covered included: metamorphic rocks (Shimron, 1974), stratigraphy and structure (Bartov, 1974; Hildebrand, 1974; Shirav, 1974), sedimentology (Levy, 1974), mineralogy and geochemistry (Gavish, 1974). However, in the early part of the 1970s there is a noticeable absence of research on the megafauna of Sinai due to the political situation. Research on megafossils in Israel was accomplished by Lewy (1972, 1973, 1976, 1977), Lewy and Samtleben (1979), Parnes (1971), Bein (1976) and increased in the middle to latter part of the 1970s (Hirsch, 1976, 1977a, 1978, 1979; Parnes, 1974, 1975, 1977), especially with regard to the molluscs, specifically the bivalves, gastropods and cephalopods. However, the brachiopod faunas remained unstudied. Work was begun on the Sinai faunas in the 1980s, facilitated by the construction of a stratigraphic section of the Jurassic rocks in Gebel El-Maghara (Goldberg et al., 1971), that allowed for the subsequent study of the brachiopods and molluscs of that important section (Feldman, 1987, Feldman and Owen, 1988; Feldman, et al., 1991; Hirsch, 1978). In addition to the research in the Sinai Peninsula, Friedman et al. (1979) described pinnacle reefs of Cretaceous age exposed along the western margin of the Dead Sea.

Hirsch (1980) described the Jurassic bivalves and gastropods from southern Israel (Hamakhtesh Hagadol) and northern Sinai (Gebel El-Maghara) in which he noted their position within the Ethiopian Province (along the southern Tethyan margin). Parnes (1980) described gastropods and a brachiopod species (*Gibbirhynchia*) from the Liassic of Makhtesh Ramon and a megafauna from the Mahmal Formation (Bajocian) of the same area. Parnes (1986) described Middle Triassic cephalopods from the Negev (Israel) and Sinai (Egypt) and Lewy (1981, 1982 and 1985) wrote a series of papers on the cephalopods and molluscs of the Middle East. Lewy and Honig (1985) also described a Late Coniacian ammonite from the lower part of the Sayyarim Formation near Elat (southern Israel). Marquez-Aliaga and Hirsch (1988) studied the migration of Middle Triassic bivalves in the Sephardic Province. Parnes et al. (1985) reported on new aspects of Triassic ammonoid biostratigraphy, paleoenvironment and paleobiogeography in southern Israel, also within the Sephardic Province.

Feldman and Brett (1997a, 1997b, 1998) reported on the paleoecology of Jurassic crinoids from Hamakhtesh Hagadol, southern Israel and described epi- and endobiontic organisms on crinoid columnals. They extended the range of the trace fossil *Tremichnus* (now known as *Oichnus*) by 100 million years. Hirsch et al. (1998) published a study on the Jurassic of the southern Levant that discussed the biostratigraphy, paleogeography and cyclic events of the region. Feldman (2002, 2005) described Triassic brachiopods from Makhtesh Ramon, southern Israel, and Lewy (1995) reported on Cretaceous rudists and ostreids (1996). Hoff (1998) described Late Cretaceous stomatopods from Israel and Jordan.

Triassic

The Triassic is exposed in Har Arif and Gebel Areif en-Naqa as well as the Ramon crater, Makhtesh Ramon, in the central Negev where a more complete section crops out ranging from Olenekian through Carnian stages. Borehole data and measurements from surface outcrops indicate that the thickness of the Triassic rocks in Israel ranges from 500-1100 m (Druckman, 1974; Feldman, 2002, 2005). The columnar section includes the Negev Group (Yamin and Zafir formations; Weisbrod, 1969, 1976) and Ramon Group (Ra'af, Gevanim, Saharonim and Mohilla formations; Zak, 1963). The section in Makhtesh Ramon consists of carbonates, sulfates, sandstones, siltstones, clays, that is largely clastic in the lower, more carbonate-rich in the middle and more evaporitic in the upper part (Druckman, 1969, 1974, 1976; Feldman, 2002, 2005) and ranges in age from Scythian (Early Triassic) to Carnian-Norian (Late Triassic). My work centers on the Middle Triassic transgressive Saharonim Formation that flooded most areas on the African-Arabian platform. Along the southern Tethyan margin there is a record of endemic taxa that characterized the Sephardic Province (Benjamini et al., 2005). The Sephardic Province (Hirsch, 1972) is represented along this margin and seems to be correlative with the western Mediterranean Muschelkalk and other strata in North Africa and the Levant (Benjamini et al., 2005; Hirsch, 1977b).

Early research on the Triassic was accomplished by the British Petroleum Company during World War II (Shaw, 1947), but data became available to workers only after oil companies and governments published the results of their drilling and exploration (Picard and Flexer, 1974). Future work in the Triassic involves the search for brachiopods in the sedimentary deposits in Makhtesh Ramon and the study of this generally neglected time period, at least in terms of brachiopod evolution and paleoecology.

Jurassic

In 1978 I first began to investigate the brachiopod faunas of Israel and the Sinai after attending the International Symposium on Sedimentology in Jerusalem. Dr. Francis Hirsch first introduced me to the Jurassic sequence at Gebel El-Maghara, northern Sinai, by organizing an expedition from the Geological Survey of Israel that included oil geologists, stratigraphers and paleontologists. The paper on *Septirhynchia* from the 2,000 m section at Gebel El-Maghara represents the first modern study of the brachiopod faunas of the region since the early twentieth century works of Douvillé (1916, 1925) and Cossmann (1925).

The goal of this research is to taxonomically study and revise the brachiopod faunas and investigate the ecological relationships in the various marine communities, particularly their structure and paleoecology. As data accumulates, the history of brachiopod species and their evolution within the Ethiopian Province and Tethyan margin will be elucidated. These data will provide a basis for the interpretation of the biogeographic history of the Ethiopian Province as well as insight into the structure and paleoecology of its marine communities (see, for example, Feldman and Brett 1998; Wilson, et al. 2008, 2010).

Endemic brachiopods taxa such as *Somalirhynchia africana, daghanirhynchia daghaniensis, Somalithyris bihendulensis, Striithyris somaliensis, Bihenithyris barringtoni,* and *B. weiri* were recognized by workers (Weir, 1925; Muir-Wood 1935) in the early to mid-twentieth century. Cooper's (1989) work on the Jurassic brachiopods of Saudi Arabia was based largely on collections made during the years 1933-1953 by field geologists of the Arabian-American Oil Company (Aramco) and the Kier-Kauffmann collections (1962) (see Feldman et al., 2001, for a more detailed discussion). His data, combined with the data collected from Sinai, Israel, and Jordan over the last several years, aid in establishing areas of endemism within the Ethiopian Province. Endemic faunas in the ammonoid Cephalopoda were recognized by Arkell (1952, 1956) and Kitchin (1912) found endemics within the trigoniacean and crassatellacean bivalves. These endemics also helped define the Ethiopian Province. Today, after decades of compiling mostly brachiopod data and revising the taxonomy of the brachiopods found within this province, we have a clearer picture of the extent of the endemism that typifies these faunas. The faunas of Israel and Jordan lie at the northernmost part of the Indo-African Faunal Realm and may therefore be related to faunas of the Tethyan Realm. Completion of a systematic revision of Israeli, Jordanian and Egyptian (Sinai) Jurassic brachiopods will enable us to define faunal- and province-realm boundaries with greater accuracy.

Very few of Cooper's (1989) species found in Saudi Arabia, collected from seven formations (Marrat, Dhruma, Tuwaiq Mountain, Hanifa, Jubaila, Arab and

Hith) and representing 1126 meters of sediment, occur in the Negev region; they are more closely related to the Sinai faunas. The Negev fauna appears to be more akin to Muir-Wood's (1935) and Weir's (1925) Somalia material. Additionally, Cooper has erected many new species and I strongly suspect that many of these are simply varieties. Shi and Grant (1993) revised some Jurassic rhynchonellids but did not deal with the distribution of genera and species, except in a general sense. They reported on taxa mostly from the United Kingdom, France, China, India, Egypt but only one from Israel. Only four genera (*Globirhynchia, Burmirhynchia, Somalirhynchia,* and *Pycnoria*) occurred within the Jurassic Ethiopian Province.

The data collected from studies on the Jurassic of Israel, Jordan and Sinai, along with data from future projects, will help determine how closely the brachiopod faunas of the Middle East can be correlated with those of other regions within the Tethyan Realm. Kitchin (1900) described a brachiopod fauna from the Jurassic limestones of the Cutch, India, in which he broadly correlated the Putchum and Charee groups with the Bathonian to Kimmeridgian of Europe. Arkell (1956) suggested that the Putchum Beds represented the Lower Callovian whereas the Charee ranged from the Upper Callovian to Oxfordian. Until now only a broad comparison with the Cutch faunas and the Ethiopian Province faunas was possible. However, with additional collecting in Israel it will be possible to correlate the stratigraphic sections and determine the taxonomic and paleogeographic relationship of the faunas. I suspect that many genera and species from Saudi Arabia, the Cutch and Israel are very closely related, but lack of sufficient material has made exact determinations impossible. For example, the rhynchonellids described by Kitchin (1900) as *Rhynchonella fornix* and *R. nobilis* are probably congeneric with the rhynchonellid *Pycnoria* described by Cooper (1989) from the Upper Bathonian to Lower Callovian of Saudi Arabia. Kitchin's *R. versabilis*, also from beds equivalent to the Upper Bajocian, are similar to those named and described by Cooper (1989), from an equivalent horizon, as *Globirhynchia crassa*. The genus *Schizoria* described by Cooper (1989) from the Dhruma Formation of Saudi Arabia shows affinities to *R. assymetrica* from the Charee Group at Jooria, India. Many terebratulids also seem to have congeneric forms on both continents. *Kutchithyris* species, similar in external morphology to those from the Cutch, have been described from both Israel and Saudi Arabia.

Some monographs (Weir, 1925, 1929; Muir-Wood, 1935), while dealing with specimens in strata ranging from Bajocian to Kimmeridgian age, give the impression of a rhynchonellid-dominated fauna of comparatively little diversity occurring in beds of Middle to Upper Jurassic age in Somalia. Beds of similar age (Dubar, 1967) from Tunisia also deal more thoroughly with the rhynchonellid rather than the terebratulid species. The conclusion drawn from such publications has produced an impression of a closer faunal and possibly ecological relationship of these areas. Likewise, areas currently being studied in beds of Bathonian to Oxfordian age in the Negev, Israel, suggest a closer comparison with the Somali fauna than with Cooper's (1989) Saudi Arabian fauna. Yet, there are a number of terebratulid species that occur in both Saudi Arabia and Israel that have not been noted from Somalia or Tunisia. Preliminary comparisons with the Israel faunas have shown that some genera from Saudi Arabia have not yet been discovered in the Negev, Somalia or Tunisia.

Paleobiogeographic data obtained from analysis of the brachiopod faunas in the Levant and Egypt will also provide insight into the sequence of rifting within the southern Tethyan Platform that led up to total isolation of the Ethiopian Province from its northern boreal counterpart by the completion of the Tethyan-North Atlantic Ocean divide. Work on the geographic distribution of brachiopod genera and species in the areas discussed above is in progress and more comparative work, especially between the faunas of Saudi Arabia and Israel, as well as those of Madagascar and the Cutch, must be done before any real attempt at closer correlation can be achieved.

Community Ecology

The Jurassic was a critical time in the evolution of marine benthic communities. The "Mesozoic Marine Revolution" (Vermeij, 1977, 1987) began then in earnest as ecological systems recovered from the devastation of the Permo-Triassic extinctions and communities took on more modern aspects with a rise in predators and a consequent infaunalization of many taxa on both soft and hard substrates. This Jurassic diversification has been described both very broadly (e.g. Sepkoski, 1977) and in numerous systematic studies of particular clades, but it is still little known at the community level. The analysis of well-preserved and well-exposed Jurassic marine invertebrate assemblages in the Negev, southern Israel, will help fill the gap in paleoecological studies and answer a set of paleogeographic and biostratigraphic questions as well. Future studies will integrate a community analysis of brachiopod-dominated and hard substrate communities within a sequence stratigraphic framework and correlate various Jurassic beds throughout the Ethiopian Province from North Africa to Saudi Arabia.

Furthermore, hard substrate communities inhabited by sclerozoan (those organisms which live on or in hard substrates such as hardgrounds, shells and other skeletons) develop under very distinct physical requirements of temperature, water depth, nutrient levels and so forth. Wilson et al. (2008) noted that marine fossil sclerozoans are commonly found on hardgrounds (synsedimentarily-cemented seafloor sediments), rockgrounds (exposed surfaces of rocks lithified much earlier), and various biotic substrates including carbonate skeletons, wood and other plant materials. Since they are almost always preserved in situ, they are excellent paleoenvironmental indicators. The proximity of individual elements in hard substrate communities are also preserved intact, including overlapping competitive relationships, commensalism, and predatory borings. These communities are easily compared with each other over time and space, so they have been very useful for evolutionary paleoecological studies (Taylor and Wilson, 2003). Sclerozoan abundance and diversity increased worldwide during the Jurassic (Taylor and Wilson, 2003; Wilson et al., 2008), probably due to the increase in carbonate hard substrates in shallow marine environments including hardgrounds (Palmer, 1982; Wilson et al., 2008) and thick carbonate skeletons such as those of oysters, sponges and corals (Stanley and Hardie, 1998; Wilson et al., 2008). The hard substrate faunas of the tropical Ethiopian Province have not yet been thoroughly described and integrated with other better known subtropical to temperate faunas in Europe and North America.

One very significant study in the Negev was that of a detailed description of bioerosion, that is, the removal of consolidated mineral or lithic substrate by the direct action of organisms as defined by Neumann (1966) and revised by Wilson et al. (2010) to signify the destruction of hard substrates by biological processes. Here numerous patch reefs and crinoids were bioeroded by various invertebrates. The significance of the discovery of these ichnospecies is that it is the first equatorial Middle Jurassic boring ichnofauna to be documented (Wilson et al. 2010).

In order to provide useful data for evolutionary, paleoecological and stratigraphic studies, the computation of "best-fit" correlation lines for several local sections within the Negev will be completed along with the construction of a composite standard (Shaw, 1964) that can also be used for intercontinental correlation (based on genera). A composite standard nearly always produces a much clearer picture of relative times of origin and extinction of species than can be provided by any single section and its use, therefore, makes recognition of evolutionary lineages and phylogenetic relationships less speculative than they would otherwise be (Raup and Stanley, 1978). In addition to biostratigraphic information, I have found other diverse data (e.g., an oolitic limestone [marker] bed that extends laterally throughout the study area in Hamakhtesh Hagadol) that would strengthen the conclusion of a strictly biostratigraphic investigation.

In future work sequence stratigraphic analysis will be used as a check on biostratigraphic correlation and as a means of integrating facies information. Fossils are the primary tools of chronostratigraphy, and biostratigraphy enables inter-regional correlation of depositional sequences; sequence stratigraphy also permits, in turn, much more detailed resolution of time within biozones (Brett, 1995). Brett (1995) believes that the sequence stratigraphic paradigm makes a number of predictions about stratigraphic pattern and its relationship to sea level, subsidence, and sedimentation processes. He notes that on the basis of sequence stratigraphy, further deductions can be made about the distribution of both lithologic and paleontologic aspects of strata. Sequences record fluctuations of a number of parameters, such as relative sea level and sedimentation rates, which are of critical importance in governing the local distribution of ancient organisms. Many of these aspects remain incompletely explored, but the sequence stratigraphic model provides a powerful heuristic tool for investigating pattern in life history (Brett, 1995). Paleoecological analysis of the fossils collected will enable me to study the many genetic relationships between fossil distributional patterns and depositional sequences because paleoecological changes are closely correlated with fluctuations in sea level and sedimentation. Brett (1998) argues that sequence stratigraphy provides a temporally constrained framework for the evaluation of ecological and evolutionary events and, for example, may permit precise evaluation of the timing of immigration, extinction or origination of new taxa in a region or on global scales.

Topics

The first section of the book deals with some Triassic brachiopods of Makhtesh Ramon, a large erosional crater in the Negev adjacent to the town of Mitzpe Ramon. A new species of the terebratulid brachiopod, *Coenothyris oweni*, is erected and described (Feldman, 2002). The specific epithet, *oweni*, was given in honor of Ellis F. Owen, for his important contributions to the study of Mesozoic brachiopods. A second paper (Feldman, 2005) describes the ecology, taphonomy (burial) and biogeography of a marine community consisting predominantly of *Coenothyris* shells and ten genera of bivalves that were apparently smothered by pulses of clay sedimentation.

The second section of this study includes papers that deal with the Jurassic brachiopods and brachiopod-dominated communities of northern Sinai, specifically Gebel El-Maghara, Gebel Engabashi and Gebel El-Minshera (Feldman et. al., 1982). The second paper in this series (Feldman, 1987) is a study of the rare Callovian brachiopod *Septirhynchia hirschi* collected from Gebel El-Maghara in 1978. Due to silicification of the shell, the specimens were extremely well-preserved and impervious to the muriatic acid in which they were prepared, resulting in the dissolution of the limestone matrix. The acid bath technique allowed me to study the interiors without resorting to the use of transverse serial sectioning. This was followed by a description of *Goliathyris lewyi*, a new genus and species of terebratulid from Gebel El-Minshera (Feldman and Owen, 1988). In the early 1990s (Feldman et al., 1991), a study of a section of the sequence (2,000 m) at Gebel El-Maghara was published. This work included 15 brachiopod species, including four new species, from an area that is critical to understanding endemic faunas of the region (e.g. Saudi Arabia, Israel and Jordan). A follow-up study on the fauna of Gebel Engabashi within the Maghara anticline (Feldman et al., 2012) provided more data on six species (one new genus and two new species) that further elucidated the paleobiogeography of the Ethiopian Province brachiopod faunas at the northern part of the Indo-African Faunal Realm.

The third section deals with research in Hamakhtesh Hagadol, another erosional crater, a short drive south of Be'er Sheva, the "capital" of the Negev. From the strata in Hamakhtesh Hagadol I collected 13 species of brachiopods that includes one new genus and 5 new species (Feldman et al., 2001). The rocks here consist of 206 m of sediments that are divided into 69 subunits (Goldberg, 1963) belonging to the Zohar and Matmor formations. In addition to the brachiopods, we found a shallow marine sclerozoan fauna (sclerozoans are organisms that live on hard substrates) in the Matmor Formation. In this community we found an encrusting fauna that lived in a shallow lagoon on the landward side of a coral reef which was surrounded by muddy sediments that contained echinoids, oysters and both rhynchonellid and terebratulid brachiopods (Wilson et al., 2008). Additional work in Hamakhtesh Hagadol resulted in the recognition of a Middle Jurassic coral-sponge reef community also in the Matmor Formation (Wilson, et al., 2010) that represents one of the first detailed studies of bioerosion in an equatorial Jurassic ecosystem. Feldman and Brett (1998) reported on epi- and endobiontic (now termed epizoozoans and endozoozoans) organisms on Late Jurassic crinoid columns from Hamakhtesh Hagadol. They were able to extend the range of *Tremichnus* (*Oichnus*) by almost 100 million years. This section ends with a summary of the biostratigraphy, paleogeography and cyclic events of the southern Levant (Hirsch et al., 1998). Here we trace events on the Gondwanian Tethys platform-shelf during the Jurassic Period and look at the rock formations, fossils (e.g. brachiopods, ammonites, ostracodes) and distribution of these organisms in the southern Levant. The section ends with a discussion of the tectono-eustatic cyclic events and a sequence stratigraphic view of these events.

The last paper is a description of rhynchonellide brachiopods from the Jordan Valley (Feldman, et al., 2012) in which are described seven brachiopod genera including two new species. This fauna was collected from the Mughaniyya Formation of northwest Jordan and inhabited a near shore environment during Jurassic times. The fauna here can be correlated with the faunas of the Aroussiah Formation in northern Sinai and the Zohar and Matmor formations in southern Israel.

REFERENCES

Ager, D.V. 1973. Mesozoic Brachiopoda. In A. Hallam (ed.), *Atlas of Paleobiogeography.* Amsterdam: Elsevier.

Arkell, W.J. 1956. *Jurassic Geology of the World*. Edinburgh: Oliver and Boyd.

------. 1952. Jurassic ammonites from Jebel Tuwayq, Central Arabia. *Philosophical Transactions of the Royal Society of London*, Series B, 236: 241-313.

Avnimelech, M. 1952. Les Mollusques du desert de Judee. *Journal de Conchyliologie* 92: 77.

Avnimelech, M. 1961a. Pachydiscid ammonite from Campanian chert of Israel. *Israel Research Council Bulletin* 10G: 7-9.

------. 1961b. Isocrinid fragment from upper Galilee of assumed Middle Cretaceous age. *Israel Research Council Bulletin* 10G: 1-4.

Avnimelech, M., A. Parnes, and Z. Reiss. 1954. Molluscs and foraminifera from the Lower Albian of the Negev (Southern Israel). *Journal of Paleontology,* 28: 835-839.

Bartov, Y. 1974. Transverse faults of the central Sinai-Negev fault zone. Abstracts of papers presented at the 1972/1973 seminar of the Geological Survey of Israel, D. Gill (ed.), 11-13.

Bein, A. 1971. Rudistid fringing reefs of Cretaceous shallow carbonate platform of Israel. *American Association of Petroleum Geologists Bulletin* 60: 258-272.

Benjamini, C., F. Hirsch, and Y. Eshet. 2005. The Triassic of Israel. In J. K. Hall, V. A. Krasheninnikov, F. Hirsch, C. Benjamini, and A. Flexer. *Geological Framework of the Levant, Volume II: the Levantine Basin and Israel,* 331-360. Jerusalem: Historical Productions-Hall.

Brett, C. E. 1995. Sequence stratigraphy, biostratigraphy, and taphonomy in Shallow marine environments. *Palaios* 10: 597-616.

-----. 1998. Sequence stratigraphy, paleoecology, and evolution: biotic clues and responses to sea-level fluctuations. *Palaios* 13: 241-262.

Cooper, G. A. 1989. Jurassic brachiopods of Saudi Arabia. *Smithsonian Contributions to Paleobiology* 65: 1-213.

Cossmann, M. 1925. Description des espéces. In H. Douvillé (ed.), Le Callovien dans le massif de Moghara, 305-328, *Bulletin Société Géologique de France,* Sér. 4, 25.

Douvillé, H. 1916. Les Terraines Secondaire dans le massif de Moghara a l'Est de l'Isthme Suez. *Academie de Science Paris, Memoire* 54: 1-184.

------. 1925. Le Callovien dans le massif de Moghara: avec description des fossiles par M. Cossmann. *Bulletin Société Géologique de France*, Sér. 4, 25: 305-328.

Druckman, Y. 1969. The petrography and environment of deposition of the Triassic Saharonim Formation and the Dolomite Member of the Mohilla Formationin Makhtesh Ramon, central Negev (southern Israel). *Geological Survey of Israel Bulletin* 49: 1-22.

------. 1974. Triassic paleogeography of southern Israel and Sinai Peninsula. Symposium, Wien, May 1973. *Schriftenreihe der Erdwissenschaftlichen Kommissionen. Oesterreichische Akademie der Wissenschaften* 2: 79-86.

Dubar, G. 1967. Brachiopodes Jurassiques du Sahara Tunisien. *Annales de Paléontologie* 53: 1-71.

Feldman, H. R. 1987. A new species of the Callovian (Jurassic) brachiopod *Septirhynchia* from northern Sinai. Journal of Paleontology, 61: 1156-1172.

------. 2002. A new species of *Coenothyris* (Brachiopoda) from the Triassic (Upper Anisian-Ladinian) of Israel. *Journal of Paleontology* 76: 34-42.

------. 2005. Paleoecology, taphonomy and biogeography of a *Coenothyris* community (Brachiopoda, Tereratulida) from the Triassic (Upper Anisian-Lower Ladinian) of Israel. *American Museum Novitates* 3479: 1-19.

Feldman, H. R. and C. E. Brett. 1997a. Paleoecology of Jurassic crinoids from Hamakhtesh Hagadol, southern Israel. Israel Geological Society Annual Meeting, 21.

------. 1997b. Epi- and endobiontic organisms on Late Jurassic crinoid columns from the Negev Desert, Israel: Implications for coevolution. Geological Society of America Northeastern Section Meeting, King of Prussia, Pennsylvania 29: 45.

------. 1998. Epi- and endobiontic organisms on Late Jurassic crinoid columns from the Negev Desert, Israel: Implications for coevolution. *Lethaia* 31: 57-71.

Feldman, H. R., F. Hirsch, and E. F. Owen. 1982. A comparison of Jurassic and Devonian brachiopods communities: trophic structure, diversity, substrate relations and niche replacement. Third North American Paleontological Convention Proceedings, Montreal, Canada, 1:169-174. (Reprinted In Y. Bartov [ed.], *Research in Sinai: Collected reprints*, vol. 3, reprint no. 81, April, 1984. *Geological Society of Israel Special Publication* no. 1 and *Journal of Paleontology* 56: 9-10, supplement to no. 2).

Feldman, H. R., and E. F. Owen. 1988. *Goliathyris lewyi*, new species (Brachiopoda, Terebratellacea), from the Jurassic of Gebel El-Minshera, northern Sinai. *American Museum Novitates* 2908: 1-12.

Feldman, H. R., E. F. Owen, and F. Hirsch. 1991. Brachiopods from the Jurassic of Gebel El-Maghara, northern Sinai. *American Museum Novitates* 3006: 1-28.

Feldman, H. R., M. Schemm-Gregory, F. Ahmad, and M. A. Wilson. 2012. Rhynchonellide brachiopods from the Jordan Valley. *Acta Palaeontologica Polonica* 57: 191–204.

Feldman, H. R., V. Radulovic, A. A. Hegab, and B. Radulovic. 2012. Brachiopods from Gebel Engabashi, northern Sinai. *Journal of Paleontology* 86: 238-252.

Freund, R. 1961. Distribution of Lower Turonian ammonites in Israel and neighbouring countries. *Israel Research Council Bulletin* 10G: 79-100.

Freund, R. and M. Raab. 1969. Lower Turonian ammonites from Israel. *Palaeontological Association, Special Paper* 4: 1-83.

Friedman, G. M., Y. Arkin, and E. Aharoni. 1979. Patch of pinnacle reefs of Cretaceous age exposed on western margin of Dead Sea (Israel). *Sedimentology* 26: 143-49.

Gavish, E. 1974. Mineralogy and Geochemistry of a coastal sabkhas near Nabek, Gulf of Eilat. Abstracts of papers presented at the 1972/1973 seminar of the Geological Survey of Israel, D. Gill (ed.), 18-19.

Grader, P. and Z. Reiss (eds.). 1959. *Geological Survey of Israel Summary of Activities.* Geological Survey of Israel, Ministry of Development.

Hildebrand, N. 1974. Structural geology of the Haimur Block, northern Gulf of Elat. Abstracts of papers presented at the 1972/1973 seminar of the Geological Survey of Israel, D. Gill (ed.), 7-9.

Hirsch, F. 1972. Middle Triassic conodonts from Israel, Southern France and Spain. *Mitteilungen Gesellschaft Geologie Bergbaustudenten* 21:811-828.

------. 1976. Sur l'origine des particularisms de la faune du Trias et du Jurassique de la plate-forme African-arabe. *Bulletin Societe Geologique France* 2: 543-552.

------. 1977a. Some conodont-, bivalve-, and gastropod-biostratigraphic results in the Triassic and Jurassic of Israel and adjacent countries. Geological Survey of Israel, Abstracts of papers presented at the 1974-76 seminar: 12-15.

------. 1977b. Note on *Neospathodus* cf. *kockeli* (Tatge) (Conodonta) from Ramon 1 well Central Negev. *Israel Journal of Earth Sciences* 26: 94-96.

-----. 1978. Biological constituents in the Callovian-early Oxfordian (Jurassic) sediments of southern Israel and northern Sinai. 10th International Congress on Sedimentology, Jerusalem, July 1978, Abstracts, 1: 309.

------. 1980. Jurassic bivalves and gastropods from northern Sinai and southern Israel. *Israel Journal of Earth Sciences* 28: 128-163.

Hirsch, F., J.-P. Bassoulett, E. Cariou, B. Conway, H. R. Feldman, L. Grossowicz, A. Honigstein, E. F. Owen, and A. Rosenfeld. 1998. The Jurassic of the southern Levant. Biostratigraphy, palaeogeography and cyclic events. *In* S. Carasquin-Soleau and É. Barrier (eds.), Peri-Tethys memoir 4: epicratonic basins of Peri-Tethyan platforms, *Mémoires du Muséum national d'Histoire Naturelle* 179: 213-135.

Hof, C. H. J. 1998. Late Cretaceous stomatopods (Crustacea, Malacostraca) from Israel and Jordan. *Contributions to Zoology* 67: 257-266.

Kitchin, F. L. 1900. Jurassic Fauna of Cutch, part. 1: The Brachiopoda. *Palaeontologica Indica*, series 3: 1-87.

Lerman, A. 1960. Triassic pelecypods from southern Israel and Sinai. *Israel Research Council Bulletin* 9G: 1-60.

Levy, Y. 1974. Sedimentary reflection of depositional environment in the Baradwil Lagoon, northern Sinai. *Journal of Sedimentary Petrology* 44: 219-227.

Lewy, Z. 1967. Some Late Campanian nostoceratid ammonites from southern Israel. *Israel Journal of Earth Sciences* 16: 165-173.

------. 1969a. Late Campanian heteromorphy ammonites from southern Israel. *Israel Journal of Earth Sciences* 18: 109-135.

------. 1969b. Upper Campanian ammonites from the Negev, southern Israel. *Israel Journal of Earth Sciences* 18: 164.

------. 1972. Xenomorphic growth in ostreids. *Lethaia* 5: 347-352.

------. 1973. Pigmentation patterns preserved on late Turonian oysters. *Israel Journal of Earth Sciences* 22: 64-65.

------. 1976. Morphology of the shell in Gryphaeidae. *Israel Journal of Earth Sciences* 25: 45-50.

------. 1977. Late Campanian *Pseudoceratites* from Israel and Jordan. *India Palaeontological Society* 20: 244-250.

------. 1981. A Late Albian *Hypengonoceras* (Ammonoidea) from the "Bentonite Bed" at Makhtesh Ramon. *Israel Journal of Earth Sciences* 30: 35-38.

------. 1982. *Gryphaeligmus* n. gen. (Bivalvia: Malleidae) from the Bathonian of the Middle East. *Journal of Paleontology* 56: 811-815.

------. 1985a. Paleoecological significance of Cretaceous bivalve borings from Israel. *Journal of Paleontology* 59: 643-648.

------. 1995. Hypothetical endosymbiotic zooxanthellae in rudists are not need to explain their ecological niches and thick shells in comparison with hermatypic corals. *Cretaceous Research* 16: 25-37.

Lewy, Z., and G. Honig. 1985. A Late Coniacian ammonite from the lower part of The Sayyarim Formation near Elat (southern Israel). *Israel Journal of Earth Sciences*, 34: 205-210.

Lewy, Z., and C. Samtleben. 1979. Functional morphology and Palaeontological significance of the conchiolin layers in corbulid pelecypods. *Lethaia* 12: 341-351.

Marquez-Aliaga, A. and F. Hirsch. 1988. Migration of Middle Triassic bivalves in the Sephardic Province. *Congress Geol. De Espana*, 2nd, Granada, 1: 301-304.

Mishnaevsky, G. 1966. Occurrence of *Ostrea Eilatensis*: A new oyster species from the Cenomanian of Israel and Egypt. *Nature* 210: 517-518.

------. 1967. Some Cenomanian ostreids from Israel. *Israel Journal of Earth Sciences* 16: 132-139.

Muir-Wood, H. M. 1935. Jurassic Brachiopoda. In W. A. Macfadyen et al., *The Geology and Palaeontology of British Somaliland*, 75-147. London: Government of the Somaliland Protectorate.

Neumann, A. C. 1966. Observations on coastal erosion in Bermuda and measurements of the boring rate of the sponge *Cliona lampa. Limnology and Oceanography* 11: 92-108.

Palmer, T. J. 1982. Cambrian to Cretaceous changes in hardground communities. *Lethaia* 15: 309-323.

Parnes, A. 1958. A Coniacian ammonite from the southern Negev (Israel): *Muniericeras lapparenti grossouvre* var. *Israel Research Council Bulletin* 7G: 167-172.

------. 1961. On the occurrence of *Pseudopygurus* Lambert in southern Israel. *Bulletin of the Research Council of Israel* 10G: 216-222.

------. 1963. The Middle Jurassic fauna of Makhtesh Ramon. *Israel Journal of Earth Sciences* 12: 91.

------. 1964. Coniacian ammonites from the Negev (southern Israel). *Geological Survey of Israel Bulletin* 39: 1-42.

------. 1965. Note on Middle Triassic ammonites from Makhtesh Ramon (southern Israel). *Israel Journal of Earth Sciences* 14: 9-17.

------. 1971. Late Lower Cambrian trilobites from the Timna area and Har Amram, southern Negev, Israel. *Israel Journal of Earth Sciences* 20: 179-205.

------. 1974. Biostratigraphic synchronization of the Middle Jurassic in Makhtesh Ramon, Gebel Maghara and Morocco. Israel Geological Society Annual Meeting, Jerusalem, Abstracts, 14-15.

------. 1975. Middle Triassic ammonite biostratigraphy in Israel. *Geological Survey of Israel Bulletin* 66: 1-35.

------. 1977. On a binodose keeled ceratitid from southeastern Spain (a representative of a line parallel to the genus *Gevanites* Parnes 1975 from the Ladinian of Israel). *Cuadernos Geologic Iberica* 4: 522-525.

------. 1980. Lower Jurassic (Liassic) invertebrates from Makhtesh Ramon (Negev, southern Israel). *Israel Journal of Earth Sciences* 29: 107-113.

------. 1981. Biostratigraphy of the Mahmal Formation (Middle and Upper Bajocian) in Makhtesh Ramon (Negev, southern Israel). *Geological Survey of Israel Bulletin* 74: 1-55.

------. 1986. Middle Triassic cephalopods from the Negev (Israel) and Sinai (Egypt), with an introduction on the Triassic in Israel by I. Zak. *Geological Survey of Israel Bulletin* 79: 9-59.

Parnes, A., C. Benjamini, and F. Hirsch. 1985. New aspects of Triassic ammonoid biostratigraphy, paleoenvironment and paleobiogeograpy in southern Israel (Sephardic Province). *Journal of Paleontology* 59: 656-666.

Picard, L., and A. Flexer. 1974. *Studies on the stratigraphy of Israel: the Triassic.* Tel Aviv: Israel Institute of Petroleum.

Raab, M. 1962. Jurassic-Early Cretaceous ammonites from the southern coastal plain, Israel. *Geological Survey of Israel Bulletin* 34: 24-30.

Raup, D. M., and S. M. Stanley. 1978. *Principles of Paleontology*. New York: W. H. Freeman & Co.

Reiner, W. 1968. Callovian gastropods from Hamakhtesh Hagadol (southern Israel). *Israel Journal of Earth Sciences* 17: 171-198.

Remy, J. M., and M. Avnimelech. 1955. *Eryon yehoachi* n. sp. et gen. *Cenomanocarcinus* cf. *Vanstraeleni stenzel* crustaces decapodes du Cretace Superieur de L'Etat D'Israel. *Bulletin de la Société Géologique de France* 5: 311-314.

Sepkoski, J. J., Jr. 1978. A kinetic model of Phanerozoic taxonomic diversity. I. Analysis of marine orders. *Paleobiology* 4: 223-251.

Shaw, A. B. 1964. *Time in Stratigraphy.* New York: McGraw-Hill Book Co.

Shaw, S. H. 1947. *Southern Palestine, geological map* on a scale of 1:250,000 with explanatory notes. Jerusalem: Government of Palestine.

Shimron, A. 1974. The metamorphic rocks of the Sinai Peninsula. Abstracts of papers presented at the 1972/1973 seminar of the Geological Survey of Israel, D. Gill (ed.), p 3.

Shi, X., and R. E. Grant. 1993. Jurassic rhynchonellids: internal structures and taxonomic revisions. *Smithsonian Contributions to Paleobiology* 73: 1-190.

Shirav, M. 1974. Stratigraphy and structure, the Quseib Graben, northern Gulf of Elat. Abstracts of papers presented at the 1972/1973 seminar of the Geological Survey of Israel, D. Gill (ed.), p 4-6.

Stanley, S. M., and L. A. Hardie. 1998. Secular oscillations in the carbonate mineralogy of reef-building and sediment-producing organism driven by tectonically forced shifts in seawater chemistry. *Palaeogeography, Palaeoclimatology, Palaeoecology* 144: 3-19.

Taylor, P. D., and M. A. Wilson. 2003. Palaeoecology and evolution of marine hard substrate communities. *Earth-Science Reviews* 62: 1-103.

Vermeij, G.J. 1977. The Mesozoic marine revolution; evidence from snails, predators and grazers. *Paleobiology* 3: 245-258.

------. 1987. *Evolution and Escalation: An Ecological History of Life.* Princeton: Princeton University Press.

Weir, J. 1925. Brachiopoda, Lamellibranchiata, Gastropoda and Belemnites, pp. 79-110. In The Collection of Fossils and Rocks from Somaliland made by Messrs. B. K. N. Wyllie and W. R. Smellie. *Monographs of the Geological Department of the Hunterian Museum, Glasgow University* 1: 1-180.

------. 1929. Jurassic Fossils from Jubaland, East Africa, Collected by V. G. Glenday. *Monographs of the Geological Department of the Hunterian Museum, Glasgow University* 3: 1-63.

Weisbrod, T. 1969. The Paleozoic of Israel and adjacent countries, part I: the subsurface Paleozoic stratigraphy of southern Israel. *Geological Survey of Israel Bulletin* 47: 1-35.

------. 1976. The Permian in the Near East. In Falke, H. (editor), *The Continental Permian in Central, West and South Europe. NATO Advanced Study Institute Series* 22: 200-214. Dordrecht, Holland: D. Reidel Publishing Group.

Wilson, M. A., H. R. Feldman, and E. B. Krivicich. 2010. Bioerosion in an equatorial Middle Jurassic coral-sponge reef community (Callovian, Matmor Formation, southern Israel). *Palaeogeography, Palaeoclimatology, Palaeoecology* 289: 93-101.

Wilson, M. A., H. R. Feldman, J. C. Bowen, and Y. Avni. 2008. A new equatorial, very shallow marine sclerozoan fauna from the Middle Jurassic (Late-Callovian) of southern Israel. *Palaeogeography, Palaeoclimatology, Palaeoecology* 263: 24-29.

Zak, I. 1963. Remarks on the stratigraphy and tectonics of the Triassic of Makhtesh Ramon. *Israel Journal of Earth Sciences* 12: 87-89.

ACKNOWLEDGMENTS

I am particularly indebted to my brilliant mentor, Dr. Arthur J. Boucot, Oregon State University, who steered me into the fascinating world of brachiopod paleontology. He provided the inspiration for my lifelong study of the Brachiopoda and was the driving force behind my love of research. Art emphasized that taxonomy is critical to progress in our field since it serves as the basis for studies in evolution, paleoecology and biogeography. I thank my professor and advisor at Rutgers University, Dr. Richard K. Olsson, who constantly challenged me with thought provoking questions. I am grateful to Dr. Niles Eldredge, the American Museum of Natural History (AMNH), for arranging my appointment as research associate in the Division of Paleontology (Invertebrates) and his constant support over the years, and Dr. Neil Landman (AMNH), for his discussions on paleoecology and cladistics. I thank the late Dr. Stephen Jay Gould, Harvard University, for general discussions on paleontology when I first began my career. I benefitted from numerous conversations, discussions and exchange of ideas from Drs. Mena Schemm-Gregory, University of Coimbra (deceased) and Mark A. Wilson, the College of Wooster. Many thanks are due to Susan Klofak and Stephen Thurston (AMNH) for specimen preparation and photographic work. Mai Reitmeyer, Research Services Librarian (AMNH), was particularly helpful to me in locating obscure references on the Mesozoic paleontology of the Middle East.

I thank Dr. Simcha Fishbane, Touro College and University System (TCUS), who first suggested putting this volume together and Dr. Michael Shmidman (TCUS), for his advice in its preparation. I thank Dr. Alan Kadish, President of the Touro College and University System, for his support and encouragement in preparing this volume. I am grateful to deans Stanley L. Boylan and Anthony Polemeni, Touro College, for providing funding from the Faculty Development Fund that enabled me to present the results of my research at various scientific venues. I acknowledge the Touro College Faculty Research Fund for supporting my field work and thank the members of its peer review panel.

My colleagues at the Geological Survey of Israel (GSI) provided field support, laboratory space, access to the library as well as critical comments and discussions over many field seasons. I am especially grateful to Drs. Francis Hirsch, Zeev Lewy and the late Abraham Parnes (GSI). Mr. Moshe Arnon (GSI) served as technician on my first trip to the Sinai Peninsula and was instrumental in making the field excursion run smoothly.

I thank my editors at Academic Press, Sharona Vedol and Deva Jasheway for critically reading the manuscript and making helpful suggestions for improvement.

It should be noted that a variety of traditional Jewish authorities and texts have dealt with the particular issues concerning the age of the universe, and have offered diverse approaches for resolving any apparent conflicts that might arise between Jewish tradition and modern science on this matter.[1] This author follows the lead of these traditional authorities in operating with the accepted principles of the science of paleontology.

1 See, e.g., Babylonian Talmud, standard editions, Hagigah 13b; Midrash Bereshit Rabba, ed. J. Theodor and C. Albeck (Jerusalem, 1996), 3, 5; Rabbi Isaac of Akko, Ozar ha-Hayyim, Ms. Moscow-Russian State Library, Guenzburg 775, 86b-87b ; Rabbi Yisrael Lifshitz, Derush Or Ha-Hayyim, in his Tiferet Yisrael on Mishnah, end of Nezikin (Danzig, 1845), 276b-279b.

Chapter One

A Comparison of Jurassic and Devonian Brachiopod Communities: Trophic Structure, Diversity, Substrate Relations and Niche Replacement

ABSTRACT

Four Jurassic (Bathonian-Callovian) brachiopod communities from Gebel El-Maghara, northern Sinai, are compared to four Devonian (Eifelian) brachiopod communities from New York. All communities recognized were examined in terms of composition, trophic structure, diversity, and relation to substrate. Conclusions reached regarding the Jurassic communities pertain only to local areas in northern Sinai and are strictly local observations. The faunas of the Jurassic communities show a close affinity with Eurasian Tethyan shelf faunas and, situated on the African continent, form an important link between the European faunas and those of Afro-Indian origin.

INTRODUCTION

During the past decade, paleobiologists have increasingly focused on trophic structure as a means of reconstructing ancient communities. One of the trends in recent years has been to compare paleocommunities and living communities (a paleocommunity is defined as a suite of preservable taxa comprising a community, as opposed to a living community, which consists of all taxa in the community, *sensu* Scott and West, 1976). In this preliminary study we compare four Devonian brachiopod communities to four Jurassic brachiopod communities and evaluate trophic structure, diversity, and substrate relations in order to assess and recognize any major trends from the mid-Paleozoic to mid-Jurassic.

The non-reef paleocommunities of the Onondaga Limestone (Devonian, Eifelian) in New York have been analyzed by Feldman (1980) and Lindemann and Feldman (1981) with respect to distribution, diversity, functional morphology, and substrate relations, but trophic structure was not studied in detail. A paper on the systematics of the Onondagan brachiopods (Feldman, 1980) includes additional distributional data and collecting localities.

The Middle Jurassic section at Gebel El-Maghara, northern Sinai, which was sampled in this study is over 600 meters thick. The brachiopod, molluscan, and echinoderm faunas were described by Douvillé (1916) but no one has studied the brachiopods in detail since Cossmann (in Douvillé, 1925) studied the Callovian bivalves and gastropods of Sinai and Hirsch (1979) and reported on the bivalves and gastropods of northern Sinai and southern Israel.

The brachiopod faunas of northern Sinai are significant in that they show a close affinity with Eurasian Tethyan shelf faunas and, situated on the African continent, form a key link between European faunas and those of Afro-Indian origin. Based upon data collected, we suspect that the Ethiopian Province (i.e., northern Sinai) was invaded by brachiopods migrating from the north in early Jurassic times which were isolated for the remainder of the Jurassic. These faunas are thought to have subsequently developed special morphological features which distinguish them from their original stock. A detailed systematic study of the Brachiopods is in preparation (Feldman, Owen and Hirsch) which will yield functional morphological data as well as result in a revision of those genera and species described before Muir-Wood (1934, 1935) and which lack internal descriptions.

TROPHIC ANALYSIS OF SPECIFIC COMMUNITIES

Atrypa-Coelospira-Nucleospira Community.—This community is found in mudstones and wackestones of the Onondaga Limestone (Devonian, Eifelian) in the Mid-Hudson Valley, southeastern New York. The environment of deposition was most likely mid-neritic with a moderately to highly agrillaceous lime mud or lime sand substrate (Feldman, 1980). Of the four Devonian communities studied, the *Atrypa-Coelospira-Nucleospira* Community shows the greatest diversity. Six major taxa were recovered (Table 1): Brachiopods (32 species), corals (13 genera), gastropods (4 species), echinoderms (indetermined number of crinoid species), trilobites (1 species) and bryozoans (2 species). The trophic nucleus of the community is composed mainly of spiriferid brachiopods, low-level suspension feeders. Note that the brachiopods show the greatest diversity in the community. The next most abundant group is the corals (high-level suspension feeders) followed by the gastropods (collectors? browsers? scavengers?), echinoderms (crinoids, passive high-level suspension feeders), trilobites (semi-infaunal burrowers? collectors? scavengers? predators?) and bryozoans (low-level and high-level suspension feeders). The *Atrypa-Coelospira-Nucleospira* Community shows excellent structure and stratification with regard to trophic levels. The high diversity appears to be indicative of a low stress environment. (See table 1)

TABLE 1: Trophic structure and diversity of the *Atrypa-Coelospira-Nucleospira* Community.

Major Taxon (in order of relative abundance)	General Morphology	Trophic Group
Brachiopods	Varied,* but predominantly dorsibiconvex, cancavo-convex, plano-convex, and biconvex spiriferids	Low-level suspension feeders
Corals	Solitary and colonial	High-level suspension feeders
Gastropods	Spinose platyceratids	Collectors? browsers? scavengers?
Echinoderms	Non-pinnulate inadunate crinoids ossicles	Passive high-level suspension feeders
Trilobites	Fragments with inflated glabellas (Phacopids?)	Semi-infaunal burrowers? collectors? scavengers? predators?
Bryozoans	Encrusting and ramose fragments	Low-level and high-level suspension feeders.

* A total of 32 species have been recovered, 17 of which are spiriferids.

Atrypa-Megakozlowskiella Community.—This community occurs in the mudstones and wackestones of the Onondaga Limestone from Cherry Valley to Clarkesville, New York. The environment of deposition was mid-neritic with a moderately argillaceous lime mud or lime sand substrate (Feldman, 1980). Diversity (Table 2) is greatest among the brachiopods (22 species), low-level suspension feeders, followed by the corals (high-level suspension feeders), echinoderms (crinoids, passive high-level suspension feeders), trilobites (semi-infaunal burrowers? collectors? scavengers? predators?), bryozoans (high-level suspension feeders), and gastropods (collectors? browsers? scavengers?). The trophic nucleus is composed mainly of spiriferid brachiopods. Here, as in the *Atrypa-Coelospira-Nucleospira* Community, there is excellent stratification of trophic levels with eight different trophic groups recognized. The high diversity appears to be indicative of a low stress environment. (See table 2).

TABLE 2: Trophic structure and diversity of the *Atrypa-Megakozlowskiella* Community.

Major Taxon (in order of relative abundance)	General Morphology	Trophic Group
Brachiopods	Varied,* but predominantly dorsibiconvex and ventribiconvex spiriferids	Low-level suspension feeders
Corals	Solitary and colonial	High-level suspension feeders
Echinoderms	Non-pinnulate inadunate crinoids ossicles	Passive high-level suspension feeders
Trilobites	Indet. Fragments (molts?)	Semi-infaunal burrowers? collectors? scavengers? predators?
Bryozoans	Trepostome? fragments	High-level? suspension feeders
Gastropods	Lenticular tropidodiscids	Collectors? browsers? scavengers?

* A total of 32 species have been recovered, 11 of which are spiriferids.

Leptaena-Megakozlowskiella Community.—This community is found in mudstones and wackestones of the Onondaga Limestone in central New York, in the vicinity of Syracuse. The environment of deposition was probably mid-neritic with a moderately to highly argillaceous lime mud or lime sand substrate (Feldman, 1980). Brachiopods (low-level suspension feeders) comprise the trophic nucleus (Table 3) with 17 species followed by corals (high-level suspension feeders), echinoderms (crinoids, passive high-level suspension feeders), trilobites (semi-infaunal burrowers? collectors? scavengers? predators?), gastropods (collectors? browsers? scavengers?), bryozoans (high-level suspension feeders) and cephalopods (predators). Again, eight different trophic groups are recognized, minimizing feeding competition in this highly stratified community. As in the two previous communities discussed, high diversity here appears to indicate a low stress environment. (See table 3)

Table 3: Trophic structure and diversity of the *Leptaena-Megakozlowskiella* Community.

Major Taxon (in order of relative abundance)	General Morphology	Trophic Group
Brachiopods	Varied,* but predominantly concavoconvex strophomenids and ventribiconvex spiriferids	Low-level suspension feeders
Corals	Solitary and colonial	High-level suspension feeders
Echinoderms	Camerate crinoid columnals	Passive high-level suspension feeders
Trilobites	Medium sized flat forms with spine-bearing pygidia, pear-shaped and inflated glabellas	Semi-infaunal burrowers? collectors? scavengers? predators?
Gastropods	Lenticular, trochiform, and discoid morphotypes	Collectors? browsers? scavengers?
Bryozoans	Ramose fragments	High-level suspension feeders
Cephalopods	Subdiscoid and lenticular morphotypes	Predators

* A total of 17 species have been recovered, 8 of which are spiriferids.

Amphigenia? Community.—This community occurs in a sandstone facies of the Onondaga Limestone in central New York. The environment of deposition was inner-neritic with a sand substrate (Feldman, 1980). Of all the Devonian communities studied, the *Amphigenia*? Community shows the least diversity (Table 4). Although only two major taxa are found in this community, they are stratified such that they feed at different trophic levels: brachiopods (low-level suspension feeders) and corals (high-level suspension feeders). The low diversity and coarse substrate are indicative of a high stress environment.

TABLE 4: Trophic structure and diversity of the *Amphigenia*? Community.

Major Taxon (in order of relative abundance)	General Morphology	Trophic Group
Brachiopods	Robust biconvex terebratulids	Low-level suspension feeders
Corals	Solitary and colonial	High-level suspension feeders

Eudesia Community.—The *Eudesia* Community (Jurassic, Upper Bathonian) is found in interbedded, in places microoncolithic, friable limestones and thin-bedded calcareous shales of the Sherif Formation, Gebel El-Maghara, northern Sinai. The general environment of deposition of the Sherif Formation in southern Israel and northern Sinai was a peritidal shelf environment indicative of alternating sequences of clastics and carbonates, representative of continually shifting river systems which drained the Arabo-Nubian shield (Goldberg and Friedman, 1974). However, locally, the environment of deposition appears to have been one of low-energy (mid-neritic) with a low rate of deposition, dominated by brachiopods and bivalves (*Eligmus, Africogryphaea*, and *Gryphaeligmus*) on a mud substrate. Brachiopod diversity is markedly reduced from the Devonian communities. In the *Eudesia* Community (Table 5) only 1 brachiopod genus (i.e. *Eudesia*) represents the trophic nucleus. There is present, however, an additional rare species of smooth terebratulid. The brachiopods (low-level suspension feeders) are closely followed in abundance by bivalves (low-level suspension feeders), rare gastropods (collectors? browsers? scavengers?) rare cephalopods (predators), and rare echinoderms (scavengers, predators). Structure and stratification of trophic levels is not as good as in the first three Devonian communities discussed above, and only five different trophic groups are recognized here. The relatively high diversity appears to indicate a low stress environment, although the presence of bivalves in the number 2 biovolume dominance position may indicate a position closer to the shore. (See table 5)

TABLE 5: Trophic structure and diversity of the *Eudesia* Community.

Major Taxon (in order of relative abundance)	General Morphology	Trophic Group
Brachiopods	Biconvex multiplicate and rare smooth terebratulids	Low-level suspension feeders
Bivalves	Ostreids, malleids	Low-level suspension feeders
Gastropods	Medium-spired morphotypes	Collectors? browsers? scavengers?
Cephalopods	Oxyconic ammonites	Predators
Echinoderms	Regular echinoids	Scavengers, predators

Ptychtothyris Community.—This community occurs in the upper part of the Sherif Formation (upper Bathonian), Gebel El-Maghara, in a shallower marine environment than the underlying *Eudesia* Community. The shale content is greater and microoncolites are present along with recrystalized limestone due to aragonite dissolution of shallow marine organisms (Z. Lewy, personal communication). Here again brachiopod diversity is reduced from that of the Devonian. The trophic nucleus (Table 6) consists of brachiopods (over 95% *Ptychtothyris,* 3% *Eudesia,* 2% indet. sp.) (low-level suspension feeders) followed in abundance by bivalves (low-level suspension feeders), gastropods (collectors? browsers? scavengers?), and echinoderms (scavengers, predators). A mid-neritic environment of deposition is assigned, although the presence of microoncolites would seem to indicate a higher energy environment. Trophic structure and stratification are similar to that of the *Eudesia* Community with only four different trophic groups recognized. Bivalves are again in the number 2 biovolume dominance position.

Certain parameters typical of opportunistic species (see Levinton, 1970; Alexander, 1977) appear to be applicable to the genus *Ptychtothyris.* Although found in adjacent strata, *Ptychtothyris* reaches overwhelming numerical abundance (95%) in some strata. There is a definite lack of size sorting within the population. The cause of this possible opportunistic explosion is not certain. It may have been related to substrate mobility and/or reduced salinity caused by the drainage of river systems from the adjacent Arabo-Nubian shield. Goldberg and Friedman (1974) report that the presence of clastic rocks within the Sherif Formation suggests runoff of sand- and mud-bearing river water from nearby land areas. (See table 6)

TABLE 6: Trophic structure and diversity of the *Ptychtothyris* Community.

Major Taxon (in order of relative abundance)	General Morphology	Trophic Group
Brachiopods	Smooth and multiplicate biconvex terebratulids	Low-level suspension feeders
Bivalves	Ostreids	Low-level suspension feeders
Gastropods	Medium-sized, medium-spired morphotypes	Collectors? browsers? scavengers?
Echinoderms	Regular echinoids, medium-spired morphotypes	Scavengers, predators

Septirhynchia Community.—The *Septirhynchia* Community (Lower Callovian?) is found in the Zohar Formation, Gebel El-Maghara, just above a series of crumbly limestones in a hard, dense, buff colored limestone with a distinct chert band at the top. The lowermost meter contains the entire fauna which is biostromal in some parts. The environment of deposition appears to have been mid-neritic. The trophic nucleus (Table 7) consists of brachiopods (over 99% *Septirhynchia*) (low-level suspension feeders), followed in abundance by bivalves (low-level suspension feeders), corals (high-level suspension feeders) and gastropods (collectors? browsers? scavengers?). No ammonites have been found associated with this community. Stratification is not well developed here, although the large size of the brachiopods may in actuality bring them into direct competition with the corals in terms of utilization of food resources. Mancenido and Walley (1979) have proposed a life position for *Septirhynchia* which would have placed the anterior commissure of gibbous forms at a level several centimeters above the sediment-water interface. With the commissure vertically oriented, the brachiopod attained a certain amount of stability on the sea floor. Juvenile forms, however, were true low-level suspension feeders since they lived attached by a pedicle accompanied by incurvature of the umbos which was typical of all gibbous forms observed in the field. Although corals are common throughout the outcrops studied, no reef mounds are present. (See table 7)

TABLE 7: Trophic structure and diversity of the *Septirhynchia* Community.

Major Taxon (in order of relative abundance)	General Morphology	Trophic Group
Brachiopods	Large, strongly costate rhynchonellids and very rare smooth terebratulids, e.g. *Ptychtothyris*	Low-level suspension feeders
Bivalves	Small, inequivalved exogyrids, and subequivalve, irregular ovate malleids	Low-level suspension feeders
Corals	Indet. ramose fragments	High-level suspension feeders
Gastropods	Medium-large sized, medium-spired morphotypes	Collectors? browsers? scavengers?

Somalirhynchia Community.—This community (Upper Callovian) occurs at the top of the Zohar Formation, Gebel El-Maghara, in a yellow, nodular, argillaceous limestone. The environment of deposition was midneritic. The trophic nucleus consists of brachiopods (low-level suspension feeders) followed in abundance by ammonites and belemnites (predators), bivalves (low-level suspension feeders), and gastropods (collectors? browsers? scavengers?). Stratification of trophic levels is fair in this community (Table 8), with five different trophic groups present. Additional data are needed before a further detailed evaluation of this community can be completed. (See table 8)

TABLE 8: Structure and diversity of the *Somalirhynchia* Community.

Major Taxon (in order of relative abundance)	General Morphology	Trophic Group
Brachiopods	Medium-large sized rhynchonellids, smooth terebratulids.	Low-level suspension feeders
Cephalopods	Ammonites: *Sowerbyceras* with acutely sigmoid constrictions on test; Involute ornate oppeliids, evolute, compressed unicarinate and pachyceratid morphotypes. Belemnite fragments; *Paracenoceras*.	Predators
Bivalves	Ostreids, pectenids	Low-level suspension feeders
Gastropods	Low-, medium-, and high-spired morphotype.	Collectors? browsers? scavengers?

DISCUSSION

Although this study deals with brachiopod communities of the Devonian and Jurassic, additional observations were made pertaining to other marine communities, especially in the Jurassic. These observations are incorporated into the discussion below. The bivalves in the Jurassic, in some cases, may have taken over the ecological niche of the Paleozoic brachiopods subsequent to the Permian crisis. We concur with Gould and Calloway (1980) that after the Permo-Triassic extinction (which affected brachiopods profoundly but clams relatively little) the clams may have been the first back after a brachiopod debacle in which clams played no causal role. They did not actively displace the brachiopods during the Permian crisis. In general, in the Jurassic, we find little evidence for a "takeover" by bivalves and often meet with distinct faunal groups of one phylum or another. It is important to note that conclusions reached here based upon data collected in the Jurassic of northern Sinai, especially regarding niche-replacement, pertain only to local areas and are strictly local observations. Further study will yield additional data which will allow more general conclusions to be drawn regarding Jurassic marine communities.

Table 9 summarizes the niche replacement and biovolume dominance of the eight communities studied. The Devonian and Jurassic communities are similar in that they are all dominated by low-level suspension feeders. However, in the vast majority of the Jurassic communities observed that were not dominated by brachiopods, bivalves invariably moved into the number 1 biovolume dominance position. Thus, the ecological niche representative of the number 1 biovolume dominance position, that is, low-level suspension feeders, was the same in both the Devonian and Jurassic. In the Devonian high-level suspension feeders (corals) were consistently in the number 2 biovolume dominance position but by Jurassic time they dropped to number 5 position. The passive high-level suspension feeders of the Devonian communities do not appear in any of the Jurassic communities. The number 3 biovolume dominance position that they had occupied in the Devonian was taken over by the predators (cephalopods) in the Jurassic. It is noteworthy that the Devonian predators (i.e. cephalopods) occupied a number 7 position indicating their increased importance as Jurassic community faunal constituents. The Devonian trilobites have no exact ecological counterpart in the Jurassic communities we observed, although a crustacean has been reported from the *Eudesia* beds (Z. Lewy, personal communication). Their extinction by the end of the Permian resulted in a vacant niche which may have been taken over, at least in part, by the echinoids. The gastropods maintained a relatively stable ecological position from Devonian through Jurassic, moving from a number 5 to number 4 biovolume dominance position. (See table 9)

TABLE 9: Niche replacement and dominance in Devonian and Jurassic brachiopod communities.

Period of occurrence and communities	BIOVOLUME DOMINANCE POSITION						
	1	2	3	4	5	6	7
Jurassic 1) *Eudesia* 2) *Ptychtothyris* 3) *Septirhynchia* 4) *Somalirhynchia*	Brachiopods (LLSF)*	Bivalves (LLSF)	Cephalopods (P)	Gastropods (C, B, S)	Corals (HLSF)	Echinoids (P, S)	—
Devonian 1) *Atrypa-Coelospira-Nucleospira* 2) *Atrypa Megakozlowskiella* 3) *Leptaena-Megakozlowskiella* 4) *Amphigenia*?	Brachiopods (LLSF)	Corals (HLSF)	Crinoids (PHLSF)	Trilobites (SB, C, S, P)	Gastropods (C, B, S)	Bryozoans (HL, LLSF)	Cephalopods (P)

* Note that the non-brachiopod dominated marine communities observed in the Jurassic of Gebel El-Maghara, not studied in detail in this report, consistently had mollusks (usually bivalves) in the number 1 biovolume dominance position. Abbreviations: LLSF = low-level suspension feeder; HLSF = high-level suspension feeder; PHLSF = passive high-level suspension feeder; P = predator; C = collector; B = browser; S = scavenger; SB = semi-infaunal burrower.

Turpaeva's (1957) well-known generalizations regarding arctic and boreal marine communities apply to three of the four Devonian communities in that they are: 1) Dominated by one trophic group (low-level suspension feeders) and, 2) Structured such that the second most dominant species belongs to a different trophic group from the most dominant species. However,

Turpaeva's (1957) third generalization, namely that a single species within a given trophic group dominates the group in terms of biomass, does not hold true here. In the *Atrypa-Coelospira-Nucleospira*, *Atrypa-Megakozlowskiella*, and *Leptaena-Megakozlowskiella* communities there is no single dominant species of low-level suspension feeder. Percentages of biovolume for the three most dominant brachiopods of these communities are: 1) *Atrypa-Coelospira-Nucleospira* Community: *Atrypa* (31%), *Coelospira* (14%), *Nucleospira* (12.5%); 3) *Leptaena-Megakozlowskiella* Community: *Leptaena* (29.9%), *Megakozlowskiella* (28.4%). Only in the *Amphigenia*? Community does a single species dominate in terms of biovolume. In the Jurassic Turpaeva's (1957) generalization that a community is normally dominated by one trophic group holds true. The low-level suspension feeders (brachiopods and bivalves) by far dominate in terms of biovolume (see tables 5-7). Also, a single genus (i.e. *Eudesia, Ptychtothyris, Septirhynchia, Somalirhynchia*) within each community dominates the group in terms of biovolume. However, Turpaeva's second generalization, and the most critical to her hypothesis that available food resources are used most economically when stratification occurs thereby minimizing competition, does not apply to three of the four Jurassic communities. In the *Eudesia*, *Ptychtothyris*, and *Septirhynchia* communities the most dominant species (brachiopods) in terms of biovolume belongs to the low-level suspension feeding trophic group and the next most dominant species (bivalves) belongs to the same trophic group.

ACKNOWLEDGEMENTS

During the course of several fieldtrips to the outcrop area in northern Sinai we were greatly assisted by M. Arnon, M. Goldberg and Z. Lewy (all of the Geological Survey of Israel). We also wish to thank Z. Lewy and P. Taylor (British Museum of Natural History) for critical review of the manuscript. Part of the field work for this project was funded by a grant to Feldman from the Division of Earth Sciences, National Science Foundation, NSF Grant EAR 76-15402.

REFERENCES

Alexander, R. R. 1977. Growth, morphology, and ecology of Paleozoic and Mesozoic opportunistic specles of brachiopods from Idaho-Utah. *Journal of Paleontology* 51(6): 1133-1149.

Douvillé, H. 1916. Les terrains secondaires dans le massif du Moghara a l'est de l'Isthme de Suez, d'apres les explorations de M. Couyat-Bartoux. Paléontologie. *Academie de Science Paris, Memoire* ser. 2, 54: 1-184.

------. 1925. Le Callovien dans le massif de Maghara, avec description des fossiles par M. Cossmann. *Bulletin de la Société géologique de France* sér. 4, 25: 303-328.

Feldman, H. R. 1980. Level-bottom brachiopod comnunities in the Middle Devonian of New York. *Lethaia* 13: 27-46.

Goldberg, M, and G. M. Friedman. 1974. Paleoenvironments and paleogeographic evolution of the Jurassic System in southern Israel. *Geological Survey of Israel Bulletin* 61.

Gould, S. J., and C. Bradford Calloway. 1980. Clams and brachiopods—ships that pass in the night. *Paleobiology* 6(4): 383-396.

Hirsch, F. 1979. Jurassic bivalves and gastropods from northern Sinai and southern Israel. *Israel Journal of Earth Sciences*, 28(4): 128-163.

Levinton, J. S. 1970. The paleoecological significance of opportunitistic species. *Lethaia* 3: 69-78.

Lindemann, R. H., and H. R. Feldman. 1981. *Paleocommunities of the Onondaga Limestone (Middle Devonian) in central New York State*. Binghamton: New York State Geological Association Guidebook for Fieldtrips in South-Central New York, 79-96.

Mancenido, M. O., and C. Walley. 1979. Functional morphology and ontogenetic variation in the Callovian brachiopod *Septirhynchia* from Tunisia. *Palaeontology* 22(2): 317-338.

Muir-Wood, H. 1934. On the internal structure of some Mesozoic brachiopoda. *Philosophical Transactions, Royal Society of London* ser. B, 223: 511-567.

------. 1935. In Macfadyen et al. *Geology and Paleontology of British Somaliland: II. The Mesozoic Palaeontology of British Somaliland*. London: Government of the Somaliland Protectorate, 75: 147.

Scott, R. W. and R. R. West. 1976. *Structure and Classification of Paleocommunities*. Stroudsburg, PA: Dowden, Hutchinson, and Ross.

Turpaeva, E. P. 1957. Food interrelationships of dominant species in marine benthic biocoenoses. In B. N. Nikitkin (ed.) *Marine Biology: Transactions of the Institute of Oceanology* 20: 137-148. (Published in U.S. by American Institute of Biological Sciences, Washington, D.C., 1959.)

Chapter Two

A New Species of the Jurassic (Callovian) Brachiopod *Septirhynchia* from Northern Sinai

ABSTRACT

Septirhynchia hirschi n. sp. is described from the Callovian (Jurassic) strata of Gebel El-Maghara, northern Sinai, Egypt. Significant ontogenetic changes from juvenile to adult include: 1) increase in height of the ventral median septum; 2) change from hypothyrid to mesothyrid pedicle foramen; 3) change from a pyriform to a gibbous outline; 4) change from a weakly defined to a strongly defined pedicle sulcus; and 5) change from a relatively straight to a strongly arched lateral commissure. All ontogenetic stages, except for neanic, possessed a pedicle tube. Muscle scars (diductor) were observed for the first time in the genus on several pedicle valve interiors; none were noted on any brachial valve interiors. Juveniles lived epifaunally, attached to the substrate by a small but functional pedicle, while adults lived semi-infaunally with the umbos buried in the mud.

INTRODUCTION

Most of the material described here was collected by the writer during several excursions to the Sinai Peninsula in recent years and represents only a small fraction of the brachiopods retrieved from those Jurassic sediments (Figures 1, 2). Most specimens of *Septirhynchia* from Sinai are silicified, which enabled their morphology to be studied and illustrated without necessitating the grinding of transverse serial sections.

Mancenido and Walley (1979) noted that it had been difficult to study the genus *Septirhynchia* due to a paucity of material. Muir-Wood and Cooper (1951) described two species of *Septirhynchia,* one of which was new (*S. pulchra*), from Harrar Province, Ethiopia (formerly Abyssinia), based on two fragmentary pedicle valves and one complete specimen. All of their material was silicified. All other descriptons (e.g., Cossmann, 1925; Dubar, 1967; Mancenido and Walley, 1979; Muir-Wood, 1935; Rousselle, 1970; Stefanini, 1932) of the genus have been based on non-silicified material which required the use of transverse serial sections in order to interpret the internal structures and consequently assign specific or generic names. In some cases the material was described even though no serial sections were made. In general, especially with respect to Mesozoic brachiopods, elucidation of the internal morphology either by etching in acid, loop excavation (sensu Cooper, 1983), or transverse serial sections is important in order to differentiate homeomorphic forms. Free valves are particularly rare in the Mesozoic, compared to the Paleozoic, because of their cyrtomatodont teeth resulting in stronger articulation between valves.

Abbreviations: L = length; W = width; T = thickness (all measurements in millimeters). AMNH = American Museum of Natural History, New York; USNM = United States National Museum, Washington, D.C.; BMNH = British Museum (Natural History), London; GSI = Geological Survey of Israel, Jerusalem; HU, Hebrew University, Jerusalem.

SYSTEMATIC PALEONTOLOGY

Family RHYNCHONELLIDAE Gray, 1848
Subfamily SEPTIRHYNCHIINAE
Muir-Wood and Cooper, 1951
Genus SEPTIRHYNCHIA Muir-Wood, 1935
SEPTIRHYNCHIA HIRSCHI n. sp.
Figures 4-11

Derivation of name. The species is named after Dr. Francis Hirsch, Geological Survey of Israel, in recognition of his outstanding contributions to the Mesozoic geology and paleontology of Israel and Sinai.

Type locality. The base (lowermost meter) of Subunit 77 (Goldberg et al., 1971) below a karstic or "rotten" limestone zone, Gebel El-Maghara, northern Sinai. Israel Grid coordinates 9978/0150 (30°43'N, 33°27' E).

Age and distribution. Most specimens were collected from a single, almost biostromal, horizon, Middle Zohar Formation (=Masajid Member of the Masajid Formation of Al Far, 1966), Gebel El-Maghara, northern Sinai (Figure 3). The horizon is traceable laterally for several hundred meters within the outcrop area. Since no ammonites were found associated with *Septirhynchia*, an exact age is difficult to determine. However, the horizon in question is bracketed above (Subunit 82, top of the Zohar Formation) by a mixed ammonite fauna consisting of *Pachyceras, Pachycardioceras, Taramelliceras, Pachyerymnoceras,* cf. *"Collotia" collotiformis, Brightia, Putealiceras,* and *Sowerbyceras,* which indicates a late Callovian to early Oxfordian age. Below (Subunit 64, Sherif Formation), a *Eudesia* sp. Zone contains the following fauna indicative of a late Bathonian age: *Bullatimorphites bullatus* and *Clydoniceras.* Consequently, the *Septirhynchia* horizon is placed in the middle to upper Callovian.

Material. There are 159 specimens in the collection (38 articulated valves, 82 pedicle valves, 39 brachial valves) which are silicified in varying degrees ranging from excellent to poor (beekite rings are present in many specimens). The holotype is in repository at the Geological Survey of Israel, Jerusalem, while the remainder of the material is deposited at the American Museum of Natural History, New York, The United States National Museum, Washington, the Hebrew University, Jerusalem, and the Geological Survey of Israel, Jerusalem.

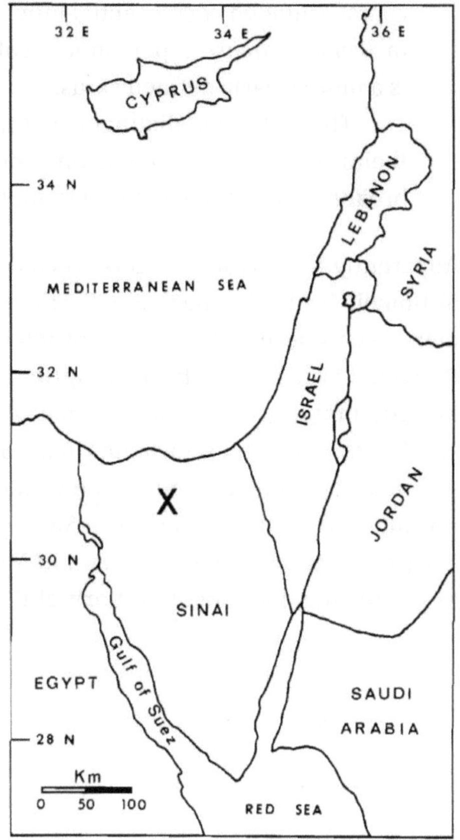

FIGURE 1: Geographic location of sample area (X).

Diagnosis. Large species of *Septirhynchia* typically with 5 chevronlike costae in well-developed ventral sulcus; dorsal fold weakly defined; an average of about 20 costae present on valve exterior each of which extends from the umbonal region to the anterior commissure; subpyriform to subovate in outline; ventral beak strongly incurved in adults; hinge teeth small, triangular; delthyrium covered by conjunct deltidial plates which often form a henidium anteriorly; pedicle foramen submesothyridid; large, bulbous cardinal process with septalium at base; crura cilifer.

Taxonomic note. Muir-Wood (1935, p. 110), in her monograph on the Jurassic brachiopoda from British Somaliland, erected a new genus and new species, *Septirhynchia mogharaensis* (not included in the "Somaliland" material), based on a single specimen from Sinai illustrated by Douvillé (1925, p. 326, Pl. vii, fig. 9a-c) called *Rhynchonella decorata* (Schlotheim). Neither Muir-Wood nor Douvillé adequately described the specimen. Muir-Wood stated:

> Although it has not been possible to examine the internal structure of this species, there appears to be no doubt from the external characters that it belongs to the genus *Septirhynchia.* It resembles species of *Septirhynchia* in its large proportions and in having areolate flanks,

coarse rounded costae, and a much incurved dorsal umbo. The umbo in *S. mogharaensis* tapers more acutely than that of *S. azaisi* and there is a more clearly defined sinus.

The species is distinguished from *Rhynchonella decorata* Schlotheim by its larger dimensions, more produced and tapering umbo, broader sinus, and coarser and more rounded costae.

Apparently Muir-Wood (1935) erected a new species based on a photograph of Douvillé's (1925) specimen. Further research into the history of this specimen reveals several taxonomic problems. 1) The type specimen upon which Douvillé (1925) based his description of *Rhynchonella decorata* from Sinai is lost. Attempts have been made to locate the specimen by writing to brachiopod specialists and various museums in Europe but to no avail (D. V. Ager, Y. Almèras, A. Prieur, personal commun.). Consequently, comparison with the specimens recently collected from northern Sinai is impossible. 2) There is a lack of exact geographic and stratigraphic data in Muir-Wood's (1935) description. She referred to "the 'Bathonian' of Darb-el-Cheikh Moghara Massif, Sinai Peninsula"

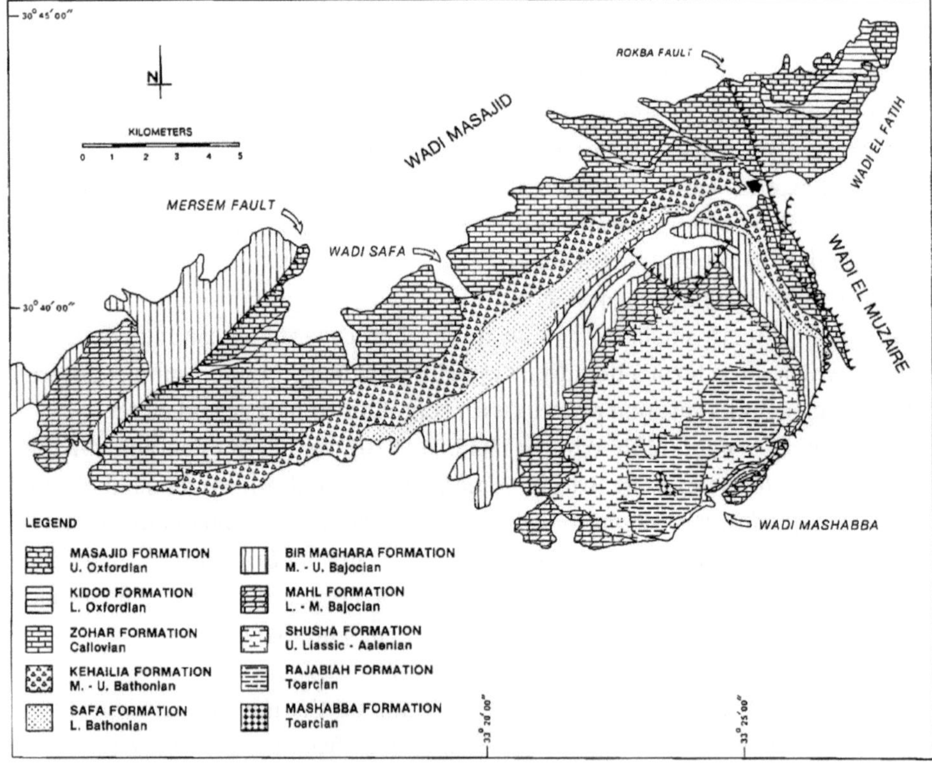

FIGURE 2: Locality map showing main area of outcrop (dark arrow) of *Septirhynchia*-bearing beds, Gebel El-Maghara, northern Sinai (modified from Al Far, 1966).

as the location of Douvillé's *R. decorata* without giving more detailed data. This information is based on Douvillé's note of 1925 in which he described how the Egyptian Geological Survey explored the Maghara region in order to find oil and in so doing collected a fauna which resembled the Bathonian *Rhynchonella decorata* in northeastern France. The collector, M. Sadek, gave the material to Cossmann, who studied most of it until his illness. At first Douvillé and Cossmann considered the fauna to be of Bathonian age but subsequently they assigned it a later age. Douvillé's (1916) description of the exact location of the material is as follows (translated from the French):

> At Darb el Cheik, on a point of the gradient north of Wadi Abu Gaza, to the east of Gebel Hmeir (on the continuation of the G. Aroussieh of Barthoux [Barthoux originally explored the region in 1913 and 1914 with apparently sensational results], G. Maghara on the map of the Geological Survey)....

The writer has explored this region on several field excursions and has been unable to locate the outcrop, although Gebel Hmeir and Wadi Abu Gaza were found. Also, Barthoux most likely explored the region on camel, as there were no paved roads there in those years. Most of the material in the writer's collection was found several kilometers to the northeast of Gebel Hmeir. The present material is silicified with internal morphologies exposed for study while Douvillé's specimen is represented only by a photograph showing three views of the external shell, which does somewhat resemble the Sinai forms. Therefore, no meaningful comparison of the two different "faunas" can be made at this time.

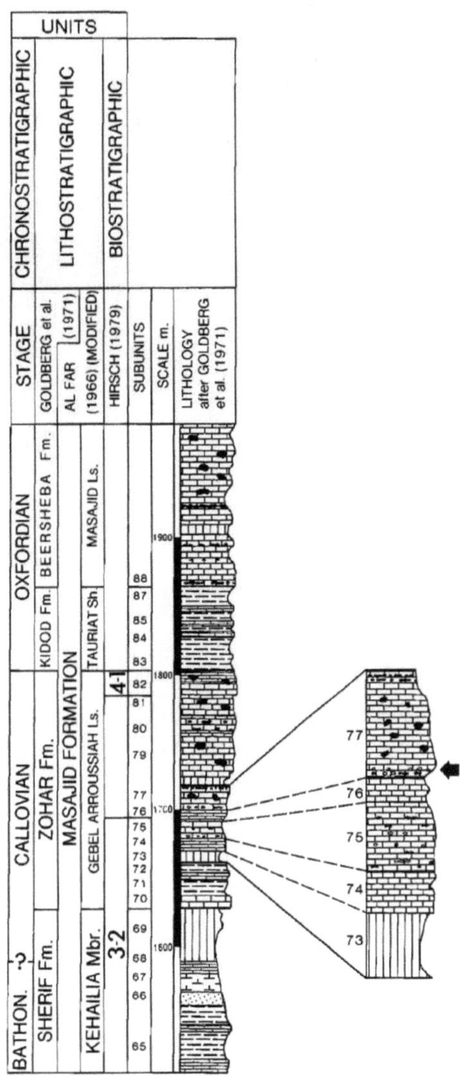

FIGURE 3: Measured stratigraphic section, Gebel El-Maghara, northern Sinai.

FIGURE 4: *Septirhynchia hirschi* n. sp., middle to upper Callovian, Gebel El-Maghara (northern Sinai), all x 1. *1-5*, holotype, GSI M6922, posterior, anterior, lateral, dorsal (oblique), and ventral views.

FIGURE 5: *Septirhynchia hirschi* n. sp., middle to upper Callovian, Gebel El-Maghara (northern Sinai), all x l. *1, 2*, paratype, HU 35560, lateral and posterior views; *3, 4,* paratype, GSI M8012, anterior and lateral views; *5-7* paratype, GSI M7193, lateral, dorsal, and anterior views; *8, 9*, paratype,

GSI M6926, lateral and anterior views; *10,* paratype, GSI M8013, pedicle valve interior; *11, 12*, paratype, AMNH 42923a, b, *11*, AMNH 42923a, view into partially dissected articulated specimen showing delicate crura partially encased in limestone matrix (compare with Figure 9 which represents the same specimen after preparation), *12*, AMNH 42923b, pedicle valve interior subsequent to etching in muriatic acid (note details of deltidial plates and henidium as illustrated in Figure 7).

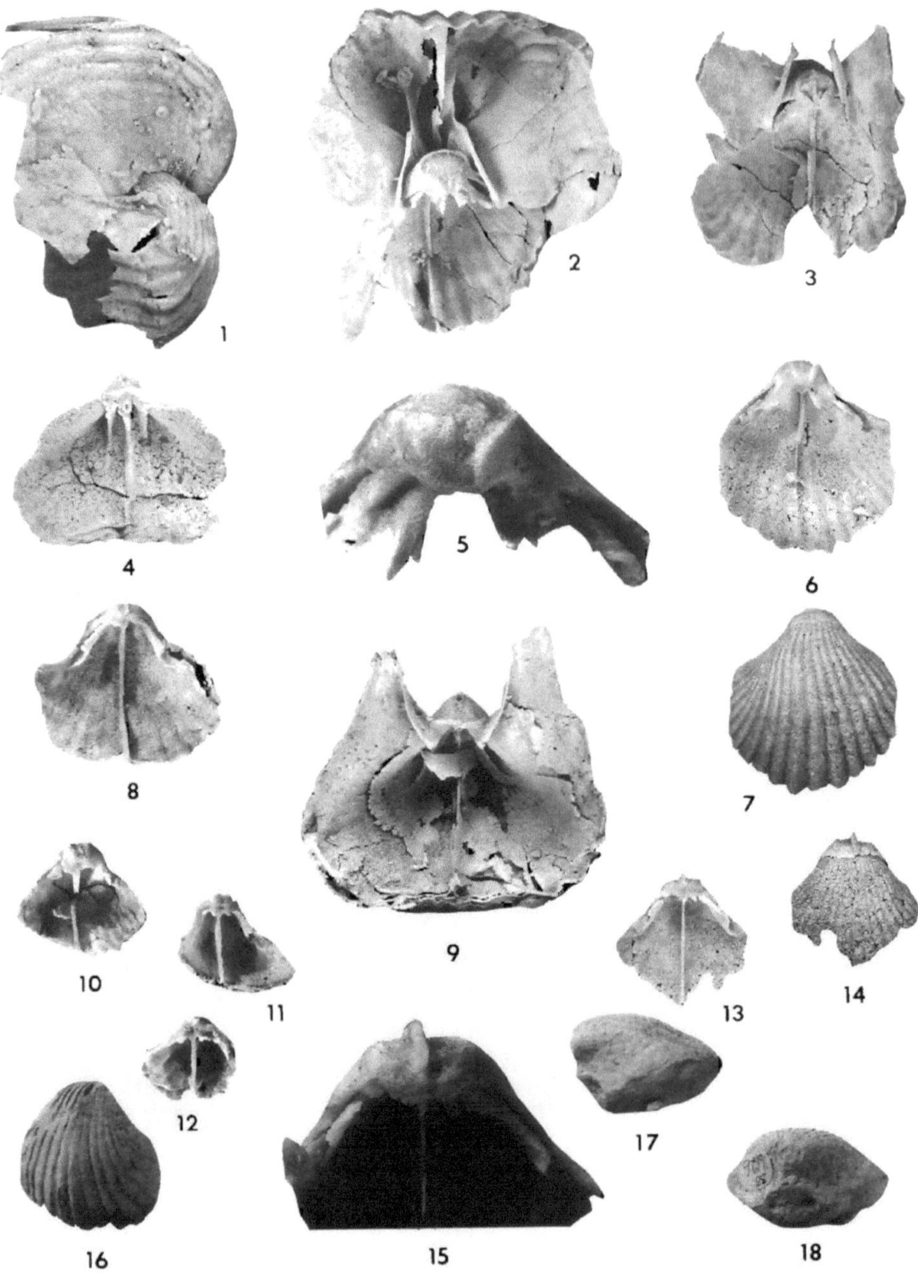

FIGURE 6: *Septirhynchia hirschi* n. sp., middle to upper Callovian, Gebel El-Maghara (northern Sinai), all x 1 unless otherwise noted. *1, 2*, paratype, USNM 401030, lateral and interior views of articulated specimen; *3*, paratype USNM 401031, interior view of articulated specimen; *4*, paratype, GSI M8014, pedicle valve interior showing pedicle tube, *5-7*, paratype, GSI M8015, *5*, brachial valve interior, posterior region, showing cardinal process, x 3, *6*, brachial valve interior, *7*, brachial valve exterior; *8*, paratype GSI M8016, brachial valve interior; *9*, paratype, GSI M8017, pedicle valve

For the above stated reasons, Muir-Wood's (1935) *Septirhynchia "mogharaensis"* (nomen inquirendum) is referred to in quotation marks.

Description. Shells large (Table 1), nontrophic and subpyriform in outline in ephebic specimens when viewed dorsally, to almost subovate in gerontic forms; maximum width reached at or near anterior commissure; both valves strongly convex with ventral valve deeper; ventral beak strongly incurved in late ephebic through gerontic forms and comes in contact with dorsal umbo; in neanic and early ephebic specimens incurvature slightly less pronounced, with no contact between ventral extremity and dorsal valves; in brephic forms beak suberect; a small, round, but distinct pedicle foramen present in all forms from brephic to early ephebic, in late ephebic pedicle foramen covered by incurved ventral beak; delthyrium includes an angle of approximately 60° in most specimens, and is covered by conjunct deltidial plates; in late neanic through middle ephebic stages pedicle foramen submesothyrid, in late ephebic it is sometimes visible and appears submesothyrid, in gerontic forms it is completely hidden.

There is a strong, distinct U-shaped sulcus on the ventral valve (weakly defined in juveniles) with a corresponding, but weaker, fold on the dorsal valve; sulcus originates deep in umbonal region but origin of fold indistinct; commonly 5 costae on sulcus and a minimum of 7-8 on each flank (Table 2); costae originate in umbonal regions of both valves and continue uninterrupted to uniplicate anterior commissure becoming wider and more prominent than on flanks; anterior commissure not preserved on any juvenile forms in collection; costae angular throughout and chevron-like in cross section anteriorly, averaging about 20 at anterior commissure in adults; costae in sulcus, and to a somewhat lesser degree on fold, tend to become asymmetrical such that the axis of each costa is tilted towards lateral margins; costae on fold more symmetrical and rounded; between costae are deep, V-shaped interspaces which become almost U-shaped posteriorly, especially in umbonal regions; planareas smooth and concave, the concavity becomes more pronounced in more advanced ontogenetic stages; ventral planareas extend sharply dorsally such that posterior segment of lateral commissure projects into dorsal valve; lateral commissure curved in adults and less so in juveniles; growth lines more numerous anteriorly and poorly defined posteriorly, in some cases entirely absent due to poor preservation.

---

interior; *10,* paratype, GSI M8018, brachial valve interior; *11*, paratype, USNM 401032, brachial valve interior; *12*, paratype, AMNH 42924, brachial valve interior; *13-15*, paratype, GSI M8019, brachial valve interior, exterior, and interior, posterior region, showing cardinal process with prominent median ridge, x 3.*16-18*, *Septirhynchia pulchra* Muir-Wood and Cooper, ?Callovian, Ego Gambo, Harrar Province, Ethiopia. *16*, BMNH BB 16125, brachial valve exterior; 17, 18, BMNH BB16l26, lateral and posterior views.

TABLE 1—Measurements of articulated specimens of *Septirhynchia* (in mm). L = length, W = width, T = thickness, NVC = number of ventral costae, NDC = number of dorsal costae, NSC number of sulcal costae, e estimated due to shell damage, i incomplete.

Specimen	L	W	T	NVC	NDC	NSC	Geographic distribution	Age	Source of data
S. hirschi n. sp.									
GSI M6922	62.3	58.3	71.1	18	17	5	N. Sinai	M. to U. Callovian	This paper
GSI M6926	37.4	43.9	71.1	22	21	5	N. Sinai	M. to U. Callovian	This paper
GSI M7193	42.5	43.9	34.5	24e	23	5	N. Sinai	M. to U. Callovian	This paper
GSI M8012	47.8	43.2e	49.8	19	18	5	N. Sinai	M. to U. Callovian	This paper
GSI M8032	51.9	—	—	20	19	5	N. Sinai	M. to U. Callovian	This paper
GSI M8033	—	—	51.1	16	15	5	N. Sinai	M. to U. Callovian	This paper
HU 35560	43.5	46.9	41.4	17e	16	5	N. Sinai	M. to U. Callovian	This paper
S. azaisi									
BMNH B46235	67.7	58.8	—	—	24	—	Somaliland	Kimmeridgian	Muir-Wood, 1935, Pl. 9, fig. 3a–c
S. pulchra									
USNM 103961	36.5i	30.0	32.0	13	12	3	Ethiopia	?Callovian	Muir-Wood and Cooper, 1951, Pl. 1, figs. 1–5, 11, 12, Pl. 2, figs. 1–6.
S. budulcaensis									
BMNH BB16128	22.2	—	15.0	14e	15e	—	Tunisia	Callovian	Mancenido and Walley coll. BMNH
S. numidiensis									
BMNH BB76533	57.0	62.0	58.0	9	8	3	Tunisia	Early Callovian	Mancenido and Walley, 1979
S. "*mogharaensis*"									
BMNH BB12360	—	—	71.2	17	—	5			
S. madashonensis									
BMNH B85631	27.5*	30.8*	28.2*	11e	10e	3	Somaliland	?Callovian	Muir-Wood, 1935, Pl. 10, fig. 11a–c.

* The holotype was remeasured due to a discrepancy with Muir-Wood's (1935, p. 109) data.

Pedicle valve interior with small, triangular hinge teeth, slightly bowed laterally, supported at juncture of inner socket ridges and dental lamellae; high, thin, curved dental lamellae extend down to valve floor and diverge at an angle of approximately 35°; a high, prominent, well-developed bladelike median septum extends from posterior region to umbonal cavity to a point at least halfway to anterior commissure; no complete anterior margins preserved in disarticulated specimens, while in few articulated shells in collection, median septum either not visible or incompletely preserved, especially towards anterior margins; the median septum and dental lamellae coalesce at posterior region of umbonal cavity; delthyrium covered by concave, conjunct deltidial plates which, late in ontogeny, form a henidium anteriorly (Figure 7), in many specimens line of juncture between deltidial plates obscured by imperfect silicification; a small submesothyridid pedicle foramen rimmed by a pedicle collar which merges internally into a narrow, well-developed pedicle tube terminating dorsally of median septum; diductor muscle scars represented by narrow, longitudinal striations, extending from base of dental lamellae and becoming shallower anteriorly until disappearing approximately one-third to one-half distance to anterior commissure; possible adductor muscle scars observed in only one specimen (GSI M8025) consisting of narrow depressions on either side of ventral median septum extending anteriorly for about 8 mm, anterior strongly crenulated internally due to impress of costae on valve floor, as posterior of valve is approached impress of costae becomes weaker.

TABLE 2: Number of costae in juvenile and adult specimens of *Septirhynchia hirschi* n. sp. PV = pedicle valve; BV = brachial valve; est. = estimated.

Juveniles			Adults		
Specimen	Valve	No. costae	Specimen	Valve	No. costae
USNM 401033	PV	25	GSI M6922	PV	10
GSI M8034	PV	24	GSI M6926	PV	12 est.
GSI M8035	PV	27	GSIM7193	PV	11
GSI M8019	BV	24	HU 35560	BV	14

In brachial valve interior dental sockets long, relatively deep and concave in cross section such that points of the bowed hinge teeth project into sockets in a medial direction; sockets extremely narrow throughout entire length, outer socket ridges smooth and diverge laterally at angle of about 85° from their point of origin at lateral margins of cardinal process; cardinal process large with a medial depression that deepens anteriorly but widens posteriorly, and is bisected by a median ridge (Figure 8); at base of cardinal process are two slightly convex delicate hinge plates divided by a narrow, simple septalium; median depression in the cardinal process grades dorsally into septalium which, in turn, is supported by a bladelike median septum; median septum composed of two knifelike vertical plates which merge and unite anteriorly and extends at least one-third of valve; at anterior margins of hinge plates are located two slightly divergent curved cilifer crura on a plane roughly parallel to hinge axis, uniformly narrow in width, projecting ventrally (Figures 9-11); no muscle scars observed; anterior commissure strongly corrugated due to costae impressed on valve floor and progressively weaken posteriorly.

Comparison.—Septirhynchia "mogharaensis" (BMNH BB 12360) from the Jurassic of Bihen, Somaliland, has 17-19 costae, slightly less than the *S. hirschi* shells from Gebel El-Maghara on average. Although both valves are articulated, the specimen is incomplete (thickness 68.5 mm, width 59.3 mm), with a tapering umbo. The presence of a ventral sulcus is questionable since the shell is somewhat crushed. Another late ephebic specimen (BMNH BB12361) has a tapering umbo and a shallow sulcus (Figure 12.14). However, it is incomplete; only the posterior portion of the pedicle valve and a fragment of the brachial valve are present. The shell has 14-15 costae and is very similar in appearance to *S. hirschi* (GSI M6922) from Gebel El-Maghara.

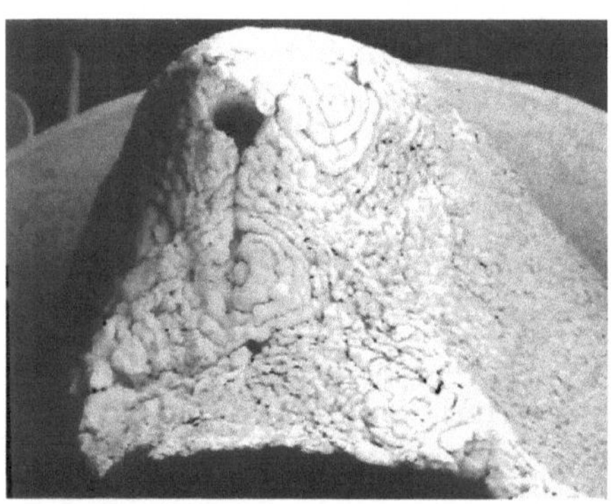

FIGURE 7: Paratype of *Septirhynchia hirschi* n. sp., OSI M8022, scanning electron micrograph showing pedicle foramen with conjunct deltidial plates and portion of henidium still intact, x 7.5.

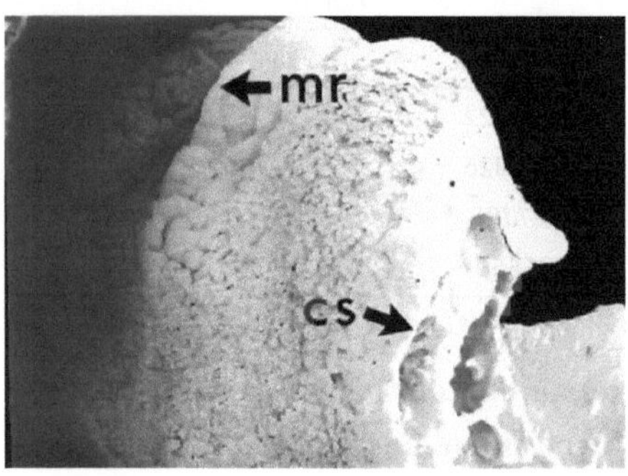

FIGURE 8: Paratype of *Septirhynchia hirschi* n. sp., OSI M8020, scanning electron micrograph showing fiat, thin, smooth recurved calcite sheet (cs) extending back over the brachial umbo representing the cardinal process (posterior oblique view); note median ridge (mr) which bisects medial depression, x8.5.

Septirhynchia "mogharaensis"? (BMNH 6618/ 50) has the same general outline as BMNH BB 12361 but is smaller with 12-13 costae. The specimen is fragmentary, possesses a shallow sulcus and tapering umbo with the dental lamellae and median septum visible through the external shell.

Septirhynchia azaisi (Cottreau) (BMNH B46235) (Figure 12.1-12.4) from the Kimmeridgian of Ida Kabeita, Somaliland, differs from *S. hirschi* in its pyriform outline of the posterior portion of the ventral valve, especially noticeable when viewed dorsally, and in its lack of a ventral sulcus. In addition, the costae in *S. azaisi* are more uniformly thick than those of *S. hirschi*, which broaden anteriorly. Both species approximate each other in all other dimensions (BMNH B46234: articulated specimen but crushed dorsoventrally with maximum length 54.9 mm, maximum width 57.2 mm; BMNH B46235: maximum length 67.6 mm, maximum width 59.6 mm, maximum thickness of pedicle valve 49.5 mm).

Muir-Wood and Cooper (1951, p. 4-5, Pl.1, figs. 6-10) described two fragmentary pedicle valves of *Septirhynchia azaisi* (Cottreau) (USNM 107066a) from Abu-uqua-Kurtcha, Ethiopia, in which the dental lamellae are less divergent and shorter than those of *S. hirschi*. Muir-Wood and Cooper's specimens, however, both possess conjunct deltidial plates as well as an anteriorly developed henidium, similar to the condition found in numerous specimens of *S. hirschi*. *Septirhynchia azaisi* may be further differentiated by its pyriform ventral umbo which seems to be characteristic of the species.

A single silicified specimen of *Septirhynchia pulchra* Muir-Wood and Cooper (USNM 103961) from the Callovian of Ego Gambo, Harrar Province, Ethiopia, is available for comparison. *Septirhynchia hirschi* is considerably larger, has a more gibbous shell, especially in the ephebic stage, and possesses more costae

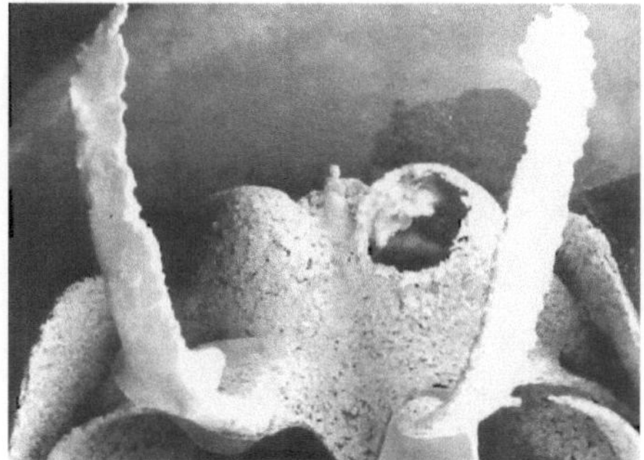

FIGURE 9: Paratype of *Septirhynchia hirschi* n. sp., AMNH 42923a, scanning electron micrograph showing cilifer crura, x5.

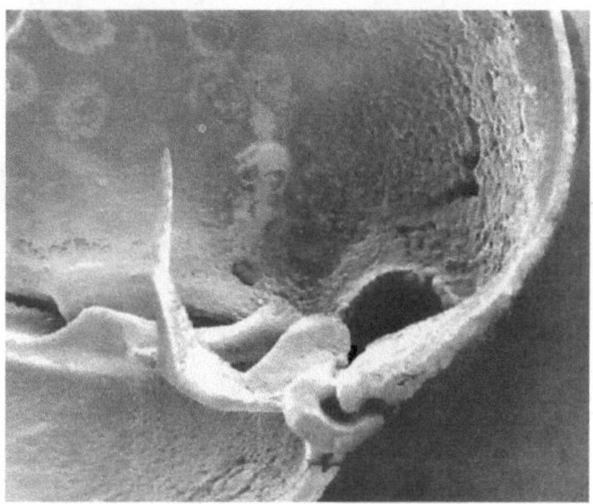

FIGURE 10: Paratype of *Septirhynchia hirschi* n. sp., AMNH 42926, scanning electron micrograph showing crus of juvenile specimen; note second crus broken at base and weakly-formed beekite rings in upper left quadrant, x6.

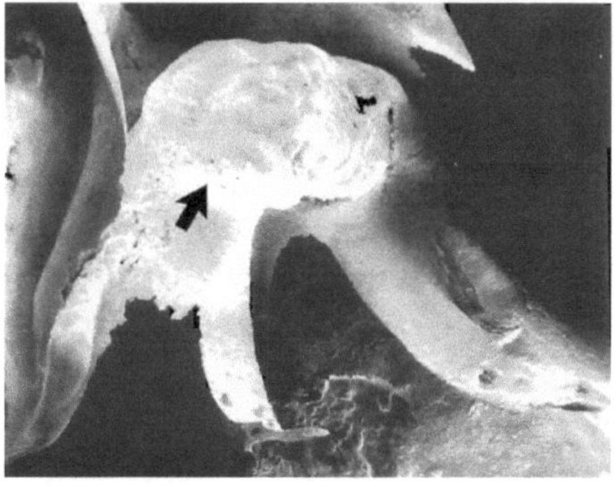

FIGURE 11: Paratype of *Septirhynchia hirschi* n. sp., AMNH 42925, scanning electron micrograph showing cardinal process (arrow) and crura, x5.

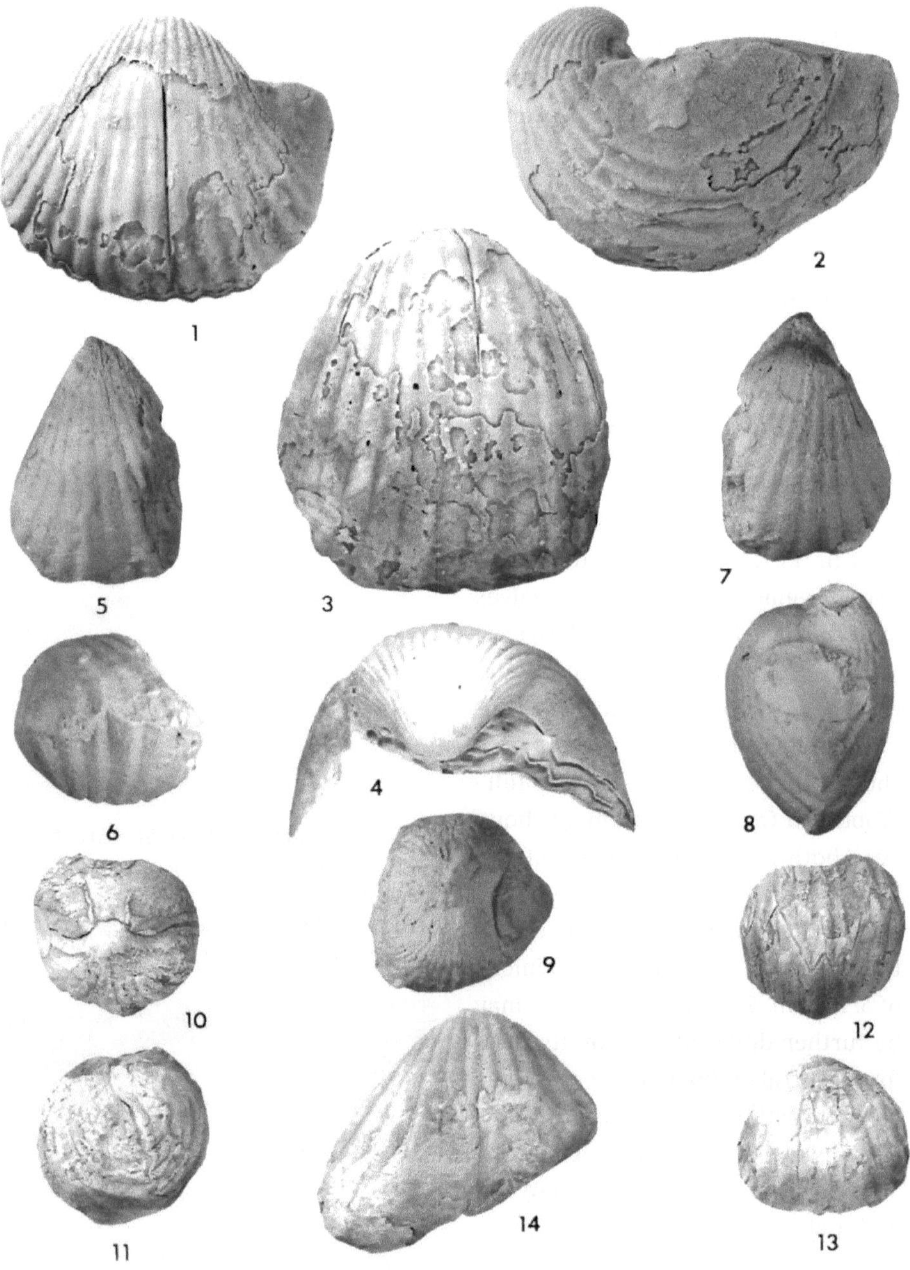

FIGURE 12: *1-4, Septirhynchia azaisi* (Cottreau, 1924), BMNH B46235, Kimmeridgian, Ida Kabeita, Somaliland; all views are of brachial valve, x1. Muir-Wood mistakenly labelled this specimen as a pedicle valve (Muir-Wood, 1935, Pl. 9, fig. 3a-c). Exterior oblique (slit represents median septum), lateral , exterior (note how slit representing median septum gradually fades anteriorly), and posteriorviews. *5-9, Septirhynchia budulcaensis* (Stefanini, 1932), BMNH BB 16128, Kimmeridgian,

(an average of 19 in *S. hirschi* as compared with 13 in *S. pulchra).* The ventral sulcus has an average of 5 costae while in *S. pulchra* there are 3 costae present in the sulcus. Internally the two species are almost identical. Both have ventrally projecting cilifer crura which emanate from the anterior ends of divided hinge plates. The hinge plates in *S. pulchra* appear to taper more anteriorly, but in both species they are divided by a septalium. The most significant difference between the two occurs in the morphology of the cardinal process. In *S. hirschi* it is a re-curved, spherical calcite sheet with a medial depression that widens and deepens anteriorly and is bisected by a median ridge, while in *S. pulchra* the cardinal process is small, knoblike, and terminates in a short spike (see Muir-Wood and Cooper, 1951, Pl. I, figs. 11, 12; Pl. 2, figs. 3-6, for comparison). In the ventral interior the dental lamellae of *S. pulchra* are subparallel, while in *S. hirschi* they are normally widely divergent. Both possess a pedicle collar. In Muir-Wood and Cooper's specimen (USNM 103961) the deltidial plates are conjunct posteriorly but form a henidium anteriorly. In the Gebel El-Maghara fauna most shells have conjunct deltidial plates while some have conjunct deltidial plates with a henidium anteriorly. It appears that the henidium develops late in the ontogeny of the species (late ephebic-gerontic).

Septirhynchia hirschi differs from *S. madashonensis* (Muir-Wood, 1935) (Figure 12.10-12.13) in its less well defined dorsal fold and greater number of costae in the ventral sulcus (5 in *S. hirschi* as compared with 3 in *S. madashonensis).* Also, *S. madashonensis* has a total of only 15 costae while *S. hirschi* has 19. In both forms, however, the costae are angular.

Septirhynchia numidiensis Mancenido and Walley may be differentiated from *S. hirschi* by the number of costae in the ventral sulcus (2-3 in *S. numidiensis* as compared with 5 in *S. hirschi)* as well as total number of costae (8-9 in adult specimens of *S. numidiensis* as compared with 19 in adult specimens of *S. hirschi).* Additionally, whereas the cardinal process in *S. numidiensis* is smooth, in *S. hirschi* it has a medial depression which widens and deepens anteriorly and is bisected by a median ridge.

Septirhynchia budulcaensis (Stefanini, 1932) (Figure 12.5-12.9) differs from *S. hirschi* in its finer costation, more numerous costae in the ventral sulcus (approximately 7 in *S. budulcaensis* as compared with 5 in *S. hirschi),* narrower planareas, and smaller size. Both shells lack a well-developed dorsal fold. Also, *S. budulcaensis* has about 24 costae on each valve. Muir-Wood (1935, p. 110)

Somaliland. Ventral, anterior, dorsal, lateral , and posterior views. *10-13, Septirhynchia madashonensis* Muir-Wood, BMNH B85631, ?Callovian, Madashon, Somaliland. Posterior (note dental lamellae shown in cross section of pedicle valve, not illustrated in Muir-Wood, 1935, Pl. 10, fig. 11a-c), lateral, anterior, and ventral views. *14, Septirhynchia "mogharaensis"* Muir-Wood, BMNH BB12361, ?Callovian, Bihen, Somaliland, pedicle valve exterior.

noted that a specimen of *S. budulcaensis* (Stefanini, 1932) from the Kimmeridgian of Somaliland lacked a median fold and sinus. However, Weir (1929, Pl. 4, fig. 5a) illustrated a specimen of *Stolmorhynchia? azaisi* (Cottreau, 1924) var. (subsequently redescribed by Stefanini as *Rhynchonella budulcaensis)* which clearly has a ventral sulcus.

Farag and Gatinaud (1962, p. 9, fig. 5a-e) described several incomplete specimens of *Rhynchonella (Stolmorhynchia?) afifi,* from a lithographic limestone at Richet Umm Werib and from a dolomitic limestone at Ouadi Gaza. The specimens closely resemble *S. hirschi* in their general outline, costation, and dorsal projection of the lateral commissure, but a definite determination of the species cannot be made at this time.

ONTOGENY AND MORPHOLOGIC VARIATION

The study of ontogenetic change and morphologic variation in *Septirhynchia hirschi* is facilitated by the availability for analysis of moderately- to well-silicified shells which reveal internal features that would have otherwise been obscured or, at best, whose structures would have been deduced from a series of transverse and median sections.

In neanic and early ephebic specimens the general outline of the shell is pyriform and the valves ventribiconvex with the pedicle valve noticeably deeper than the brachial valve. Ephebic specimens are subpyriform in outline and biconvex with the brachial valve slightly deeper than the pedicle valve; gerontic specimens are consistently gibbous.

The ventral sulcus is weakly defined in juveniles and originates just anterior to the beak. In adult forms, however, the ventral sulcus is strongly defined. There is a decrease in the number of costae from juvenile to adult stage (Table 2) due to the relative expansion of the planareas at the expense of the small lateral costae as noted by Mancenido and Walley (1979, p. 320) in their description of *Septirhynchia numidiensis* from Tunisia.

The lateral commissure is relatively straight in juveniles but becomes strongly arched dorsally in adults. In early neanic forms the lateral commissure is only slightly arched dorsally but the arch increases dramatically in ephebic and gerontic specimens.

A ventral median septum is present in all ontogenetic stages, becoming progressively higher in older specimens. Mancenido and Walley (1979, p. 322) noted the absence of a ventral median septum in a juvenile specimen of *Septirhynchia numidiensis,* approximately 25 mm long, and reported that upon examination of other fragmentary forms there was no trace of a ventral median septum in juveniles of this species. *Septirhynchia azaisi* has a low ventral median septum, described as a knifelike ridge (Muir-Wood and Cooper, 1951, p. 5, Pl. 1, figs. 6, 9), which increases gradually in height anteriorly. The ventral median septum of *Septirhynchia pulchra* (Muir-Wood and Cooper, 1951, Pl. 2, fig. 1) is lower than that found in *S. azaisi, S. numidiensis,* and *S. hirschi* and barely extends beyond the base of the dental lamellae. Due to the extremely delicate nature of the structure, a perfectly preserved ventral median septum has yet to be found among the shells from Sinai, thus necessitating deferral of further speculations on its ontogeny and variation until additional material is collected.

TABLE 3: Measurements (in mm) of pedicle tube and pedicle foramen in *Septirhynchia hirschi* n. sp.

Specimen	Ontogenetic stage	Diameter of pedicle tube	Diameter of pedicle foramen
GSI M8013	neanic	0.8	0.8
GSI M8014	early ephebic	0.7	0.8
USNM 401034	ephebic	0.8	0.9
USNM 401035	ephebic	0.8	0.9
GSI M8036	ephebic	0.7	0.7
GSI M8017	gerontic	0.9	0.9
AMNH 42928	gerontic	0.9	concealed
AMNH 42929	gerontic	0.9	0.9
GSI M8026	gerontic	0.9	0.9*

* Damaged.

The pedicle foramen is located at the apex of the delthyrium in early neanic forms and in some cases does not come in contact with the delthyrial margins, but lies entirely within the delthyrium proper, anterodorsally of the beak ridges (GSI M8034). Occasionally, however, the pedicle foramen does come in contact with the delthyrial margins. In both cases the deltidial plates are conjunct (Figure 7) but it is sometimes difficult to observe the line of fusion between the plates which, if absent, would be equivalent to the symphytium of the Terebratulacea. In neanic forms the pedicle foramen is hypothyrid; it lies anterodorsal to the beak ridges. In ephebic shells the pedicle foramen migrates further away from the base of the delthyrium and lies on each side of the beak ridges with the larger portion of the foramen still within the delthyrium (that is, within the interarea) and the smaller portion in the umbo resulting in a submesothyridid condition. As ontogenetic development progresses, late ephebic forms become mesothyrid while in gerontic specimens the delthyrium and pedicle foramen are covered by the incurvature of the ventral umbo and the relative position of the pedicle foramen cannot be determined accurately.

TABLE 4: Measurements (in mm) of diductor muscle scars in *Septirhynchia hirschi* n. sp. PV = pedicle valve.

Specimen	Valve	Length of diductor scar	Ontogenetic stage
USNM 401033	PV	4	neanic
GSI M8028	PV	5	early ephebic
GSI M8021	PV	9	ephebic

In well-preserved specimens a narrow, cylindrical, internally directed pedicle tube (see Figure 6.4, GSI M80 14) is often present. Since this structure is quite delicate and easily broken, it is often missing, even in well-silicified shells. The pedicle tube consists of secondary shell material and functioned as a supportive casing for the proximal portion of the pedicle. The diameter of the pedicle foramen was measured in specimens in which the pedicle tube was well preserved and compared to the diameter of the pedicle tube (Table 3). Based on the data collected several interesting trends may be noted.

1) The pedicle tube is not present in early neanic forms but appears first in the late neanic stage because it either did not develop at that ontogenetic stage (i.e., GSI M8037) or was not preserved.

2) The pedicle tube does not increase significantly after the animal attained the late neanic to early ephebic stage (the observed range of the pedicle tube diameter varies between 0.7 and 0.9 mm).

3) Therefore, the pedicle itself does not increase significantly in diameter once the brachiopod reached the late neanic to early ephebic stage.

4) The pedicle tube is either narrower than, or equal to, the diameter of the pedicle foramen, which suggests that the pedicle became smaller (narrower) within the umbonal cavity and expanded externally. This suggests further that the pedicle was most likely not very robust and may well have served as a tether, in the manner of *Cryptopora,* rather than as a supportive structure which kept the animal suspended above the sediment-water interface.

Muscle scars were observed for the first time in *Septirhynchia* due to the availability of silicified material which yielded well-preserved internal structures. Diductor scars were preserved in 10 specimens, all pedicle valves (AMNH 42927, USNM 401033, GSI M8021, M8023, M8024, M8027, M8028, M8029, M8030, M8031), and possible adductor scars were noted in one pedicle valve (GSI M8025). There were no muscle scars observed on the brachial valves in the collection. The diductor scars are narrow, longitudinal grooves, beginning just anterior to the base of the dental lamellae and extending anteriorly. The grooves are a bit wider and deeper posteriorly but one specimen (GSI M8030) has a single, shallow subellipsoidal scar preserved. Based on the morphology of an articulated specimen (AMNH 42927), these markings are interpreted as diductor scars due to their position on the pedicle valve. They are situated such that any muscles which articulated into the grooves would have been unable to bypass the cardinal process in order to attach themselves onto the brachial valve floor as would an adductor muscle. One specimen (GSI M8025) has shallow grooves on either side of the ventral median septum which may represent adductor scars since they are situated more anteriorly than the diductor markings. Also, muscles articulating into these grooves would not rub against the cardinal process since they are located at approximately midlength; the adductors would easily reach the brachial valve floor without obstruction.

In neanic specimens (Table 4) the diductor scars are short, longitudinal, narrow grooves which, as the brachiopod grew, became proportionately longer and wider. There does appear to be some variation in ephebic forms typified by a wider posterior region of the scar in several cases (GSI M8021, M8027, M8031) to a subellipsoidal shape in one case (GSI M8030).

The cardinal process in neanic specimens of *Septirhynchia hirschi* consists of a flattened, recurved sheet of calcite bisected by a low median ridge. As the brachiopod passed into the ephebic stage, the cardinal process became bulbous and the median depression deepened anteriorly and widened posteriorly (Figures 8, 1 I). The median ridge is often more pronounced here than in the neanic stage, but this may be due to erosion of the surrounding shell material. In only one specimen (GSI M8019) was the median ridge found to be unusually elevated. In the gerontic stage the cardinal process (GSI M8026) lost the median ridge and became relatively flat, extending well back onto the dorsal umbo. The cardinal process in *S. pulchra* (Muir-Wood and Cooper, 1951, p. 3, Pl. 1, fig. 3, Pl. 2, figs. 3-6) is similar to that of *S. hirschi* but is smaller, knoblike and terminates in a short spike.

It seems clear that there was a change in the relationship of *Septirhynchia hirschi* to the substrate during ontogeny, although the change was not a radical one similar to that displayed by the productid brachiopods. Grant (1963) noted that all known productid species except *Linoproductus angustus* normally remained suspended by their spines to cylindrical objects such as crinoid columnals only in their earliest stages and soon broke away either to live free on the sea floor or attached in different ways to different objects. As the brachiopod became larger it reached a critical weight above which the spines were unable to provide support, resulting in the shell breaking off and dropping to the sea floor, thus effecting a radical change in substrate. Similar morphological features present in progressive ontogenetic stages in *Septirhynchia hirschi* and *S. numidiensis*, as described by Mancenido and Walley (1979), indicate that the mode of life of the two species was very similar. Specifically, juveniles of both species lived attached to the substrate by a small but functional pedicle. Whether the pedicle actually supported the animal above the sediment-water interface in early stages of development is questionable; however, the presence of a pedicle tube is at least evidence of an early attached stage (Boucot, 1981, p. 39). During later ontogenetic stages, both species adopted a semi-infaunal mode of life with the umbos buried in soft sediment and the lateral commissure almost vertical. Mancenido and Walley (1979, p. 325) argued that nearly all features of the adult morphology can be interpreted as adaptations to this mode of life and went on to discuss particular adaptive features. Their discussion is relevant to the life mode of *Septirhynchia hirschi* which shares with *S. numidiensis* the following conditions in the adult (ephebic and gerontic) stages.

1) The ventral planareas projected into the dorsal valve, thereby allowing the valves to gape without seepage of mud through the buried posterior portion of the commissure.

2) Fused deltidial plates (Figure 7) in adult stages prevented mud from entering over the enclosed dorsal umbo.

3) Stability on the sea floor was increased by the vertical orientation of the lateral commissure, symmetrical balance of the approximately equal valves, and gibbous shape of the adult, as well as strongly incurved pedicle beak which would have allowed both umbos to lie at approximately the same level within the mud.

4) The fractional volume of the brachiopod occupied by skeleton decreased as ontogeny progressed, thus decreasing the bulk density of the organism.

5) The loss of a functional pedicle would have facilitated the free movement upwards of the brachiopod as the valves gaped and levered the shell upwards in the sediment.

CONCLUSIONS

The following morphologic and ontogenetic changes are found in *Septirhynchia hirschi* collected from Gebel El-Maghara, northern Sinai.

1) Juvenile shells are pyriform in outline while adults tend to be gibbous.

2) The ventral sulcus became more strongly defined as the animal grew.

3) The lateral commissure is relatively straight in juveniles but is strongly arched dorsally in adults.

4) A ventral median septum is present in all ontogenetic stages, becoming progressively higher as the brachiopod matured.

5) The pedicle foramen ranges from hypothyridid in juveniles to mesothyridid in adults.

6) A narrow, cylindrical, internally directed pedicle tube extends into the umbonal cavity in all ontogenetic stages except for early neanic.

7) Narrow, longitudinal diductor muscle scars are present on some pedicle valves in the collection; no muscle scars were observed on any brachial valve.

8) The cardinal process consists of a thin, recurved calcite sheet with a medial depression bisected by a median ridge.

9) Juveniles lived attached to the substrate by a small but functional pedicle; adults lived semi-infaunally with the umbos buried in the mud and lateral commissures almost vertical.

ACKNOWLEDGMENTS

This research was funded by grants from the National Geographic Society and the Explorers Club Exploration Fund for work which was carried out during my tenure as Visiting Scientist at the Geological Survey of Israel, Jerusalem. Special thanks are due to Y. Druckman, Head, Stratigraphy, Mapping and Oil Division, Geological Survey of Israel (GSI), for providing office space and laboratory and library facilities; his hospitality was much appreciated. M. Arnon, of the same institution, acted as a field assistant and technician during the course of two field seasons. I thank D. V. Ager (University College, Swansea), G. A. Cooper (USNM), F. Hirsch (GSI), Z. Lewy (GSI), and E. F. Owen (BMNH) for critical review and providing valuable suggestions for improvement. C. D. Walley (University College, Swansea) kindly loaned me unpublished material which proved useful, and F. Collier (USNM), E. F. Owen (BMNH), and S. Rothmann (HU) deserve thanks for the loan of fossil specimens from their respective institutions. D. V. Ager sent me casts of *Septirhynchia numidiensis* and *S. budulcaensis* from his personal collection which were invaluable in comparing those species to *S. hirschi* as well as other species of *Septirhynchia.* I also acknowledge the help of L. Duffy, Scientific Technician, American Museum of Natural History, who assisted in working with the Museum's Cambridge Stereoscan 250 scanning electron microscope on which the photomicrographs illustrated in this paper were taken. P. Harries assisted with the photography.

REFERENCES

Al Far, D. M. 1966. *Geology and coal deposits of Gebel El-Maghara (Northern Sinai)*. Geological Survey of Egypt, Paper 37: 1-59.

Boucot, A. J. 1981. *Principles of Benthic Marine Paleoecology*. New York: Academic Press.

Cooper, G. A. 1983. The Terebratulacea (Brachiopoda), Triassic to Recent: a study of the brachidia (loops). *Smithsonian Contribution to Paleobiology* 50: 1-445.

Cossman, M. 1925. Description des espèces. In H. Douvillé (ed.), Le Callovien dans le massif de Moghara. *Bulletin Société Géologique de France* Ser. 4, 25: 305-328.

Douvillé, H. 1916. Les Terrains Secondaire dans le massif de Moghara a l'Est de l'Isthme Suez. *Academie de Science Paris, Memoire* 54.

------. 1925. Le Callovien dans le massif de Moghara: avec description des fossiles par M.Cossman. *Bulletin Société Géologique de France* sér. 4, 25: 305-328.

Dubar, G. 1967. Brachiopodes Jurassiques du Sahara Tunisien. *Annales de Paléontologie* (Inv.), 53: 33-101.

Farag, I. A. M, and W. Gatinaud. 1962. Six espèces nouvelles du genre *Rhynchonella* dans le roches Jurassiques d'Egypte. *Journal of Geology of the United Arab Republic* 1960, 4: 81-88.

Goldberg, M., et al. 1971. Preliminary columnar section of the Jurassic of Gebel Maghara. *Geological Survey of Israel*, Report No. *MMI* 117 1.

Grant, R. E. 1963. Unusual attachment of a Permian linoproductid brachiopod. *Journal of Paleontology* 37: 134-140.

Mancenido, M. O., and C. D. Walley. 1979. Functional morphology and ontogenetic variation in the Callovian brachiopod *Septirhynchia* from Tunisia. *Paleontology* 22: 317-337.

Muir-Wood, H. M. 1935. *The Mesozoic paleontology of British Somaliland: Jurassic Brachiopoda*, 7:75-147. London: Government of the Somaliland Protectorate.

Muir-Wood, H., and G. A. Cooper. 1951. A new species of the Jurassic brachiopod genus *Septirhynchia. Smithsonian Miscellaneous Collections* 116: 1-6.

Rouselle, L. 1970. Sur une "faunule" de Rhynchonelles du Jurassique supérieur de la region de Laghouat (Sud Algérien). *Bulletin Société Géologique de France* Ser. 7, 12: 573-578.

Stefanini, G. 1932. Echinodermi, Vermi, Briozoi e Brachiopodi del Giur-Lias della Somalia. *Palaeontographica Italica*, n.s. 2, 32(1931): 81-130.

Weir, J. 1929. Jurassic fossils from Jubaland, East Africa. Collected by V. G. Glenday. *Monographs of the Geological Department of the Hunterian Museum, Glasgow University*, 3: 1-63.

Chapter Three

Goliathyris lewyi, New Species (Brachiopoda, Terebratulacea) from the Jurassic of Gebel El—Minshera, Northern Sinai

ABSTRACT

Goliathyris Lewyi, new genus and species is described from the *Lamberticeras Lamberti* Zone, Jurassic (Upper Callovian) of Gebel El-Minshera, northern Sinai. Homeomorphic with *Aulacothyris* spp., G. *Lewyi* is questionably assigned to the family Dyscoliidae based on its internal resemblance to *Trigonithyris,* but resembles a zeilleriid externally.

INTRODUCTION

The present study is part of a preliminary investigation of the brachiopod faunas of the northern Sinai undertaken by us and colleagues in the Geological Survey of Israel. Our long-range goal is to complete a taxonomic revision of the brachiopod faunas of northern Sinai as well as those of Arabia which will help us establish the early history of brachiopod species and their evolution within the "Ethiopian" Faunal Province. Analysis of present data supports our contention that this province was invaded by brachiopods migrating from the north in Early Jurassic times which were isolated for the remainder of the Jurassic. These faunas are thought to have subsequently developed special morphological characteristics which distinguish them from their original stock. In addition, we are investigating the distribution of brachiopod species across faunal realm and province boundaries, specifically the Indo-African Faunal Realm which is now widely dispersed on various continental fragments. The Sinai is situated at the northern part of this realm and consequently the brachiopods are very likely to include species belonging to the equatorial Tethyan Realm.

Taxonomic revision of the Sinai brachiopods will also enable us to define, with greater accuracy, faunal realm and province boundaries. The "Ethiopian" Faunal Province, for example, is recognizable from early in the Jurassic until the middle and possibly the end of the Cretaceous by the presence of endemic taxa at the species, genus, and family level. These endemics have been recognized in the ammonoid cephalopoda (Arkell, 1952, 1956), in the trigoniacean and crassatellacean bivalves (Kitchin, 1912), and particularly in the brachiopods (Weir, 1925; Muir-Wood, 1935). This same province has been recognized in India, East Africa, and Madagascar and at the end of the Jurassic in South America; it may also extend eastward as far as New Caledonia. Its first occurrence seems to be in the shallow seas following rifts formed during the breakup

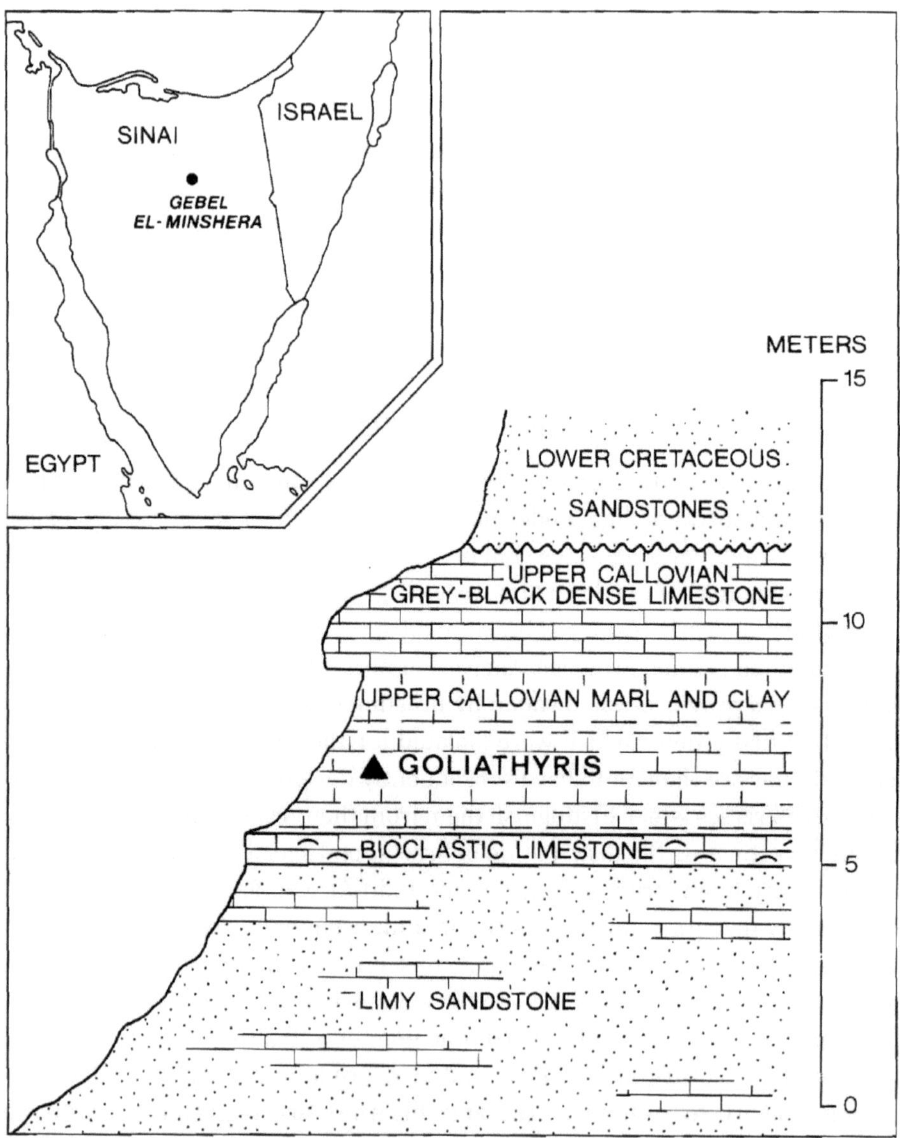

FIGURE 1: Locality map of northern Sinai (inset) and detailed lithologic section of outcrop area. The triangle represents approximate strata from which *Goliathyris lewyi*, new species, was collected.

of Gondwanaland but is apparently limited at an unknown southern margin, as none of its species are known in the geosynclinal contemporaneous deposits of Antarctica or New Zealand. During much of the Early Jurassic, ubiquitous genera such as *Tetrarhynchia, Lobothyris,* and *Zeilleria,* found in the faunas of most known outcrops, became so generalized that it is difficult to plot them in terms of distribution (Ager, 1973). The description and revision of specialized forms in particular will facilitate the delineation of province boundaries as well as the recognition of faunal realms and distribution of genera.

Table 1: Stratigraphic Distribution of Some Jurassic Ammonites at Gebel El-Minshera, Northern Sinai, and Correlation with Ammonite Faunas in Central Saudi Arabia

Unnamed stratigraphic units, Gebel El-Minshera	Ammonites (this report)	Ammonites (Imlay, 1970)	Ammonite Faunas (Arkell, 1952)
Cretaceaous			
Sandstone	—	—	—
UNCONFORMITY			
Jurassic (European Upper Callovian)			
Gray-black dense limestone	*Lamberticeras lamberti* Zone *Peltoceras athleta* Zone		
Marl and clay (with *G. lewyi*)	"*Clydoniceras*" *Quenstedtoceras* *Pachyerymnoceras*	*Pachyerymnoceras*	
Bioclastic limestone	*Paracenoceras*	*Pachyceras*	
Limy sandstone	—	*Erymnoceras*	*Erymnoceras*

The *Goliathyris* described herein as *G. lewyi,* new genus and species, was collected from loose debris on a slope of marl and clay (fig. 1) of Upper Callovian age along with *Pachyerymnoceras, Quenstedtoceras,* and *"Clydolliceras" pseudodiscus* Arkell (Cephalopoda). The marly unit is overlain by gray-black dense limestone, also of Upper Callovian age, containing within its matrix specimens of *Peltoceras trifidum* (Quenstedt), "*Clydoniceras,*" and *Pachyerymnoceras* (Cephalopoda); *Putealiceras, Pseudomelania* (Gastropoda), and unidentified corals. The top of the unit lies unconformably under a series of Lower Cretaceous sandstones. A thin, bioclastic limestone unit containing *Pachyerymnoceras* and *Paracenoceras prohexagonum* (Cephalopoda) directly underlies the marl and clay unit from which the *Goliathyris* specimens were collected.

The age of the upper part of the Jurassic strata at Gebel El-Minshera appears to be Upper Callovian (table 1) on the basis of ammonites of the *Lamberticeras lamberti* Zone (Z. Lewy, personal commun.) with a possible Upper Bathonian to Upper Callovian unconformity.

ABBREVIATIONS

AMNH = American Museum of Natural History, Department of Invertebrates
GSI = Geological Survey of Israel, Paleontology Division
USNM = United States National Museum of Natural History, Department of Paleobiology, Smithsonian Institution

SYSTEMATIC PALEONTOLOGY

SUBORDER TEREBRATULIDINA WAAGEN, 1883
SUPERFAMILY TEREBRATULACEA GRAY 1840
FAMILY UNCERTAIN
Goliathyris, new genus

TYPE SPECIES: *Goliathyris lewyi,* new species.

INCLUDED SPECIES: Type species only.

GENERIC DIAGNOSIS: Extremely large, strongly sulcate, nonstrophic, permesothyridid with incurved dorsal umbo and massive zeilleriid beak. Broadly pentagonal in dorsal view. Hinge plates horizontal, becoming concave toward floor of brachial valve. Cardinal process massive and bilobate.

CHARACTER ANALYSIS: Based on a cladogram (fig. 6) it is postulated that *Goliathyris* is more closely related to *Trigonithyris* than to either *Dyscolia* or *Goniobrochus,* all three genera belonging to the family Dyscoliidae (Cooper, 1983; Muir-Wood, 1965). In support of this hypothesis two synapomorphies are presented (table 3): (1) noncapillate radial ornamentation, and (2) a massive cardinal process. *Goniobrochus* is more closely related to *Dyscolia* than either is to *Goliathyris* or *Trigonithyris* based on the following two synapomorphies: (1) ventrally directed hinge plates and (2) square loops. *Dyscolia* differs from *Goliathyris* in its sulcate anterior commissure, permesothyrid pedicle foramen, and broadly pentagonal outline, all of which are considered plesiomorphic characters at this (generic) level of analysis. Cooper (1983) noted that *Trigonithyris* does not belong with the Dyscoliidae because of its well-developed outer hinge plates. *Goliathyris* resembles a zeilleriid externally in its beak and pedicle foramen (symplesiomorphies) but differs from the family Zeilleriidae in having no dental plates or median septum. Based on analysis of the following characters the authors believe that the present classification (Cooper, 1983; Muir-Wood, 1965) in which *Dyscolia, Goniobrochus,* and *Trigonithyris* are placed in the family Dyscoliidae must be reconsidered. Clearly, the four genera do not share any apomorphic characters. Additional data, such as the exact morphology of the loop in *Trigonithyris* and *Goliathyris,* will provide information in support or rejection of the proposed cladogram (fig. 6). Characters used in the above analysis and their phylogenetic significance are described below.

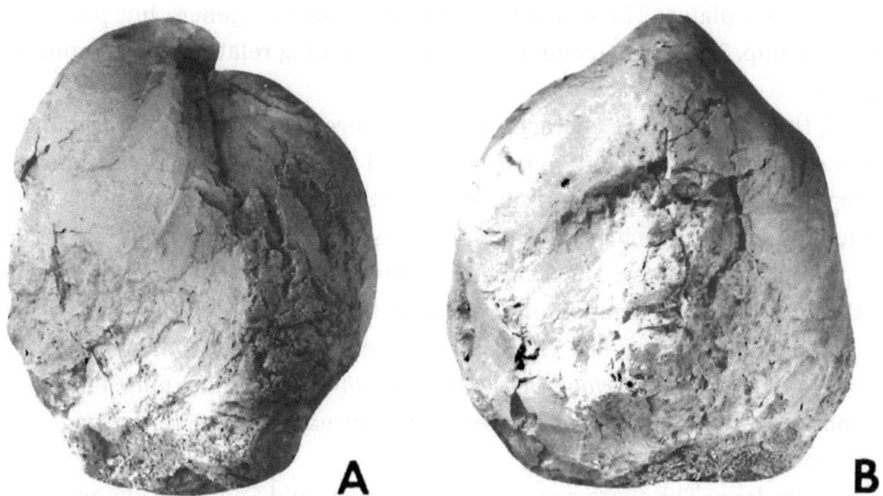

FIGURE 2: *Goliathyris lewyi,* new species. A, B. Lateral, ventral views, holotype GSI M7261 a. x1.1.

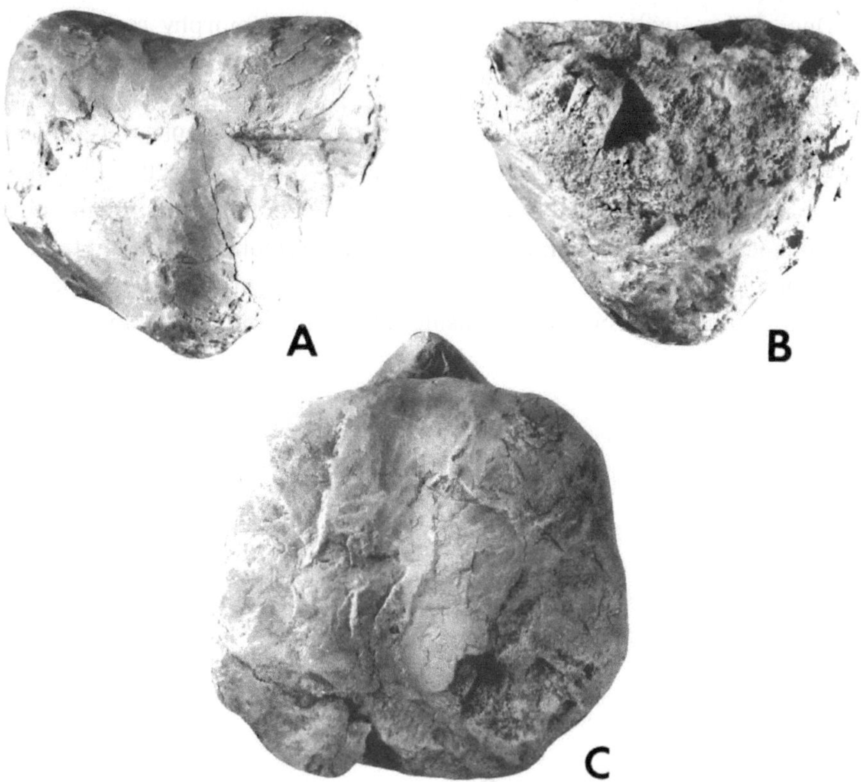

FIGURE 3: *Goliathyris lewyi,* new species. A, B, C. Posterior, anterior, dorsal views, holotype GSI M7261a. x 1.1.

1. Dental plates: The character is absent in all four genera but present in the outgroup. This gives us no information regarding relationships among the genera.

2. Hinge plates: The character (ventrally directed hinge plates) is found in the outgroup, *Dyscolia,* and *Goniobrochus* and thus represents a plesiomorphy at that level of analysis while its alternative character (concave hinge plates directed toward the brachial valve floor) represents a synapomorphy.

3. Cardinal process: The character is plesiomorphic in *Dyscolia* (found also in the outgroup) but synapomorphic in *Trigonithyris* and *Goliathyris.*

4. Ornamentation: The character (noncapillate radial ornamentation) is found in the outgroup, *Trigonithyris,* and *Goliathyris,* and thus represents a plesiomorphy at that level of analysis while its alternative character (zigzag capillae) represents a synapomorphy.

5. The loop: The character is found in *Dyscolia* and *Goniobrochus* and not in the outgroup, thus representing a synapomorphy uniting the two genera; it is unknown in *Trigonithyris* and *Goliathyris.* If a zeilleriid loop is eventually found in these genera it would represent a plesiomorphy.

One of the auxiliary criteria of phylogenetic apomorphy recognized by Hennig (1966) for distinguishing apomorphic from plesiomorphic characters is used in the character analysis discussed above. The criterion used here is that of geological character precedence which, according to Wiley (1981), rests on the rule that when one character is found entirely in the geologically older members of a monophyletic group while the alternative homology is found only in the geologically younger members of the same monophyletic group, then the older homolog is the plesiomorphic character. In accordance with the view that stratigraphic position not be considered as a priori evidence for character

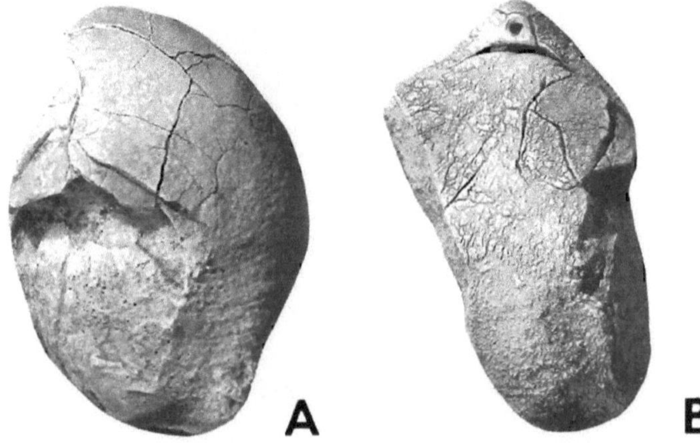

FIGURE 4: *Goliathyris lewyi,* new species. A, B. Lateral, dorsal views, paratype GSI M7261b. x 1.

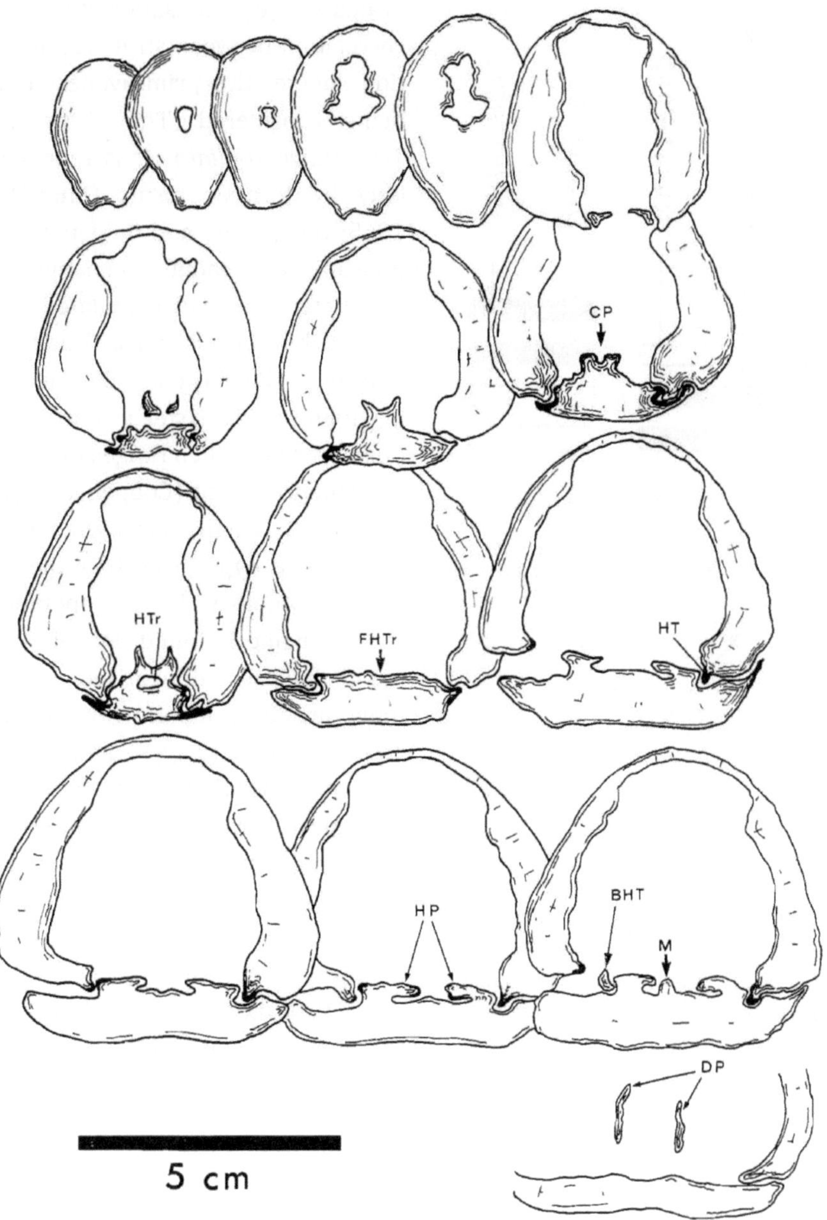

FIGURE 5: Transverse serial sections of *Goliathyris lewyi,* new species. GSI M7261b. Due to the nature of the matrix it was impossible to accurately record distances of serial sections from beak. However, the distance from the last transverse serial section to the umbo is 38 mm. Abbreviations: BHT = broken hinge tooth; CP = cardinal process; DP = descending processes; FHTr = flattened hinge trough; HP = hinge plates; HT = hinge tooth; HTr = hinge trough; M = myophragm (= euseptoidum). Note how hinge plates change from almost horizontal to slightly concavoconvex, thicken, and point toward the floor of the brachial valve.

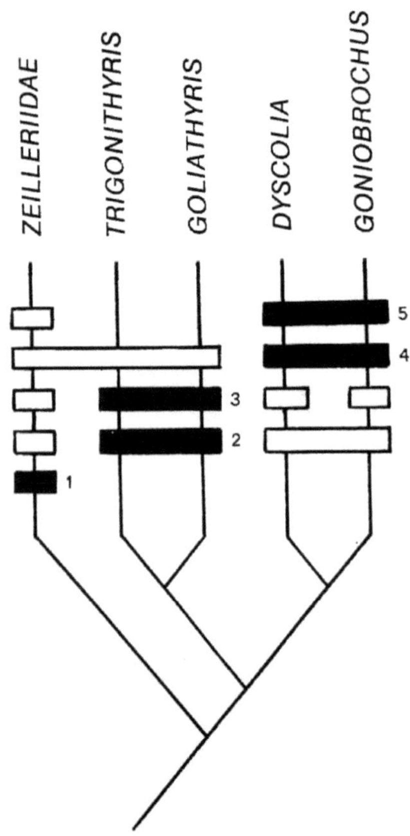

FIGURE 6: A phylogenetic hypothesis of relationships among Jurassic, Pliocene, and Recent terabratulid brachiopods. See text for discussion. Filled boxes = apomorphic condition; empty boxes = plesiomorphic condition.

evaluation (e.g., Schaeffer et al., 1972), correlation between stratigraphic position and relative primitiveness is now being considered. The sister group *Trigonithyris-Goliathyris* is represented only by Jurassic forms (Muir-Wood, 1965; this paper) and its genera share noncapillate radial ornamentation (character 4, above) postulated as plesiomorphic since it is also found in the outgroup (Zeilleriidae). Zigzag capillae, found in the sister group *Dyscolia-Goniobrochus,* which ranges in age from Pliocene to Recent (Cooper, 1983), is a postulated synapomorphy, thus supporting the hypothesis that the alternative homology is the apomorphic character and is found entirely within the geologically younger members of a monophyletic group. However, upon examination of characters 2 and 3, it is evident that when applying the criterion of geological character precedence there is an incongruity in that the apomorphic homologies are found entirely within the geologically older group. Additional data will aid in resolving this problem.

ETYMOLOGY: Named after Goliath, the Philistine giant from Gath, slain by [King] David.

Goliathyris lewyi, new species
Figures 2-5

DIAGNOSIS: Only known species of the genus; same as for genus.

DESCRIPTION: The shell is extremely large (table 2 figs. 2-4), smooth, rostrate, nonstrophic, and broadly pentagonal when viewed dorsally. A small, circular pedicle foramen is present just posterior to a poorly defined beak ridge on the ventral umbo, resulting in a permesothyrid condition. The interarea is obscured by the incurved ventral umbo. A wide, moderately shallow, dorsal sulcus originating slightly anterior to the concealed dorsal umbo, widens and

TABLE 2
Measurements of *Goliathyris lewyi* Compared with Homeomorphic Species of *Aulacothyris* from the "European" and "Ethiopian" Faunal Provinces
(all measurements in millimeters)

Species	(L)	(W)	(T)	Locality	Faunal province	Age
Goliathyris lewyi, sp. nov.						
GSI M7261a	74.0*	61.2	49.5	Gebel El-Minshera	EP	JUR
GSI M7261b	72.6*	—	43.5	Gebel El-Minshera	EP	JUR
Aulacothyris jubaensis (Weir, 1929, pl. IV, fig. 15. XI)	21.9	19.8	10.2	Somaliland	EP	JUR
A. somaliensis (Stefanini, 1931, pl. VI, fig. 6. XI)	25.0	20.0	15.0	Somaliland	EP	JUR
A. resupinata						
USNM 88703	22.8	19.7	15.0	Morocco	EP	JUR
A. impressa						
AMNH 1514a	19.4	16.5	10.5	Bavaria	EUP	JUR
AMNH 1514b	15.2	13.5	8.3	Bavaria		
AMNH 1514c	16.0	14.8*	8.0	Bavaria	EUP	JUR
A. carinata						
AMNH 1509a	21.7	19.4	12.8	Bavaria	EUP	JUR
AMNH 1509b	19.7	17.1	11.1	Bavaria	EUP	JUR
AMNH 1509c	21.8	19.0	13.2	Bavaria	EUP	JUR
A. pala						
AMNH 1523a	16.6	12.1	10.0	Vils in the Tyrol	EUP	JUR
AMNH 1523b	19.7	12.9	11.5	Vils in the Tyrol	EUP	JUR
AMNH 1523c	16.9	12.0	9.5	Vils in the Tyrol	EUP	JUR
A. bernardina						
USNM 30949a	18.6	16.5	10.7	Villers-sur-mer, France	EUP	JUR
USNM 30949b	15.8	13.9	8.3	Villers-sur-mer, France	EUP	JUR
Aulacothyris sp.						
USNM 89083	21.5	16.3	11.2	England	EUP	JUR
A. blakei						
USNM 104730a	14.6	13.2	7.9	England	EUP	JUR
USNM 104730b	12.8	12.9	7.7	England	EUP	JUR
USNM 104730c	12.0	11.9	6.6	England	EUP	JUR
A. meriani						
USNM 66932a	27.8	19.1	13.8	England	EUP	JUR
USNM 66932b	21.7	16.3	12.7	England	EUP	JUR
A. gregalis						
AMNH 19303a	10.2	10.8	6.1	Sarajevo, Bosnia	EUP	TRI
AMNH 19303b	9.0	7.5	4.9	Sarajevo, Bosnia	EUP	TRI
AMNH 19303c	9.4	9.3	4.8	Sarajevo, Bosnia	EUP	TRI
A. angusta						
AMNH 1932a	7.0	6.4	4.0	Germany	EUP	TRI
AMNH 1932b	7.1	7.1	4.4	Germany	EUP	TRI

* Damaged.

deepens anteriorly. Although much of the ventral exterior is poorly preserved, it is evident that a strong ventral fold opposed the dorsal sulcus. The height of the fold appears to have been significantly greater than the depth of the sulcus. The shell is concavoconvex with the ventral valve almost carinate when viewed posteriorly. The anterior commissure is not preserved but, based on shell morphology; a reconstruction seems to indicate that it is strongly sulcate. One lateral commissure is preserved extending about two-thirds of the shell length and, although partially concealed by debris anteriorly, appears to be uniformly straight. The lateral slopes of the ventral valve are quite steep, while on the dorsal valve they flare out from the sulcus, rise dorsally, and descend at approximately a 90° angle to the lateral commissure.

Although radial ornament is absent on the smooth shell, there are some distinct, irregularly spaced growth lines on the anterior half of the specimens.

INTERNAL CHARACTERS: The following description has been obtained with great difficulty from a series of transverse serial sections (fig. 5) made through the umbonal region of a silicified specimen from the type locality at Gebel El-Minshera, northern Sinai.

TABLE 3: Data Used to Analyze the Phylogenetic Relationships of Four Genera of Terebratulid Brachiopods (Apomorphies are in italics).

Character	Outgroup (Zeilleriidae)	*Trigonithyris*	*Goliathyris*	*Dyscolia*	*Gonio-brochus*
1. Dental Plates	Present	Absent	Absent	Absent	Absent
2. Hinge Plates	Ventrally directed	*Concave toward brachial valve floor*	*Concave toward brachial valve floor*	Ventrally directed	Ventrally directed
3. Cardinal Process	Small	*Massive*	*Massive*	Small	Absent
Ornamentation	Noncapillate	Noncapillate	Noncapillate	*Zigzag capillae*	*Zigzag capillae*
Loop	Long	Unknown	Unknown	*Square*	*Square*

The umbonal cavity of the ventral valve is elongate-oval in transverse outline in the early stages and is partially filled with dense callous. The cavity gradually develops a more triangular outline, meeting and articulating early with the dorsal valve. A massive, bilobate cardinal process develops and persists until the strong, peg like hinge teeth are well inserted into the dorsal sockets. Horizontal, elongate hinge plates, showing little or no differentiation from the inner socket ridges, develop a ventrally convex transverse outline and are deflected dorsally.

No evidence was obtained of crural bases and all that remains of the obviously broken brachial loop is seen as two subparallel, inwardly curving descending processes (or possibly crural processes). Due to difficulty in sectioning and incompleteness of the shell, the last transverse serial section (fig. 5) was taken at 38 mm from the beak.

ETYMOLOGY: The species is named for Dr. Zeev Lewy, Paleontology Division, Geological Survey of Israel, Jerusalem, in recognition of his valuable contributions to the Mesozoic paleontology of the Middle East.

TYPES: The holoptype (GSI M7621a) and paratype (GSI M7621b) have been placed in repository in the paleontological collection of the Geological Survey of Israel, Jerusalem.

OCCURRENCE AND AGE: Upper Callovian, *Lamberticeras lamberti* Zone, Gebel El-Minshera, northern Sinai (latitude and longitude: 30°19'N, 33°43'E; Israel grid coordinates 9690/0219).

REMARKS: This genus internally resembles *Trigonithyris* described by Muir-Wood (1935: 131) from the ?Argovian of British Somaliland. At the time, she suggested that the genus *Trigonithyris* was allied to *Pygope* and *Pygites* but later (Muir-Wood, 1965: H807) assigned it to the family Dyscoliidae, giving its stratigraphic position as Upper Jurassic (?Oxfordian).

The transverse serial sections of the type species *Trigonithyris eruduwensis* given by Muir-Wood (1935: 132) were produced before the invention of the Croft serial grinding apparatus. Although poor by modern standards, her sections clearly indicate the well-developed cardinal process and the almost horizontal, ventrally convex, and extensive hinge plates which show little or no differentiation from the inner socket ridges—morphological characters which distinguish this genus from other terebratulid genera so far examined from the "Ethiopian" Faunal Province. *Goliathyris lewyi* has slightly concave to horizontal hinge plates (see fig. 5) which are very similar to *Trigonithyris eruduwensis* Muir-Wood. However, *G. lewyi* is sulcate while *T. eruduwensis* is rectimarginate with an erect beak and large pedicle foramen.

In *Goliathyris lewyi,* new species, the inner socket ridges appear to be more extensive and the cardinal process considerably more developed than is seen in the serial sections of Muir-Wood's genus. This may be due to the difference in relative size of the two specimens representing the species *G. lewyi* and *T. eruduwensis.* Muir-Wood's serial sections were from a small, adult form, whereas those of the very large *G. lewyi* specimen are of an old individual showing some evidence of gerontic thickening of the valves and cardinalia.

Although in its external morphology, especially its zeilleriid beak, *Goliathyris lewyi* is suggestive of some terebratulidae, such as *Rugitela impressa* (Von Buch) and *R. bernardina* (D'Orbigny), it differs from these species in its less lobate outline and deeper sulcus. *R. impressa* has marked mesothyridid beak ridges and a distinct median septum in the dorsal valve, both of which are absent in *G. lewyi.*

R. bernardina (D'Orbigny) also differs in this way from *G. lewyi* and has shallower valves, an acuminate to subpyriform outline, and a shallower sulcus.

Weir (1929) described a new terebratulid species as *Aulacothyris jubaensis* Weir (which is also figured in Stefanini, 1931), from the Juba Limestone of Dakatch, Somaliland, ranging in age from Callovian to Corralian. Although assigned to the terebratellacean genus *Aulacothyris,* Weir's species *jubaensis* shows little of the generic characters associated with that genus apart from a similarity in general morphological outline, as is seen in the original description:

> Shell obovate-subpentagonal in outline, somewhat acuminate towards the anterior extremity, which is regularly rounded and not re-entrant. In the region of greatest width, which is situated rather nearer to the posterior than to the anterior extremity of the shell, the lateral margins are subangular. Dorsal valve sulcate, the depression becoming more pronounced towards the anterior end; lateral surfaces of the dorsal valve convex, depressed to the level of the commissures. Ventral valve deep, the keel somewhat acutely rounded. Beak. short, thick, closely pressed to the dorsal umbo, concealing the symphytium.

Weir (op. cit.) contended that this species was closely related to "*A. curvifrons*" a species which is currently assigned to the terebratulid genus *Pseudoglossothyris.* In general outline it has much in common with *Goliathyris lewyi,* new species, but differs from this species in its more acuminate anterior, having the dorsal valve more depressed toward the lateral margins and the pedicle foramen closer to the dorsal umbo concealing the symphytium. Also, its point of greatest width is near the posterior extremity and the lateral margin subangular.

DISCUSSION: Brachiopods from the "Ethiopian" Faunal Province are in need of study from both a taxonomic and paleobiogeographic point of view. Muir-Wood's (1934, 1935) studies on Middle Jurassic brachiopods, and especially those from the "Ethiopian" Faunal Province, need extensive revision before significant paleobiogeographic analysis can be undertaken. Evidence from East Africa suggests that there was a general faunal gradient in a north-south direction. In Callovian times the bivalve *Neocrassina unilateralis* was a southerly occurring species which was replaced northward by *Neocrassina scytalis.* These changes are believed to be partly temperature controlled and are supported by the occurrence of coral-*Diceras* deposits in the Middle Jurassic of Ethiopia but not further south.

The distribution of brachiopods follows these trends. An increased diversity of species occurs in the Callovian of Somaliland compared with beds of the same age from Tanzania while the brachiopods of Kutch, India, are somewhat intermediate in morphology. Further study of the Jurassic brachiopods of northern Sinai, East Africa, Madagascar, and India will aid in reconstructing the early history of the "Ethiopian" Faunal Province since the breakup of Gondwanaland. Since northern Sinai is situated at the tip of this province, the Sinai faunas are likely to include species belonging to the area within the equatorial Tethyan Realm which serves as a link between the European faunas and those of Afro-Indian origin.

ACKNOWLEDGMENTS

We thank Drs. C. H. C. Brunton, British Museum (Natural History), London, G. A. Cooper, National Museum of Natural History, Washington, D.C., and Z. Lewy, Paleontology Division, Geological Survey of Israel, Jerusalem, for discussion, comments, and critical review of the manuscript. Feldman is grateful to Dr. Y. Druckman, Head, Stratigraphy, Mapping and Oil Division, Geological Survey of Israel, for providing laboratory facilities and office space during his tenure as Visiting Scientist in Israel where a portion of the research for this paper was conducted. Feldman also acknowledges the assistance of Dr. F. Hirsch, Mapping Division, Geological Survey of Israel, for organizing field expeditions to the Sinai Peninsula and assisting in the interpretation of the complex regional stratigraphy.

The research was funded by EARTHWATCH and the Center for Field Research, Watertown, Mass., the Explorers Club Exploration Fund, and the National Geographic Society under grant No. 2868-84 to Feldman.

REFERENCES

Ager, D. V. 1973. Mesozoic Brachiopoda. In A. Hallam (ed.), *Atlas of paleobiogeography*, 413-436. Amsterdam: Elsevier.

Arkell, W. J. 1952. Jurassic ammonites from Jebel Tuwayq, central Arabia. *Philosophical Transactions, Royal Society of London* ser. B, 236: 241-313.

------. 1956. *Jurassic geology of the world*. Edinburgh: Oliver and Boyd.

Cooper, G. Arthur. 1983. The Terebratulacea (Brachiopoda), Triassic to Recent: A study of the brachidia (loops). *Smithsonian Contributions Paleobiology* 50: 1-445.

Hennig, W. 1966. *Phylogenetic systematics*. Urbana: University of Illinois Press.

Imlay, R. 1970. *Some Jurassic ammonites from central Saudi Arabia*. U.S. Geological Survey Professional Paper 643-D: DI-DI7.

King, W. 1850. *A monograph of the Permian fossils of England.* Palaeontographic Society, Monograph, 3: 1-258.

Kitchin, F. L. 1912. *Palaeontological work: England and Wales: summation of programme for* 1911. *Memoir of the Geological Survey of Great Britain and Museum of Practical Geology*, 59-60.

Muir-Wood, H. M. 1934. On the internal structure of some Mesozoic Brachiopoda. *Philosophical Transactions, Royal Society of London* ser. B, 223: 511-567.

------. 1935. *The Mesozoic palaeontology of British Somaliland. Pt. 7, Jurassic Brachiopoda*, 75-147, pls. 8-13. London: Government of the Somaliland Protectorate.

------. 1965. Mesozoic and Cenozoic Terebratulidina. In R. C. Moore (ed.), *Treatise on invertebrate paleontology part H, Brachiopoda*, H762-H816. Lawrence, KS: Geological Society of America and University of Kansas Press.

Schaeffer, B., M. K. Hecht, and N. Eldredge. 1972. Paleontology and phylogeny. *Evolutionary Biology* 6: 31-46.

Stefanini, Guiseppe 1931. Echinodermi, Vermi, Briozoi e Brachiopodi del Giuralias della Somalia. *Palaeontographia Italica* 32: 81-141.

Weir, J. 1925. Brachiopoda, Lamellibranchiata, Gastropoda and Belemnites. In The collection of fossils and rocks from Somaliland made by Mssrs. B. K. N. Wyllie and W. R. Smellie, *Monographs of the Geological Department of the Hunterian Museum, Glasgow University* 1(6): 79-110, pls. 11-14.

------. 1929. Jurassic fossils from Jubaland, East Africa. Collected by V. J. Glenday, and the Jurassic geology of Somaliland III. *Monographs of the Geological Department of the Hunterian Museum, Glasgow University*, 3: 1-63, pls. 1-5.

Wiley, E. O. 1981. *Phylogenetics: The theory and practice of phylogenetic systematics.* New York: Wiley.

CHAPTER FOUR

Brachiopods from the Jurassic of Gebel El-Maghara, Northern Sinai

ABSTRACT

This study is part of a taxonomic revision of the brachiopod faunas of the Middle Eastern "Ethiopian" Faunal Province, specifically Egypt (northern Sinai) and Israel. As a result of these studies we will be able to establish a biogeographic history of the brachiopods along the Tethyan margin, gain insight into the structure of the various brachiopod-dominated marine communities, and study the evolution of different brachiopod stocks through the Middle Jurassic.

Brachiopods were collected from approximately 2000 m of Jurassic (Dogger) limestones, shales, and sandstones at Gebel El-Maghara, northern Sinai. The sample area lies at the junction of the Indo-African and Tethyan faunal realms and consists of breached anticlines similar to those found throughout the Negev, Transjordan, the Lebanon, and the Antilebanon. No modern study of the brachiopods of this area has been undertaken within the last 65 years. Fifteen brachiopod species (11 rhynchonellids, 4 terebratulids) are reported, including the following new species: *Burmirhynchia cooperi, Avonothyris variabilis, Bihenithyris pyriformis,* and *Kutchithyris parnesi.*

INTRODUCTION

This project is a continuation of work undertaken by the present authors in northern Sinai (Feldman, 1987; Feldman and Owen, 1988; Feldman et al., 1982) in order to complete a taxonomic revision of the brachiopod faunas which will help us establish the early history of brachiopod species and their evolution within the "Ethiopian" Faunal Province. We expect to investigate the distribution of brachiopod species across faunal realm boundaries (e.g., the Indo-African Faunal Realm) and study the biogeographic history of the area as well as the structure of its marine communities.

The brachiopod, molluscan, and echinoderm faunas of Sinai were studied by Douvillé (1916), the Callovian gastropods and bivalves by Cossmann (in Douvillé, 1925) and Hirsch (1979), and the ammonites by Arkell (1952) and Parnes (1974). Structural, stratigraphic, and mapping studies were carried out by Range (1920), Moon and Sadek (1921), Hoppe (1922), Farag (1959), Al Far (1966), and Goldberg et al. (1971). Additional detailed work was done at

FIGURE 1: Geographic location of sample area (X).

Gebel El-Minshera by Farag and Shata (1954), Farag and Omara (1955), and Bartov and Freund (1968).

Farag (1957, 1959) and Farag and Gatinaud (1960a, 1960b) studied the brachiopods of northern Sinai, and their work is noted herein and revised as necessary. Farag's (1959) faunal list, to a large extent, is taken from the works of Douvillé (1916) and Arkell (1952). Feldman (1987) described a new species of the rare genus *Septirhynchia* from the Callovian of Gebel El-Maghara and Feldman and Owen (1988) erected a new terebratulid genus and species (*Goliathyris lewyi*) from the Callovian of Gebel El-Minshera. Cooper's (1989) monograph on the Jurassic brachiopods from Saudi Arabia was invaluable in understanding the taxonomy of the northern Sinai faunas.

STRATIGRAPHIC SETTING

Gebel El-Maghara (figs. 1, 2), a classic Pliensbachian-Oxfordian section approximately 2000 m thick, lies at the junction of the Indo-African and Tethyan faunal realms in the Sinai Peninsula (Picard and Hirsch, 1987). Strata were exposed for study by erosion funnels uncovering the cores of anticlines and major upwarps which can be found throughout the Sinai, Negev, Samaria, Transjordan, the Lebanon, and the Antilebanon.

The stratigraphy of Gebel El-Maghara used here is based on the work of Al Far (1966), Goldberg et al. (1971), Picard and Hirsch (1987), and our own field observations (fig. 3). In the discussion below, subunit numbers refer to those of Picard and Hirsch (1987).

Pliensbachian to Toarcian: Goldberg et al. (1971) measured and lithologically described subunits 1-6, which correlate with Al Far's (1966) Mashabba Formation. The strata outcrop south of Shushet el Maghara, in Wadi (Sadd el-) Mashabba, and are divided into two parts, each about 40-50 m thick. The lower part (subunits 1-4) consists of poorly fossiliferous reddish, coarse-to-medium grain crossbedded sandstone interdigitated with fossiliferous shale and limestone. The upper part (subunits 5 and 6) consists of algal or oncolitic pelmicrites with some shale that contain occasional brachiopods and corals.

The Rajabiah Formation overlies the Mashabba Formation and, in the lower part (subunits 7-9), consists of organogenic coralgal pelmicritic to oolitic-oncolitic limestones with abundant solitary corals, brachiopods, and pelecypods. Subunit 10 is mainly sandstone, shale, and minor limestone. The middle part (subunits 11-13) is lithologically similar to the lower part but differs in that subunit 13 forms a conspicuous morphological scarp (22 m thick). The upper

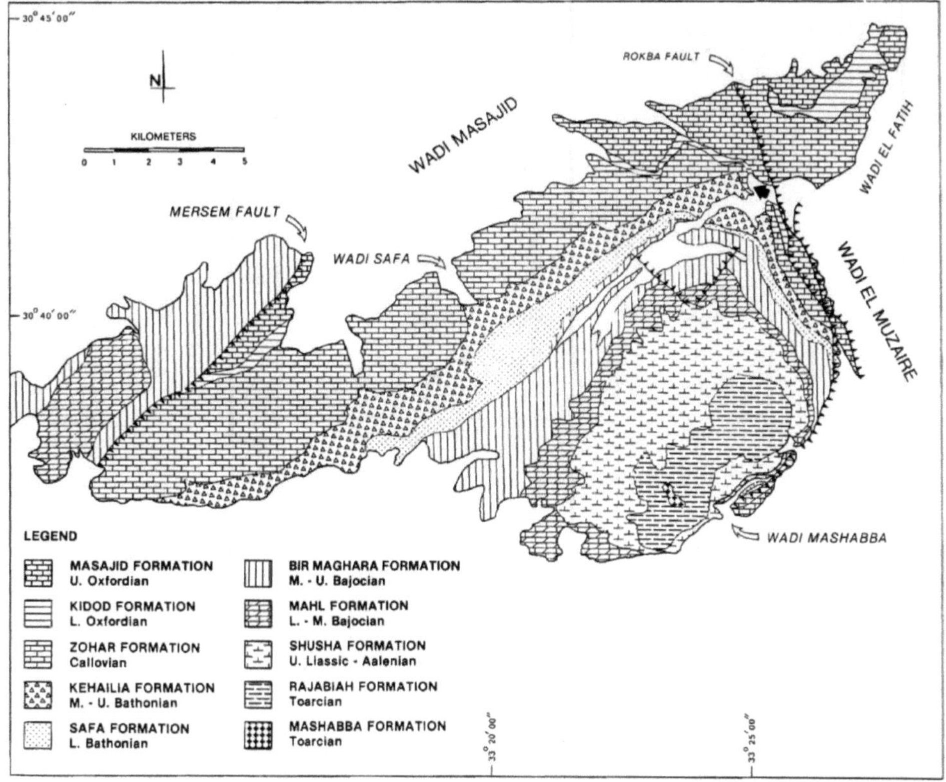

FIGURE 2: Locality map showing the breached anticlinal domes at Gebel El-Maghara, northern Sinai. Black arrow represents location of *Septirhynchia*-bearing beds of Feldman (1987).

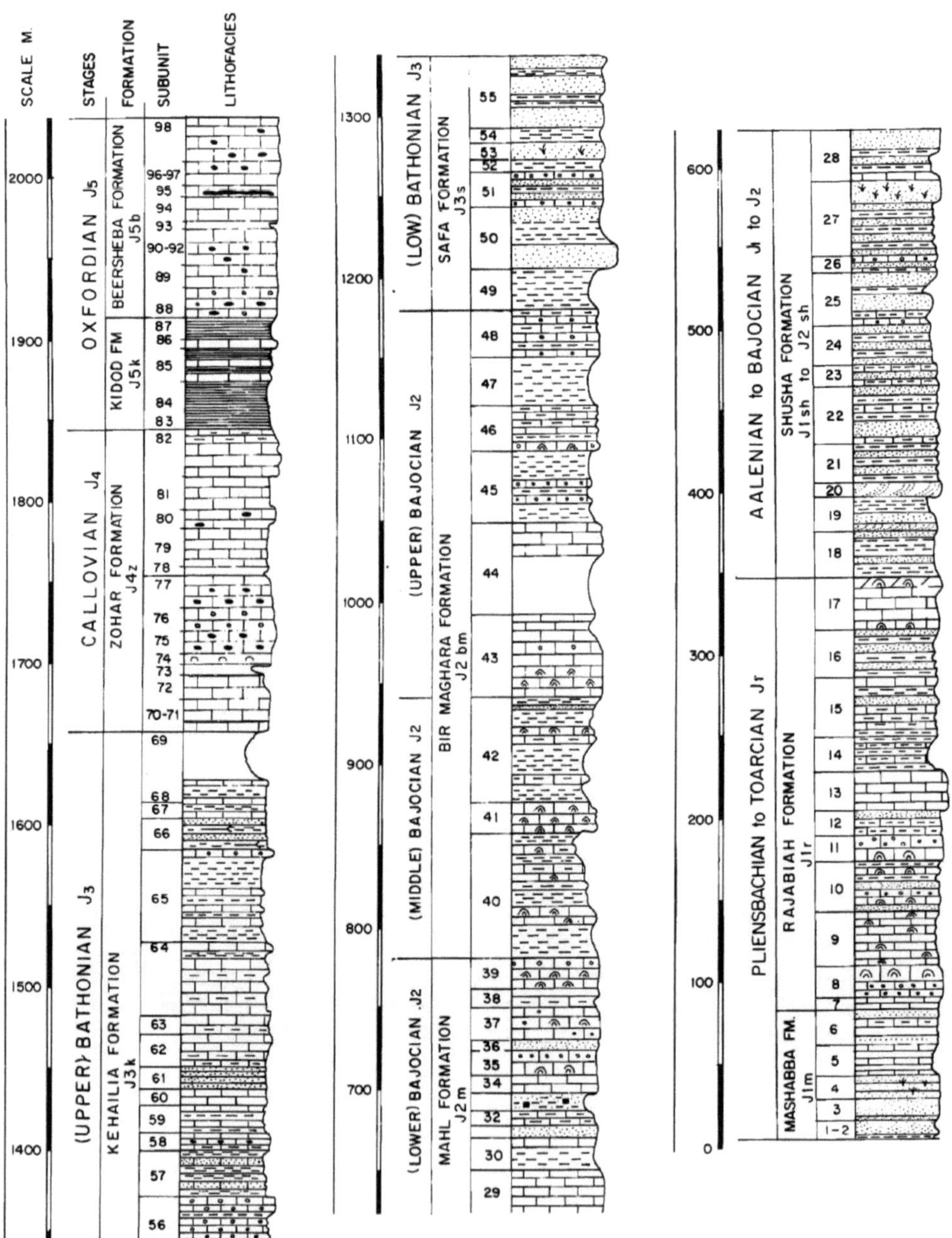

FIGURE 3: Columnar section of Gebel El-Maghara, northern Sinai (modified from Picard and Hirsch, 1987, and used with permission of the Israel Academy of Sciences and Humanities).

part (subunits 14-17) contains less carbonate and more of a sand-shale sequence with abundant brachiopods and bivalves.

Aalenian to Lower Bajocian: The Shusha Formation (subunits 18-28) overlies the Rajabiah Formation. It consists of coarse grained, often crossbedded, hematitic sandstones which are occasionally interlayered with oncolitic-oolitic, calcarenitic-dolomitic strata with some plant-bearing shales, sandy limestones, and silty mudstones. Goldberg et al. (1971) noted the occurrence of plant remains in nearly all subunits of the Shusha Formation.

Middle to Upper Bajocian: This sequence was divided by Al Far (1966) into three members (Mahl, Mowerib, Bir) of the Bir Maghara Formation but revision by Picard and Hirsch (1987) resulted in several changes noted below. The Mahl Member was raised to the rank of formation and the Mowerib and Bir members were combined into the Bir Maghara Formation.

The Mahl Formation (subunits 29-39) consists of a lower part of massive oolitic-oncolitic coralline limestone with a sandstone unit (subunit 36) and shale unit (subunit 38). Throughout the formation, relatively diverse faunas of brachiopods, pelecypods, gastropods, ammonites, corals, and plant remains are found.

The Bir Maghara Formation (subunits 40-48) begins with the Middle Bajocian *Dorsetensia* beds (subunit 40) and continues to the top of the Upper Bajocian *Ermoceras* beds (subunit 48). The lower part consists of oncolitic limestones, occasionally dolomitic and sandy, with some interbedded variegated shale, salt, and gypsum crusts grading into a dense, brown-gray organogenic lime unit. Above this lies a gray-green shale unit with interbedded limestone, followed by two limestone units and another shale sequence, which is capped by a brown oolitic-oncolitic limestone with interbedded shale. The Bir Maghara Formation outcrops in the Shushet el Maghara upwarp and near the Mersem Fault where it comes into direct contact with Callovian and Oxfordian strata.

Bathonian: The Lower Bathonian Safa Formation is an alternating sequence of sandstone and shale (subunits 49-55) with crossbedded, hematitic sands alternating with limonitic-stained sandy shales and several lenticular coal seams. The fauna is relatively poor, perhaps because of the nonmarine conditions of deposition as indicated by an absence of marine sediments (except for two 2-3 m thick oolitic limestone beds in subunit 54). The Upper Bathonian Kehailia Formation (subunits 56-69) consists, in the lower part (subunits 56-61) of shales and oolitic carbonates with some sandstone interbeds. Subunits 62-64 have no sandstone but are composed of carbonates and some shale, while subunits 65-68 consist of shales with some thin oolitic-oncolitic) limestone beds and occasional thin sandstone interbeds. Stromatoporoids were first recognized by Goldberg et al. (1971) in subunit 67, thus indicating proximity to the Callovian-Oxfordian facies typical of this group.

Callovian: The Zohar Formation (subunits 70-82) consists of massive bedded lithographic and calcarenitic, occasionally oolitic-oncolitic limestones showing karstic features. Silicification is prevalent as indicated by nodular and layered cherts and flints as well as a silicified fauna of stromatoporoids, molluscs, and brachiopods, particularly the large rhynchonellid *Septirhynchia* (Feldman, 1987). Lewy (198la, 1981b) recognized an unconformity between earliest and late Callovian strata based on the absence of Middle Callovian ammonites. Picard and Hirsch (1987) noted a hardground contact surface between the Callovian (Zohar) Limestone and the overlying Oxfordian Kidod shales exposed in Wadi Abu Gaza and Gebel Aroussiah as well as in Mount Hermon's Majdal Shams section. They interpreted this as a short depositional break due to epeirogenic movements.

Oxfordian: The Lower Oxfordian Kidod Formation (subunits 83-87) outcrops on the northeast flank of the Maghara upwarp between the Mersem Fault and Wadi Tauriat and consists of greenish to reddish-brown shales, calcareous at the base and limonitic, with concretions at the top. Subunit 85 is composed of alternating greenish-brown, soft shale and brown, argillaceous, glauconitic thin-bedded limestone. There is one gray, argillaceous limestone unit (subunit 86) bearing a molluscan fauna. The Upper Oxfordian Beersheba Formation (subunits 88-98), exposed in Wadi Abu Gaza, Gebel Rokba, and north of Wadi Tauriat consists of light brown to light gray limestones with lenticular flint and layered chert. As in the Callovian Zohar Formation, there is silicification of faunal constituents (stromatoporoids, corals, molluscs, echinoid spines), although fossils are absent from the middle subunits of the formation and more numerous at the top. North of Wadi Abu Gaza and northeast of Tauriat, yellow marls are interbedded with flinty micrites.

ABBREVIATIONS

Institutions and Localities

AMNH = American Museum of Natural History
BMNH = British Museum of Natural History
GSI = Geological Survey of Israel
USNM = Smithsonian Institution, United States National Museum

Measurements

(L) maximum length of shell
(W) maximum width of shell
(T) maximum thickness of shell

SYSTEMATIC PALEONTOLOGY

PHYLUM BRACHIOPODA
CLASS ARTICULATA HUXLEY, 1869
ORDER RHYNCHONELLIDA KUHN, 1949
SUPERFAMILY RHYNCHONELLOIDEA GRAY, 1848
FAMILY RHYNCHONELLIDAE GRAY, 1848
SUBFAMILY TETRARHYNCHIINAE AGER, 1965

GENUS *SOMALIRHYNCHIA* WEIR, 1925
Somalirhynchia africana Weir, 1925
Figures 4A-F

Somalirhynchia africana Weir, 1925: 80, pl. 12, figs. 20-23. Muir-Wood, 1935: 94, pl. 10, figs. 7a-c, text figs. 7, 8. Ager, 1965: H614, figs. 497, 9a, b. Dubar, 1967: 30, pl. 2, figs. 5a, b. Abbate et al., 1974: 439, pl. 39, fig. 4. Cooper, 1989: 58, pl. 12, figs. 37-41.

TYPE SPECIES: *Somalirhynchia africana* Weir, 1925, emended Muir-Wood, 1935.

LECTOTYPE: Selected by Muir-Wood, 1935: Wylie Coll. Hunterian Museum, Glasgow University (Ll393a).

HORIZON: Subunits 33, 43, 47, 53, 54, 64, 68, 82, Gebel El-Maghara, Sinai.

GEOLOGICAL OCCURRENCE: Lower Bajocian to Oxfordian.

REMARKS: Complete series of transverse serial sections of the genus *Somalirhynchia* are rarely given in systematic descriptions and interpretations of the internal structures of the type species are often taken from Muir-Wood's (1935: 95, fig. 8) original series which, in our opinion, falls short of present day standards. A more complete and up-to-date series of a recognizable species belonging to the genus is shown in a series of transverse serial sections figured by Dubar (1967: 33, fig. 7) featuring *Somalirhynchia smelliei* Weir from the Oxfordian of Tunisia and illustrating the well-developed septalium and high median septum in the dorsal valve and more clearly defined hinge plates and crural bases.

Somalirhynchia africana var. *smelliei* Weir is one of several varieties or variants which are sometimes found with the type species, ranging from a more transversely oval general outline to more produced umbonal features, more numerous and less deeply incised costae, and broader and lower arcuate anterior fold and sulcus.

A good example of the type species *Somalirhynchia africana* Weir collected from subunit 82 (GSI M6922), Gebel El-Maghara, is figured here (fig. 4A-C) and is one of 25 specimens from that horizon and locality showing the full range of variation in costation, convexity, and general outline (table 1). Other examples which fall within this range have been collected from Upper Callovian subunits 33 and 43 at Hamakhtesh Hagadol in the Negev, southern Israel, and one is figured here (fig. 4D-F). The species also occurs in a similar horizon at Mount Hermon, northern Israel.

Specimens figured by Dubar (1967) as *Somalirhynchia* cf. *africana* Weir (figs. 6-11) from the Oxfordian of Tunisia are also considered here to be within the morphological range of the type species and can be compared to eight specimens collected by Feldman from subunit 68, Gebel El-Maghara. An additional 21 specimens belonging to this series are housed in the collections of the Geological Survey of Israel, Jerusalem.

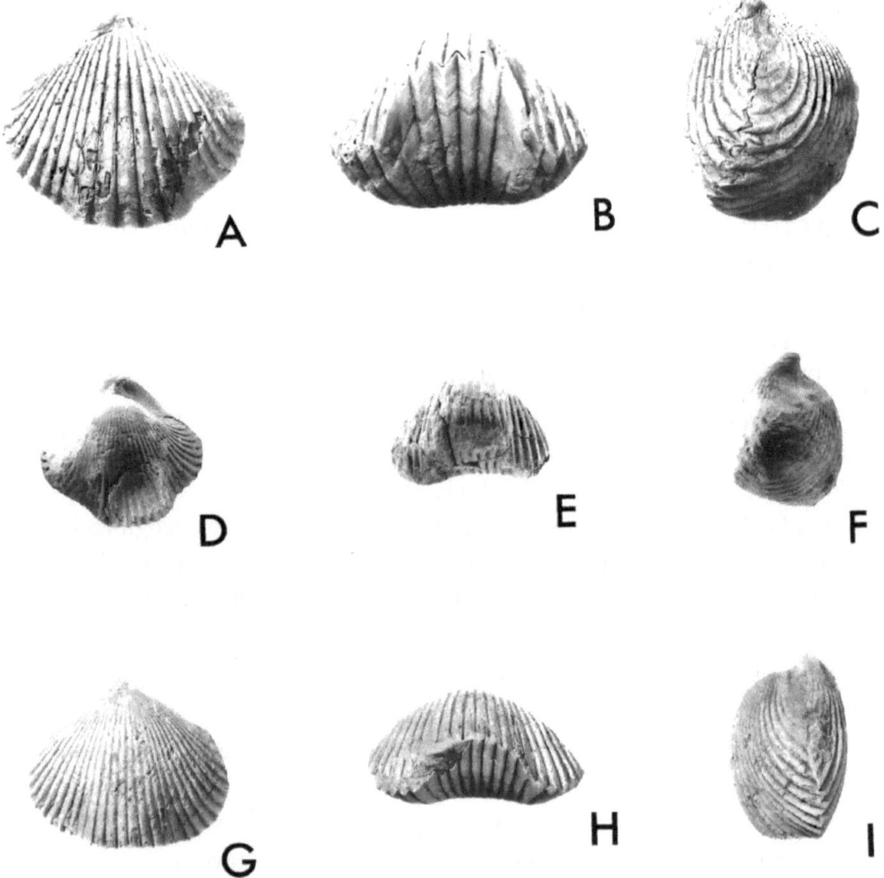

FIGURE 4: *Somalirhynchia africana* Weir, 1925. A-C, Dorsal, anterior, lateral views, GSI M6922, x1; D-F, dorsal, anterior, lateral views, AMNH 44183, x1. *Somalirhynchia bihenensis* Muir-Wood, 1935. G-I, Dorsal, anterior, lateral views, AMNH 44184, x1.

TABLE 1: Measurements (mm) of *Somalirhynchia africana* Weir, 1925

Specimen	(L)	(W)	(T)	Subunit
B. 78955[a]	32.5	34.7	23.5	
GSI M6922[b]	38.1	39.0	29.1	82[c]
AMNH 44178	39.6	44.5	30.6	82
AMNH 44179	40.0	41.1	31.7	82
AMNH 44180	41.4	39.1	32.4	82
AMNH 44181	39.9	39.5	32.8	82
AMNH 44182	37.0	41.1	24.7	82
AMNH 44183[b]	25.5	27.5	19.2	53, 54
BMNH BB.86880	35.1	38.0	27.2	48
BMNH B8.86881	32.0	33.5	23.0	48
BMNH B8.86882	34.0	37.5	30.0	48
BMNH BB.86883	35.0	44.5	32.0	48
BMNH BB.86884	33.0	38.0	29.1	48
BMNH B8.86885	33.1	34.5	28.4	48

a. Species emended by Muir-Wood (1935) who figured this specimen from Bosti, N.W. of Bihendula, Somalia. No holotype selected by Weir (1925, 1929, 1930).

b. Figured specimen.

c. Subunit numbers refer to the section at Gebel El-Maghara unless otherwise noted.

The geological range for the genus *Somalirhynchia* given by Ager (1965) is Upper Jurassic (Oxfordian). Cooper (1989: 3) gave an extended range for the genus from Lower Bathonian to Upper Kimmeridgian, but in northern Sinai the genus ranges from Lower Bajocian to Upper Callovian. Species assigned to this genus by Makridin (1955) and Childs (1969) from Upper Jurassic horizons are still in need of further investigation before these records can be established.

Somalirhynchia bihenensis
Muir-Wood, 1935
Figure 4G-I

Somalirhynchia bihenensis Muir-Wood, 19350: 101, pl. 10, figs. 5a-c.

HOLOTYPE: Natural History Museum, London, BMNH B.85409 from Bihen Pass, Somalia (formerly British Somaliland).

HORIZON: Subunit 76, Gebel El-Maghara, Sinai.

GEOLOGICAL OCCURRENCE: Lower Callovian.

REMARKS: Differing from the type species, *Somalirhynchia africana* Weir, in general outline, this species has an average of 30 rounded to subangular and less deeply incised and radiating costellae, flatter valves with a less posteriorly inflated dorsal valve, lower or less clearly defined median dorsal fold but maintaining a similar anterior profile to the type species with a broad arcuate uniplication and moderately long linguiform extension.

TABLE 2: Measurements (mm) of *Somalirhynchia bihenensis* Muir-Wood, 1935.

Specimen	(L)	(W)	(T)	Comments
BMNH B.85409	27.8	30.1	19.5	Holotype, Muir-Wood, 1935: pl. 10, fig. 5a-c Bihen Pass, Somalia.
AMNH 44184[a]	28.9	31.5	18.0	Subunit 82, Gebel El-Maghara, Sinai, Egypt.
AMNH 44185	17.1	21.3	10.9	Subunit 47, Hamakhtesh Hagadol, Negev, Israel.

a. Figured specimen.

The specimen figured here (fig. 4G-I) was collected from subunit 82 at the top of the Zohar Formation (Upper Callovian), Gebel El-Maghara. A smaller specimen, probably a juvenile, was collected from subunit 47 (Late Callovian) at Hamakhtesh Hagadol in the Negev. The shell figured herein differs slightly from the original figured by Muir-Wood (1935: pl. 10, figs. 5a-c) in having a more broadly transversely oval dorsal outline or slightly greater width (table 2). In all other aspects the species are identical.

The specimen described and figured by Dubar (1967: 29, pl. 2, fig. 16) from the Oxfordian of Tunisia has a great deal in common with the specimens figured here in both general outline and costation but differs slightly in having a more acutely inflated dorsal umbo.

Somalirhynchia arabica Cooper, 1989
Figures 5, 6

Somalirhynchia africana var. *jordanica* (Noetling, 1887) Muir-Wood, 1935: 97, pl. 10, figs. 8a, b.

Somalirhynchia cf. *jordanica* (Noetling, 1887) Dubar, 1967: 27, pl. 2, figs. 2a-c, 3a-c, text figs. 5a, b.

Somalirhynchia arabica Cooper, 1989: 58, pl. 14, figs. 1-5, pl. 15, figs. 25-35.

LOCALITY: Subunit 48, Gebel El-Maghara, Sinai.

GEOLOGICAL OCCURRENCE: Upper Bajocian.

REMARKS: The broadly triangular specimen figured here agrees with the original description and figure of the species (Cooper, 1989: 58, pl. 14, figs. 1-5) particularly in general dorsal outline. Cooper (p. 59) distinguished his species from that figured by Muir-Wood as *Somalirhynchia africana* var. *Jordanica* as having a more highly developed or stronger dorsal fold. Muir-Wood's figured specimen (1935: pl. 10, fig. 8a, b) is slightly more triangular in general outline and is shown without lateral or anterior views. The specimen figured here (figs. 5, 6; table 3) is, in our opinion, nearer in morphology to both the original specimen figured by Cooper and the specimens figured by Dubar (1967: pl. 2, figs. 2a-c, 3a-c) than to the specimen figured by Muir-Wood (1935: pl.

FIGURE 5: *Somalirhynchia arabica* Cooper, 1989. A-C, Dorsal, lateral, anterior views, GSI M4349, x1.

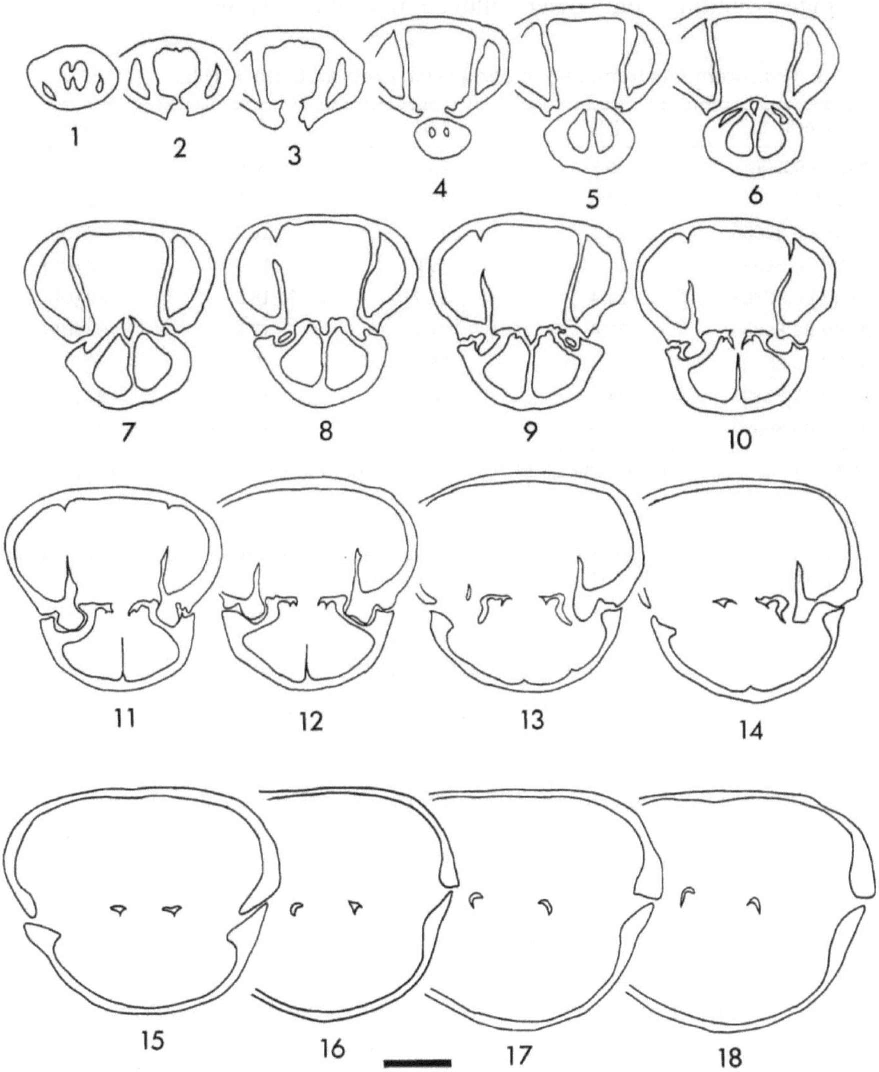

FIGURE 6: *Somalirhynchia arabica* Cooper, 1989 (numbers show distance in mm between sections and [in parentheses] distance from beak): **1,** 1.8 (1.8); **2,** 0.3 (2 . 1); **3,** 0.4 (2.5); **4,** 0.3 (2.8); **5,** 0.3 (3 . 1); **6,** 0.4 (3.5); **7,** 0.4 (3.9); **8,** 0.3 (4.2); **9,** 0.3 (4.5); **10,** 0.4 (4.9); **11,** 0.2 (5.1); **12,** 0.3 (5.4); **13,** 0.4 (5.8); **14,** 0.2 (6.0); **15,** 0.4 (6.4); **16,** 0.3 (6 .7); **17,** 0.2 (6 .9); **18,** 0.2 (7.1). Scale bar equals 5 mm; GSI M4349.

10, fig. 8a, b). Transverse serial sections given by Dubar (1967: 28, fig. 6) of *Somalirhynchia* cf. *Jordanica* have much in common with those figured by Cooper for *Somalirhynchia arabica* (1989: 59, fig. 32) particularly in the slightly ventrally convergent dental lamellae and in the position and shape of the hinge plates and crural bases.

Cooper's type and paratypes come from the Upper Dhruma Formation (Hisyan Member) and the Tuwaiq Mountain and Hanifer formations, a geological range extending from Lower Callovian to Kimmeridgian.

TABLE 3: Measurements (mm) of *Somalirhynchia arabica*, Cooper, 1989.

Specimen	(L)	(W)	(T)	Subunit
USNM 380206[a]	26.5	23.0	30.0	
GSI M4349[b]	22.0	25.0	16.5	48[c]
BMNH BB.86886	19.6	25.7	16.2	82 (UP)
BMNH BB.86887	24.7	29.0	19.9	82 (UP)
BMNH BB.86888	24.4	32.7	18.5	82 (UP)
BMNH BB.86889	21.8	27.4	17.2	82 (UP)
BMNH BB.86890	25.3	28.5	20.5	82 (LP)
BMNH BB.86891	24.2	28.8	17.7	82 (LP)
BMNH BB.86892	23.9	30.4	19.4	82 (LP)

a. Holotype, figured in Cooper, 1989, pl. 14, figs. 1-5, Upper Dhruma Formation and Tuwaiq Mountain Formation, Saudi Arabia.

b. Figured specimen.

c. Subunit numbers refer to the section at Gebel El-Maghara unless otherwise noted. (UP) refers to the upper part of subunit 82, while (LP) refers to the lower part.

GENUS *SCHIZORIA* COOPER, 1989

Schizoria elongata Cooper, 1989

Figures 7, 8

Schizoria elongata Cooper, 1989: 55, pl. 14, figs. 17-27, text fig. 31.

HORIZON: Subunits 38-40, Gebel El-Maghara, Sinai.

GEOLOGICAL OCCURRENCE: Lower to Middle Bajocian.

REMARKS: The specimen figured here (fig. 7) is a medium to large elongate rhynchonellid and is larger than the holotype (table 4). The general outline, type of costellation, and poorly defined folding of the dorsal valve are similar to features noted by Cooper (1989: pl. 14, figs. 17-27).

INTERNAL STRUCTURES: The general transverse outlines shown in the transverse serial sections (fig. 8) conform with Cooper's (1989: 55, fig. 31), differing only in the less persistent median septum in the dorsal valve and less persistent, subparallel dental lamellae in the ventral valve. A fairly deep, well-developed septalium is noted in the specimen sectioned here from Gebel El-Maghara, a structure not seen in the series of species sectioned by Cooper (fig. 31). In the description, however, Cooper states (p. 56) that a moderately large septalium is developed in the brachial valve, presumably between sections 9 and 10. It should be noted that the Sinai specimens are larger than Cooper's.

Cooper's holotype and paratypes of *Schizoria elongata* come from the Lower Dhruma Formation (between the *Dorsetensia* and *Ermoceras* zones) of Saudi Arabia, equivalent to a stratigraphic range of Lower to Upper Bajocian.

TABLE 4: Measurements (mm) of *Schizoria elongata*, Cooper, 1989.

Specimen	(L)	(W)	(T)	Subunit
USNM 380260b[a]	12.7	10.7	10.2	
GSI M6823[b]	26.5	21.0	18.4	39[c]
AMNH 44186	25.5	18.2	17.8	39
AMNH 44187	25.2	20.3	18.0	39
AMNH 44188	22.1	16.4	15.7	39
AMNH 44189	26.1	21.7	16.7	39
AMNH 44190	24.1	21.4	18.8	39
AMNH 44191	25.0	20.9	15.0	39
AMNH 44166	23.7	19.3	15.5	40
AMNH 44167	27.6	24.1	15.1	40
AMNH 44168	25.7	21.0	17.5	40
BMNH BB.86893	25.7	20.2	17.6	40
BMNH BB.86894	25.7	19.4	15.4	40
BMNH BB.86895	23.7	18.1	18.2	40
BMNH BB.86896	27.2	23.5	16.9	40

a. Holotype, figured by Cooper, 1989, pl. 14, figs. 17-21, from Lower Dhruma Formation, Saudi Arabia.
b. Figured specimen.
c. Subunit numbers refer to the section at Gebel El-Maghara unless otherwise noted.

Schizoria elongata differs from *Sphenorhynchia plicatella,* the type species of that genus, in its more elongate-oval general outline, more evenly biconvex valves, shorter, more massive umbo, and less numerous costellae.

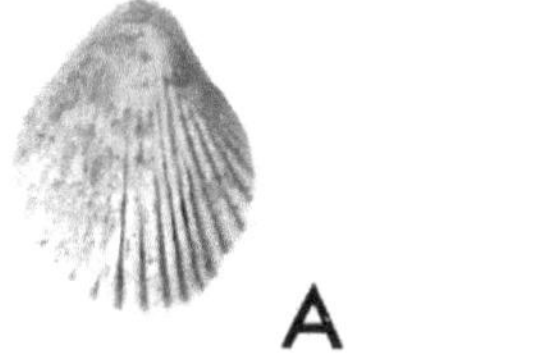

A B C

FIGURE 7: *Schizoria elongata* Cooper, 1989. A-C, Dorsal, lateral, anterior views, GSI M6823, x1.

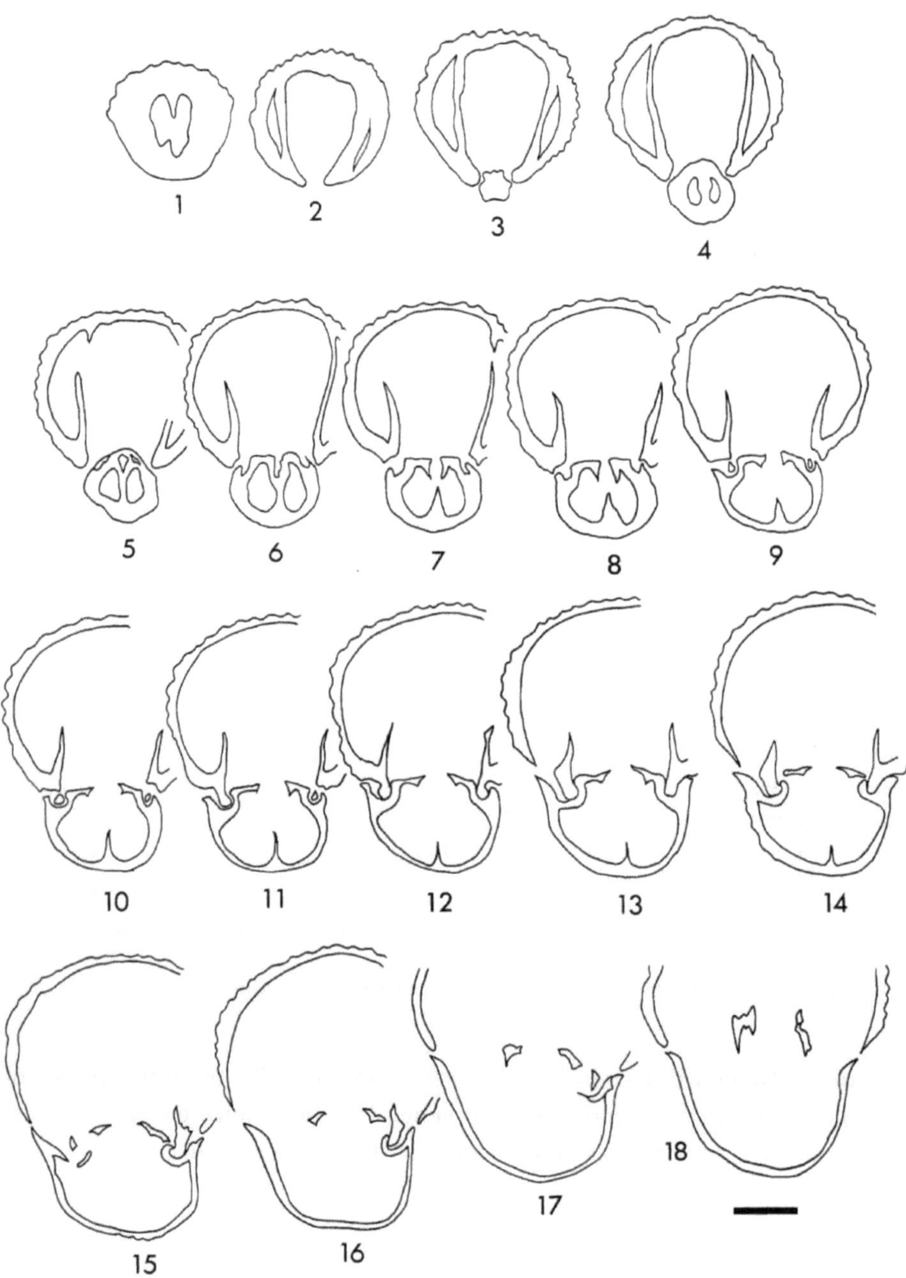

FIGURE 8: *Schizoria elongata* Cooper, 1989 (numbers show distance in mm between sections and [in parentheses] distance from beak): **1**, 2.5 (2 .5); **2**, 0.4 (2.9); **3**, 0.6 (3.6); **4**, 0.4 (3.9); **5**, 0.3 (4.2); **6**, 0.5 (4.7); **7**, 0.4 (5 . 1); **8**, 0.4 (5.5); **9**, 0.3 (5.8); **10**, 0.3 (6.1); **11**, 0.4 (6.5); **12**, 0.3 (6.8); **13**, 0.4 (7.2); **14**, 0.3 (7 .5); **15**, 0.4 (7 .9); **16**, 0.4 (8.3); **17**, 0.3 (8 .6); **18**, 0.2 (8.8); all structures gone after 8.8 mm; scale bar equals 5 mm; GSI M6823.

GENUS *PYCNORIA* COOPER, 1989

Pycnoria compacta Cooper, 1989
Figures 9, 10

Pycnoria compacta Cooper, 1989: 51-53, pl. 12, figs. 6-10.

LOCALITY: Subunit 59, Gebel El-Maghara, Sinai.

GEOLOGICAL OCCURRENCE: Upper Bathonian.

REMARKS: Cooper described two species for his new genus *Pycnoria, P. compacta* and *P. magna.* The specimen figured here is a large (table 5), coarsely costate rhynchonellid. It was collected from subunit 59, considered to be an Upper Bathonian equivalent by European standards. Cooper's specimens are stated (1989: 52) to have come from the Dhruma Formation (zone not specified) and are probably of similar Upper Bathonian age.

TABLE 5: Measurements (mm) of *Pycnoria compacta*, Cooper, 1989.

Specimen	(L)	(W)	(T)	Subunit
USNM 380453[a]	19.0	18.0	16.3	
GSI M6891[b]	21.0	19.6	22.1	59[c]
AMNH 44192	20.1	20.0	20.2	59
AMNH 44193	27.8	19.6	20.9	59
BMNH BB.86897	23.9	25.4	19.3	57, 58
BMNH BB.86898	18.5	19.6	15.4	57, 58
BMNH BB.86899	17.8	18.9	13.7	57, 58
BMNH BB.86900	17.6	17.4	11.9	57, 58
BMNH BB.86901	22.7	23.8	19.8	57, 58

a. Holotype, figured by Cooper, 1989, pl. 12, figs. 6-10, from Dhruma Formation, Saudi Arabia.
b. Figured specimen.
c. Subunit numbers refer to the section at Gebel E1-Maghara unless otherwise noted.

Our specimen (fig. 9) is slightly more acutely triangular in dorsal outline with a more highly developed median dorsal fold and subsequent longer linguiform extension. The 12 costae on each valve are sharp, angular, and deeply incised with two on the fold and three in the ventral sulcus.

A

B

C

FIGURE 9: *Pycnoria compacta* Cooper, 1989. A-C, Dorsal, lateral, anterior views, GSI M6891, x1.

After making a permanent cast, a series of transverse serial sections (fig. 10) was made of the specimen figured here (fig. 9) that allows comparison with Cooper's (1989: figs. 29, 30) specimens of *Pycnoria magna,* the type species. Although Cooper's specimens lack details of hinge plates and crural bases, it seems from the general transverse outline, the nature of the dental lamellae, height, persistence of the median septum, and subquadrate shape of the hinge

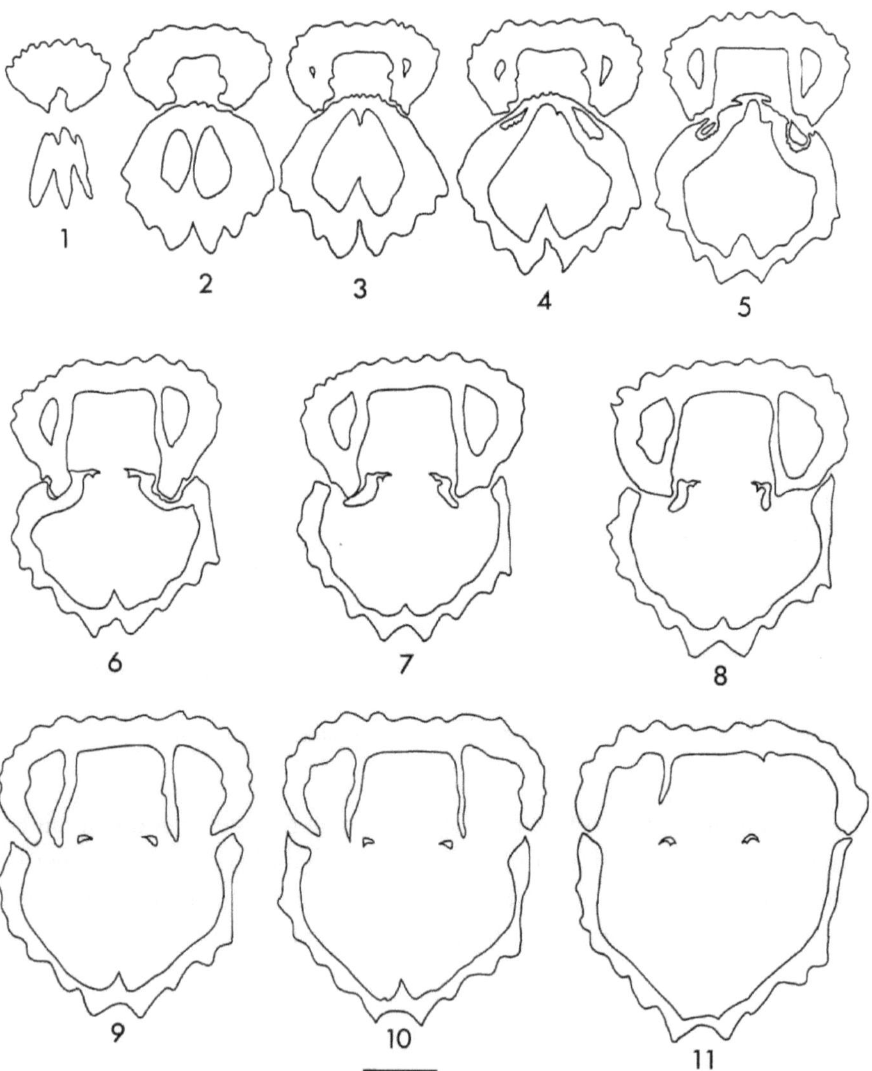

FIGURE 10: *Pycnoria compacta* Cooper, 1989 (numbers show distance in mm between sections and [in parentheses] distance from beak): **1,** 2.4 (2.4); **2,** 0.6 (3 .0); **3,** 0.4 (3.4); **4,** 0.4 (3.8); **5,** 0.3 (4.1); **6,** 0.4 (4 .5); **7,** 0.3 (4.8); **8,** 0.5 (5.3); **9,** 0.4 (5.7); **10,** 0.3 (6 .0); **11,** 0.4 (6.4). Scale bar equals 5 mm; GSI M6891.

teeth (fig. 30) that the two forms agree in details shown in Cooper's transverse serial sections.

Although our specimen is assigned here to the species *P. compacta*, it is possible that it could fall within the parameters of variation described for the species *P. magna*.

Pycnoria magna Cooper, 1989
Figure 11 A-C

Pycnoria magna Cooper, 1989: 53, pl. 12, figs. 11-36; pl. 18, figs. 26-36.

LOCALITY: Subunit 57, Gebel El-Maghara, Sinai.

GEOLOGICAL OCCURRENCE: Upper Bathonian.

REMARKS: This is the larger of the two species described by Cooper for the genus *Pycnoria* and is the type species for the genus (table 6). It is diagnosed as large, wide, and strongly costate.

The specimen figured here has 13 strong, angular, deeply incised costae with four on a well-defined arcuate median dorsal fold and three in a shallow, comparatively narrow sulcus. It is one of several specimens collected from Gebel Maghara which include most of the minor variations depicted by Cooper in his illustrations of the species (1989: pls. 12, 18). The specimen departs slightly from the typical form in having a more highly developed median dorsal fold and a slightly wider and more broadly arcuate anterior commissure. We consider these points to be minor morphological variations seen within the full range for the species which should not be allowed to separate it from the type species that we consider stratigraphically important.

INTERNAL STRUCTURES: As for the species *Pycnoria compacta* Cooper, here described.

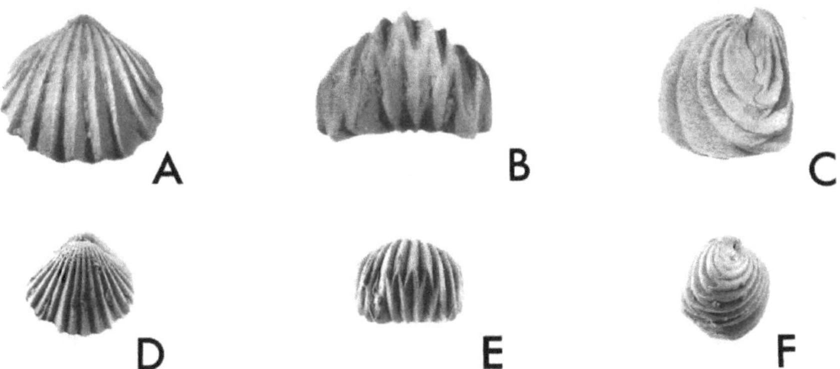

FIGURE 11: *Pycnoria magna* Cooper, 1989. A-C, Dorsal, anterior, lateral views, AMNH 44194, x1; *Globirhynchia dubia* Cooper, 1989. D-F, Dorsal, anterior, lateral views, GSI M6900, x1.

TABLE 6: Measurements (mm) of *Pycnoria magna* Cooper, 1989.

Specimen	(L)	(W)	(T)	Subunit
USNM 380215a[a]	24.0	24.2	11.2	
AMNH 44194[b]	25.9	28.0	20.7	57[c]
AMNH 44195	25.4	28.0	20.7	57
AMNH 44196	23.9	25.7	21.3	57
AMNH 44169	24.9	27.3	21.2	57
AMNH 44170	26.4	29.8	20.8	57
BMNH BB.86902	26.8	27.7	23.7	57
BMNH BB.86903	25.8	28.6	21.2	57
BMNH BB.86904	25.0	27.6	21.8	57

a. Holotype, figured by Cooper, 1989, pl. 12, figs. 11-36; pl. 18, figs. 26-36, from Dhruma Formation, Saudi Arabia.

b. Figured specimen.

c. Subunit numbers refer to the section at Gebel El-Maghara unless otherwise noted.

TABLE 7: Measurements (mm) of *Globirhynchia dubia* Cooper, 1989; *G. crassa* Cooper, 1989; *G. subtriangulata* Cooper, 1989.

Specimen	(L)	(W)	(T)	Subunit
Globirhynchia dubia Cooper, 1989				
GSI M6900[a]	17.0	18.6	14.9	59, 60[b]
BMNH BB.86905	21.0	20.1	15.5	59, 60
BMNH BB.86906	18.2	18.8	14.9	59, 60
BMNH BB.86907	17.0	18.1	13.9	59, 60
BMNH BB.86908	16.1	16.0	11.9	59, 60
Globirhynchia crassa Cooper, 1989				
AMNH 44197[a]	19.2	19.0	15.1	59, 60
BMNH BB.86909	14.9	14.7	11.0	59, 60
BMNH BB.86910	15.2	15.1	12.0	59, 60
BMNH BB.86911	14.2	14.0	10.1	59, 60
Globirhynchia subtriangulata Cooper, 1989				
AMNH 44198[a]	18.9	17.6	12.8	59, 60[b]
BMNH BB.86912	17.3	19.9	10.3	59, 60
BMNH BB.86913	17.4	16.9	9.5	59, 60
BMNH BB.86914	18.9	20.1	12.7	59, 60
BMNH BB.86915	19.5	20.9	14.9	59, 60
BMNH BB.8916	15.9	17.5	16.1	59, 60
BMNH BB.86917	18.1	19.3	17.9	59, 60
BMNH BB.86918	17.5	17.9	13.7	59, 60
BMNH BB.86919	19.5	20.9	14.7	59, 60

a. Figured specimen.

b. Subunit numbers refer to the section at Gebel El-Maghara unless otherwise noted.

GENUS *GLOBIRHYNCHIA* BUCKMAN, 1915

Globirhynchia dubia Cooper, 1989

Figures 11 D-F, 12

Globirhynchia dubia Cooper, 1989: 40, pl. 10, figs. 31-36.

LOCALITY: Subunits 59, 60, 64, Gebel El-Maghara, Sinai.

GEOLOGICAL OCCURRENCE: Upper Bathonian.

DESCRIPTION: Medium size (table 7), coarsely costate rhynchonellid; biconvex with dorsal valve more acutely inflated posteriorly producing subquadrate to broadly oval lateral profile; dorsal valve well developed with five angular, deeply incised costae among 17 ornamenting valves; ventral valve has similar number of costae, four of which occupy broad, moderately extensive, linguiform extension anteriorly, producing shallow sulcus.

INTERNAL CHARACTERS: Transverse serial sections (fig. 12) made from specimens identified as *Globirhynchia? dubia* Cooper exactly match those given by the author (Cooper, 1989: pl. 41, fig. 21) for *G. subtriangulata* and also compare favorably with those of the type species *G. subobsoleta* from the Upper Inferior Oolite (Bathonian) of Gloucestershire, England.

REMARKS: The pattern of variation which we recognize within the specimens collected from Gebel El-Maghara, Sinai, under the generic name of *Globirhynchia* include some of the species described by Cooper (1989). We can find little justification for the separation of these forms into new taxa and have, therefore, selected the most typical of those which we acknowledge to be morphologically different. All of Cooper's species—*G. concinna, G. crassa, G. dubia, G. subtriangulata,* and *G. triangulata*—were collected from zones within the Lower to Middle Dhruma Formation of Saudi Arabia, whereas the specimens figured here from Gebel El-Maghara were collected from beds equivalent to the Upper Bathonian of Europe.

Globirhynchia crassa Cooper, 1989

Figure 13A-C

Globirhynchia? crassa Cooper, 1989: 39, pl. 10, figs. 7-12.

LOCALITY: Subunits 59, 60, Gebel El-Maghara, Sinai.

GEOLOGICAL OCCURRENCE: Upper Bathonian.

REMARKS: The specimen figured here is one of 38 in the collection (table 7). The diagnosis given for this species by Cooper (1989: 39) would fit many of the specimens figured by him as both *Globirhynchia* and *Gibbirhynchia* (pl. 10). The species can be distinguished by their generic characters as seen in transverse serial sections.

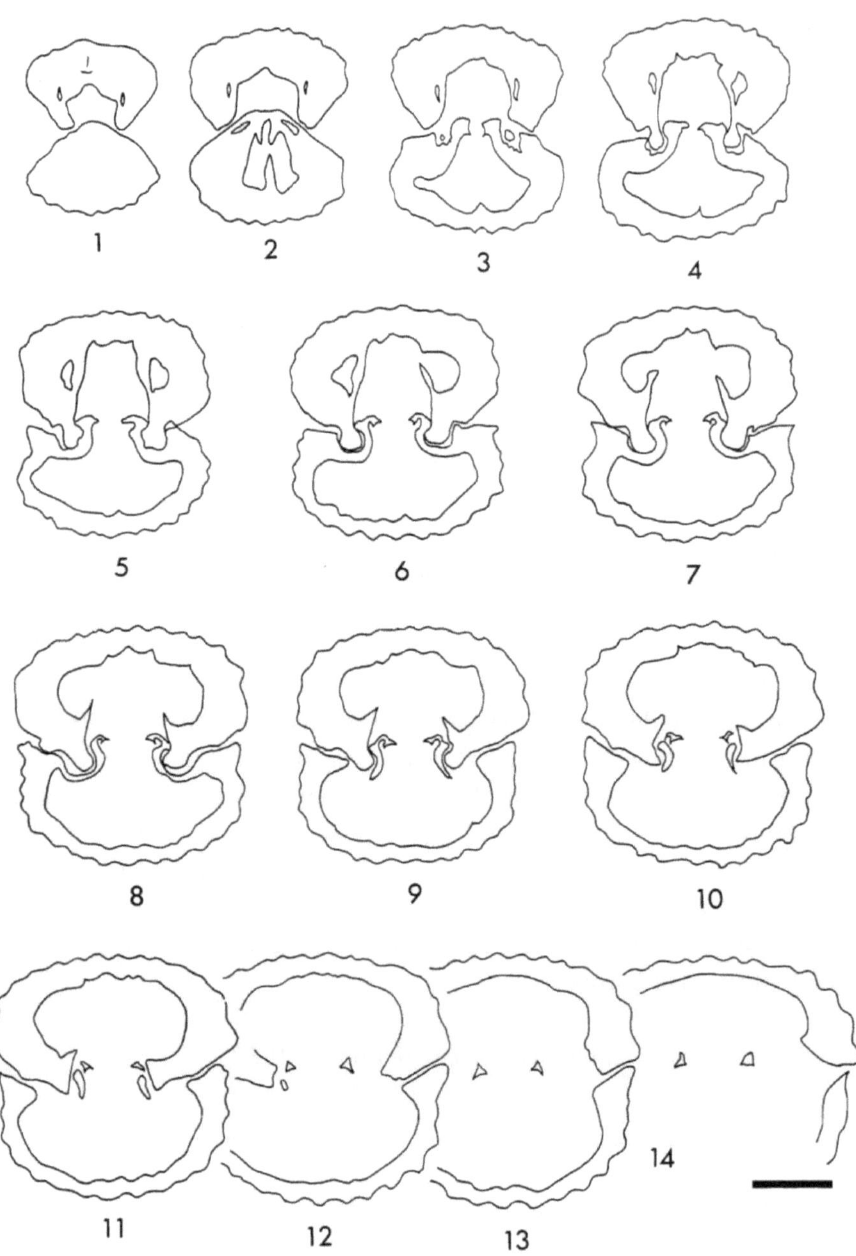

FIGURE 12: *Globirhynchia dubia* Cooper, 1989 (numbers show distance in mm between sections and [in parentheses] distance from beak): **1**, 1.8 (1.8); **2**, 0.6 (2.4); **3**, 0.3 (2.7); **4**, 0.25 (2.95); **5**, 0.41 (3.36); **6**, 0.3 (3 .66); **7**, 0.3 (3.96); **8**, 0.2 (4 . 16); **9**, 0.6 (4.76); **10**, 0.45 (5.21); **11**, 0.4 (5 .61); **12**, 0.35 (5.96); **13**, 0.45 (6.41); **14**, 0.4 (6.81). Scale bar equals 5 mm; GSI M6900.

Globirhynchia subtriangulata
Cooper, 1989, Figure 13D-F

Globirhynchia subtriangulata Cooper, 1989: 40, pl. 10, figs. 45-55; pl. 17, figs. 28-37.

LOCALITY: Subunits 59, 60, Gebel El-Maghara, Sinai.

GEOLOGICAL OCCURRENCE: Upper Bathonian.

REMARKS: As with *Globirhynchia crassa* and *G. dubia*, the specimen figured here is one of 38 specimens collected from Gebel El-Maghara and compares favorably with Cooper's figured specimens (pl. 10, figs. 45-55; pl. 17, figs. 28-37), agreeing in general size (table 7) and outline, number and type of costae, and convexity of valves.

GENUS *BURMIRHYNCHIA* BUCKMAN, 1918
Burmirhynchia cooperi, new species
Figure 13G-I

DIAGNOSIS: Small, elongate-oval, strongly costate *Burmirhynchia.*

LOCALITY: Subunits 46-48, Bir Maghara Formation, Gebel El-Maghara, Sinai.

GEOLOGICAL OCCURRENCE: Upper Bajocian.

ETYMOLOGY: In honor of Dr. G. A. Cooper, formerly Paleontologist Emeritus, Department of Paleobiology, Smithsonian Institution, Washington, D.C.

DESCRIPTION: Shells small (table 8), acutely biconvex; dorsal valve has 16 well-defined, sharp or acutely triangular (in cross section) and deeply incised, costae; corresponding number of costae on ventral valve with four in sulcus and four on almost imperceptible median fold on the dorsal valve; umbo massive, slightly elongate, with somewhat incurved beak; foramen large, interarea broad and extensive; conjunct deltidial plates well exposed; anterior commissure subquadrate in profile with extensive trapezoidal linguiform extension.

INTERNAL CHARACTERS: Unknown.

REMARKS: In general outline, beak features, and type of folding, this species closely resembles *Burmirhynchia*? *bicostata* which Cooper (1989: 16, pl. 2, figs. 50-57) described from the Lower Dhruma Formation *(Ermoceras* Zone) of Saudi Arabia. He figured a small specimen with a narrow sulcus, less clearly defined than the specimen figured as *Burmirhynchia cooperi.* Cooper seemed uncertain about assigning his species *B. bicostata* to the genus *Burmirhynchia* because of its unknown internal structures. We feel more confident about assigning the species described and figured here to the genus *Burmirhynchia.*

The difference between Cooper's species and our species here described is mainly in the number of costae in the sulcus.

TABLE 8: Measurements (mm) of *Burmirhynchia cooperi,* New Species.

Specimen	(L)	(W)	(T)	Subunit
GSI M8070[a]	24.0	20.9	19.4	46-48[b]
BMNH BB.86921	23.8	22.2	17.1	46-48
BMNH BB.86922	20.1	17.7	14.2	46-48
BMNH BB.86923	23.9	22.1	17.9	46-48
BMNH BB.86924	19.4	18.9	17.0	46-48
BMNH BB.86925	21.5	18.4	16.9	46-48
BMNH BB.86926	21.3	18.7	16.3	46-48
BMNH BB.86927	23.2	21.3	18.7	46-48
BMNH BB.86928	23.89	21.5	20.1	46-48
BMNH BB.86929	21.2	17.5	15.8	46-48

a. Holotype, figured specimen.

b. Subunit numbers refer to the section at Gebel El-Maghara unless otherwise noted.

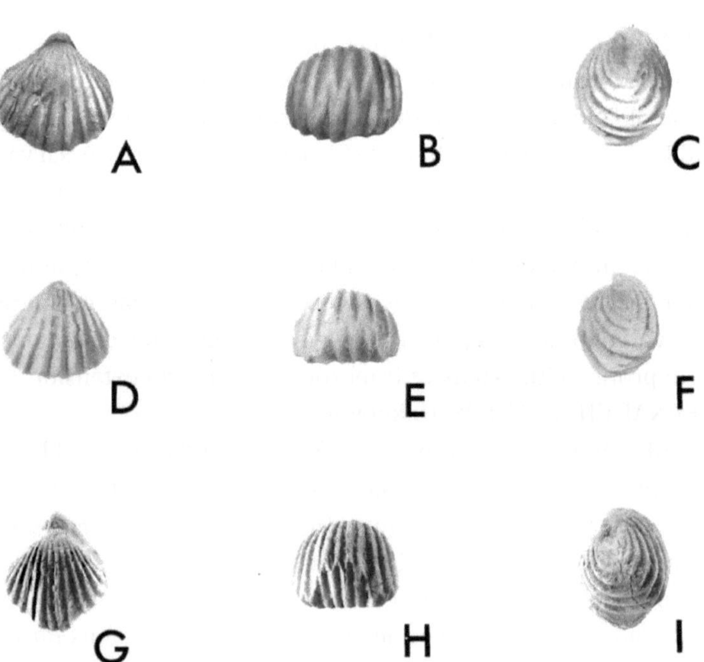

FIGURE 13: *Globirhynchia crassa* Cooper, 1989. A-C, Dorsal, anterior, lateral views, AMNH 44197, x1; *Globirhynchia subtriangulata* Cooper, 1989. D-F, Dorsal, anterior, lateral views, AMNH 44198, x1; *Burmirhynchia cooperi.* new species. G-I, Dorsal, anterior, lateral views, AMNH 44199, x1.

ORDER TEREBRATULIDA WAAGEN, 1883
SUBORDER TEREBRATULIDINA WAAGEN, 1883
SUPERFAMILY TEREBRATULIDOIDEA GRAY, 1840
FAMILY TEREBRA TULIDAE GRAY, 1840
GENUS *KUTCHITHYRIS* BUCKMAN, 1918
Kutchithyris parnesi, new species
Figure 14A-C

DIAGNOSIS: Medium size, pentangulate *Kutchithyris.*

LOCALITY: Subunit 64, Gebel El-Maghara, Sinai.

GEOLOGICAL OCCURRENCE: Upper Bathonian.

ETYMOLOGY: In honor of the late Dr. Abraham Parnes, Geological Survey of Israel.

DESCRIPTION: Biconvex pentagonal terebratulid with width almost equal to length; shells medium size (table 9); maximum width attained at midlength; ventral valve has short, massive umbo and slightly incurved beak; foramen large, circular, and permesothyrid; symphytium not exposed; dorsal valve inflated posteriorly; two well-defined carinae originate posteriorly and diverge anteriorly bordering shallow sulcus which deepens toward anterior margin; anterior commissure sulciplicate; both valves have numerous, clearly delineated, concentric growth lamellae.

REMARKS: Cooper (1989: 98, pl. 26, figs. 7-9; pl. 27, figs. 28-30) recognized two terebratulid species which he broadly assigned to *Kutchithyris?* without providing specific data. The tentative assignment is rather surprising in view of the characters seen in his species 1 (pl. 27, figs. 7-9); they concur with those described for the genus by Moore (1965: H781, H783, figs. 5a-d).

Kutchithyris parnesi, new species, differs from the type species of *K. acutiplicata* (Kitchin) in its less inflated dorsal valve and less acute anterior sulciplication. It differs from *Kutchithyris*? species 2 of Cooper (1989: pl. 26, figs. 28-30) in its distinct pentagonal outline and less acute anterior plication, but agrees in general outline and degree of sulciplication seen in *Kutchithyris*? species 1 (Cooper, 1989: pl. 27, figs. 7-9), regarded as of Bathonian age.

TABLE 9: Measurements (mm) of *Kutchithyris parnesi,* New Species.

Specimen	(L)	(W)	(T)	Subunit
GSI M8071[a]	28.6	25.7	18.5	64[b]
BMNH BB.86930	29.7	26.9	16.6	82
BMNH BB.86931	28.4	25.1	15.0	82
BMNH 88.86932	31.8	31.1	19.5	82

a. Holotype, figured specimen.

b. Subunit numbers refer to the section at Gebel El-Maghara unless otherwise noted.

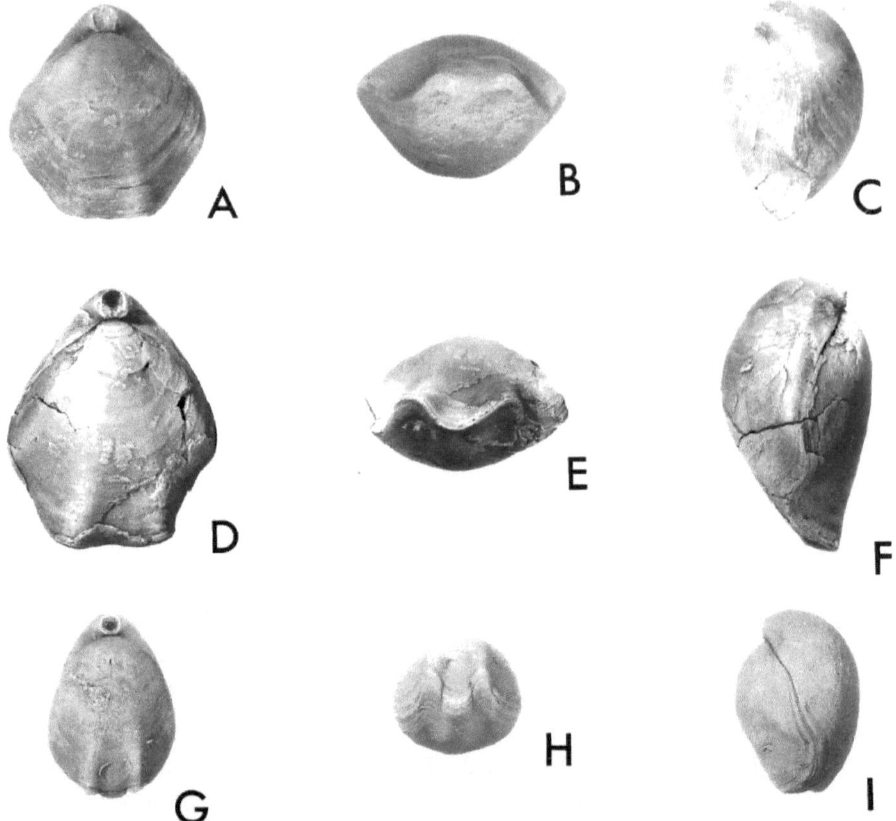

FIGURE 14: *Kutchithyris parnesi,* new species. A-C, Dorsal, anterior, lateral views, GSI MS071, x1; *Avonothyris variabilis,* new species. D-F, Dorsal, anterior, lateral views, GSI MS072, x1; *Bihenithyris pyriformis,* new species. G-I, Dorsal, anterior, lateral views, AMNH 44215, x1.

TABLE 10: Measurements (mm) of *Avonothyris variabilis,* New Species.

Specimen	(L)	(W)	(T)	Subunit
GSI M8072[a]	42.4	34.4	23.0	82[b]
AMNH 44202	39.7	30.8	21.2	82
AMNH 44203	34.8	26.4	20.5	82
AMNH 44204	34.4	26.7	20.6	82
AMNH 44205	23.5	19.2	14.9	82
AMNH 44206	35.2	28.9	22.4	82
GSI M4538	40.4	31.6	23.3	82
GSI M4361	35.2	27.8	20.8	82
GSI M4523	37.5	30.6	21.6	82

a. Holotype, figured specimen.

b. Subunit numbers refer to the section at Gebel El-Maghara unless otherwise noted.

GENUS *AVONOTHYRIS* BUCKMAN, 1915
Avonothyris variabilis, new species
Figure 14D-F

DIAGNOSIS: Broadly subpentagonal *Avonothyris.*

LOCALITY: Subunit 82, Gebel El-Maghara, Sinai.

GEOLOGICAL OCCURRENCE: Upper Callovian.

ETYMOLOGY: Latin *Varius,* in allusion to the different morphotypes typical of the genus.

DESCRIPTION: Shell biconvex, varying in general from subpentagonal to oval with maximum width usually attained at mid-length; average dimensions of the specimens studied (table 10) are as follows: length 31.4 mm; width 28.6 mm; thickness 22.4 mm; dorsal valve flatter than ventral valve but occasionally has marked umbonal inflation; two faint carinae develop from just anterior to mid length, bordering wide, shallow sulcus, forming low fold which becomes more acute anteriorly; ventral valve more evenly convex with slight carination of umbonal area umbo short, beak suberect, with large, often labiate, pedicle foramen; beak ridges epithyrid, symphytium obscure; anterior commissure varies from almost uniplicate to biplicate, but commonly biplicate; ornamentation consists of well-defined, concentric growth lamellae.

REMARKS: The specimen figured here (fig. 14D-F) represents a group of large, variable terebratulids sometimes referred to by authors as *Cereithyris wylliei* (Weir) which, in our opinion, they do not resemble. Further transverse serial sections are required before the generic status of *A. variabilis* is established, but there can be little doubt that its present taxonomic position will be maintained.

A specimen, somewhat similar in general outline and lateral profile to *Avonothyris variabilis,* new species, was figured by Dubar (1967: pl. 3, fig. 20a, b) who referred it to *Charltonithyris bihenensis* (Weir). Dubar's specimen is stated to have been collected from the Callovian *Septirhynchia* bed between Tazerdunet and Ksar Kedima, Tunisia.

GENUS *BIHENITHYRIS* MUIR-WOOD, 1935
Bihenithyris pyriformis, new species
Figures 14G-I, 15

DIAGNOSIS: Broadly oval to pear-shaped, acutely biconvex *Bihenithyris.*

LOCALITY: Subunit 64, Gebel El-Maghara, Sinai.

GEOLOGICAL OCCURRENCE: Upper Bathonian.

ETYMOLOGY: Latin *pirum (pyrum),* in allusion to the shell's pear-shaped outline.

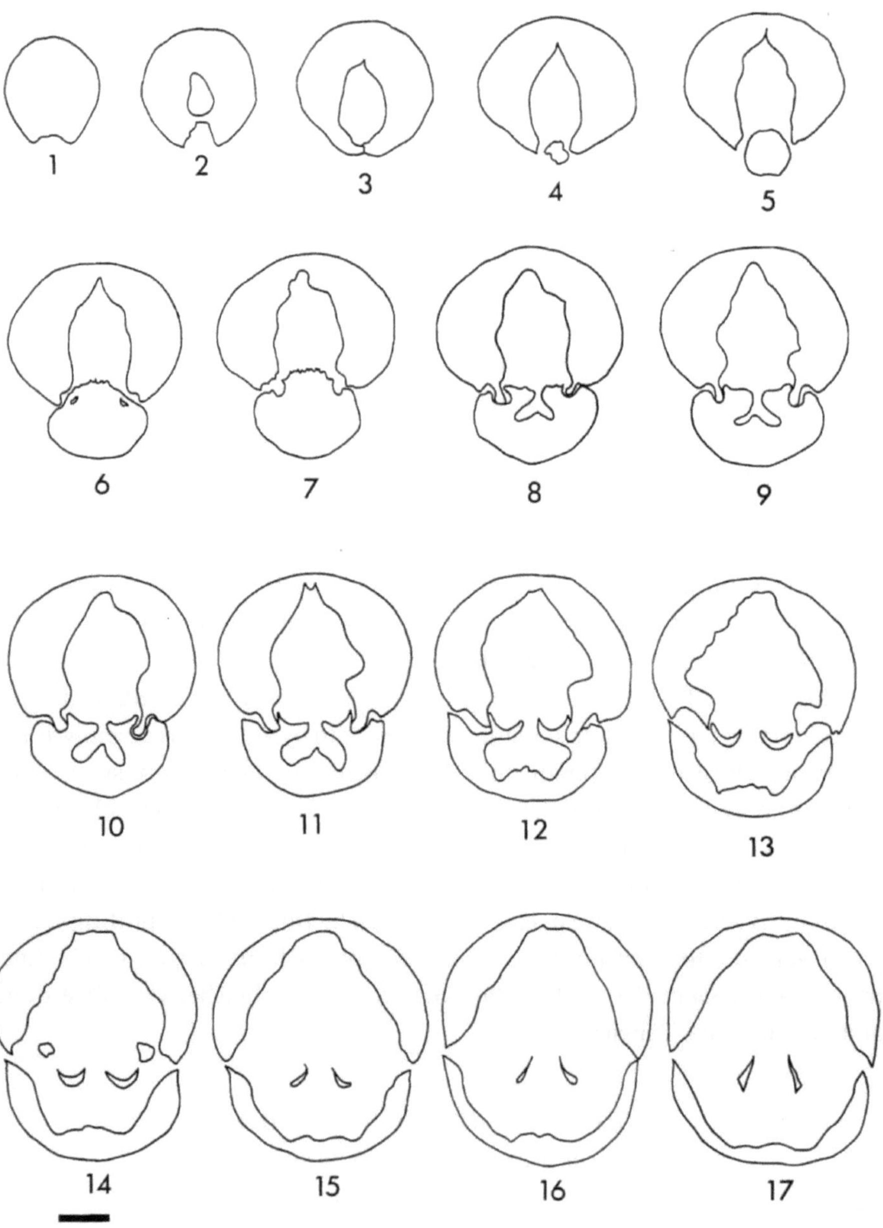

FIGURE 15: *Bihenithyris pyriformis,* new species (numbers show distance in mm between sections and [in parentheses] distance from beak): **1,** 1.4 (1.4); **2,** 0.4 (1.5); **3,** 0.3 (2.1); **4,** 0.3 (2.4); **5,** 0.3 (2.7); **6,** 0.6 (3.3); **7,** 0.4 (3.7); **8,** 0.5 (4 .2); **9,** 0.3 (4.5); **10,** 0.3 (4 .S); **11,** 0.3 (5.1); **12,** 0.4 (5 .5); **13,** 0.3 (5 .S); **14,** 0.2 (6.0); **15,** 0.3 (6.3); **16,** 0.3 (6 .6); **17,** 0.2 (6.S). Scale bar equals 5 mm; AMNH 44215.

DESCRIPTION: Shells medium size; dorsal aspect pyriform with greatest width attained at about two-thirds shell length; lateral profile shows posterior inflation of dorsal umbo; two well-defined carinae border narrow median sulcus which begins at approximately half the length of dorsal valve and deepens anteriorly; ventral umbo massive with large circular foramen and rounded mesothyrid beak ridges; anteriorly, ventral valve develops two shallow sulci matching carinae of dorsal valve and, with a degree of lateral constriction, forms narrow sulciplicate anterior margin.

REMARKS: In many ways, the general outline, morphological features, and size of the specimen figured here (fig. 14G-I) agree with those of a specimen described and figured by Cooper (1989: pl. 29, figs. 23-25) as *Stenorina paralella* but, as the name suggests, that species has almost parallel flanks as opposed to the pyriform outline of *Bihenithyris pyriformis.* It also differs from our species in having a more acutely convex ventral valve and a slightly more elongate subcarinate ventral umbo.

Bihenithyris pyriformis resembles a specimen figured by Muir-Wood (1935: pl. 12, fig. 5a-c) as *Bihenithyris weiri* but differs from that species in its shorter umbo, more oval outline, and less anterolateral constriction.

B. pyriformis, new species, also resembles a specimen described and figured by Muir-Wood from the Jordan Valley (1925: pl. 15, fig. 5a-c) as *Heimia jabbokensis.* But it differs in its smooth or rounded beak ridges; less oval general outline; and broader, less acutely sulciplicate anterior margin; and less umbonal inflation of the dorsal valve. Maximum shell width in *B. pyriformis* occurs more anteriorly than in *H. jabbokensis.*

The transverse serial sections illustrated here (fig. 15) are from the figured specimen after permanent casting. They compare favorably with those given by Muir-Wood (1935: 112, fig. 13) for the type species *Bihenithyris barringtoni.*

Dimensions of holotype (BMNH BB.86933): (L) = 32.7; (W) = 25.5; (T) = 17.6.

GENUS *PTYCTOTYHRIS* BUCKMAN, 1918

Ptyctothyris sinaiensis, new species

Figures 16, 17

DIAGNOSIS: Large, oval to subtriangular *Ptyctothyris.*

LOCALITY: Subunit 48, Gebel El-Maghara, Sinai.

GEOLOGICAL OCCURRENCE: Upper Bajocian.

ETYMOLOGY: The name refers to the Sinai Peninsula.

DESCRIPTION: Shells medium to large (table 11), evenly biconvex; dorsal valve with well-marked concentric growth lamellae and two faint radiating

TABLE 11: Measurements (mm) of *Ptyctothyris sinaiensis,* New Species.

Specimen	(L)	(W)	(T)	Subunit
GSI M8073[a]	42.2	34.5	23.2	48[b]
BMNH BB.86935	39.7	30.8	21.3	48
BMNH BB.86936	45.5	37.1	27.3	64
BMNH B8.86937	40.0	36.7	20.4	64
BMNH BB.86938	40.4	31.6	23.1	64

a. Holotype, figured specimen.

b. Subunit numbers refer to the section at Gebel El-Maghara unless otherwise noted.

carinae developing anteriorly and bordering shallow sulcus which deepens slightly toward anterior margin of shell; very slight lateral constriction of anterior part of valve present producing incipient sulciplicate anterior commissure; ventral valve short and massive with suberect beak, labiate pedicle foramen and well-developed epithyrid beak ridges; symphytium not exposed.

INTERNAL STRUCTURES: From the series of transverse serial sections figured here (fig. 17) it is possible to see that a pedicle collar has developed within the posterior portion of the ventral umbo. Broad, flat cardinal process develops early and remains until hinge plates begin to appear; strong hinge teeth articulate well with large, deep sockets in dorsal valve; brachial loop given off ventrally, beginning with gently curving and ventrally directed hinge plates which, in turn, give rise to elongate and inwardly inclined crural processes.

REMARKS: The specimen figured here as *Ptyctothyris sinaiensis,* new species, differs from the type species *P. stephani* (Davidson) in its narrower dorsal outline, labiate foramen, and incipient episulcation of the anterior margin.

The internal morphology as shown in the transverse serial sections (fig. 17), while agreeing with that of the type species, differs from that given for *Ptyctothyris*? *daghaniensis* Muir-Wood (1935: 123, fig. 20). The shapes of the hinge plates of *P.*? *daghaniensis* are more elongate and geniculate than those of our specimen.

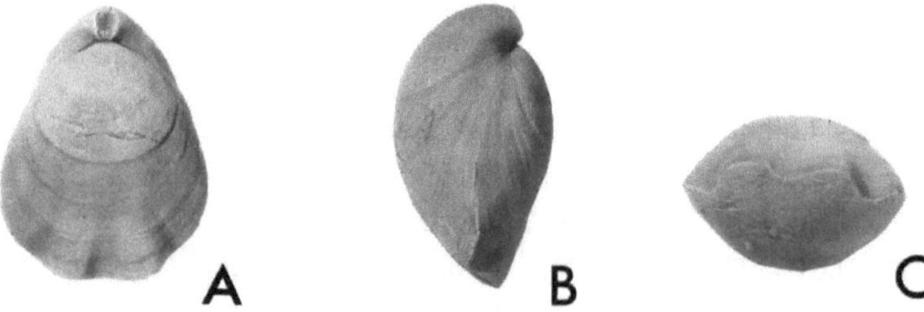

FIGURE 16: *Ptyctothyris sinaiensis,* new species. A-C, Dorsal, lateral, anterior views, GSI MS073, x1.

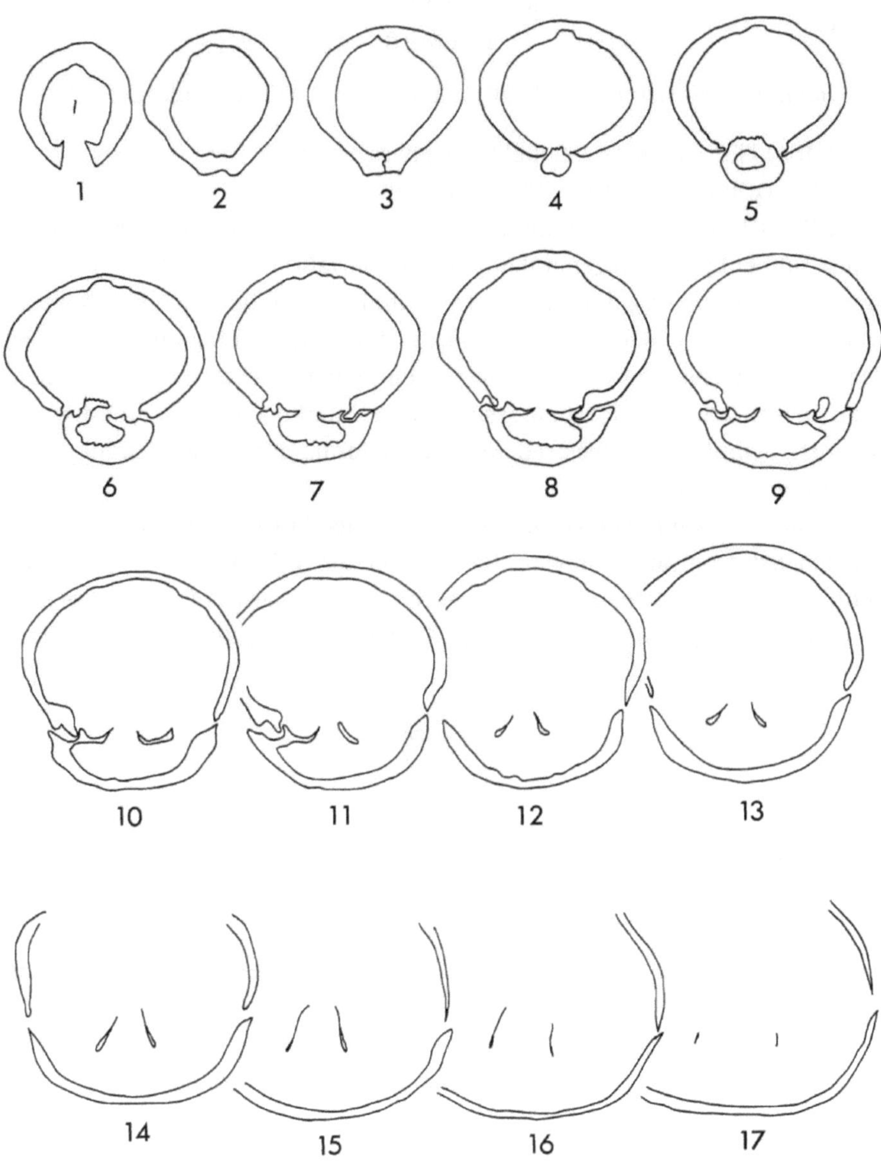

FIGURE 17: *Ptyctothyris sinaiensis,* new species (numbers show distance in mm between sections and [in parentheses] distance from beak): **1**, 1.95 (1.95); **2**, 0.8 (2.75); **3**, 0.4 (3.155); **4**, 0.3 (3.45); **5**, 0.6 (4.05); **6**, 0.4 (4.45); **7**, 0.4 (4.85); **8**, 0.3 (5.15); **9**, 0.4 (5.55); **10**, 0.3 (5.85); **11**, 0.6 (6.45); **12**, 0.4 (6.85); **13**, 0.5 (7.35); **14**, 0.3 (7.65); **15**, 0.3 (7.95); **16**, 0.3 (8.25); **17**, 0.4 (8.65). Scale bar equals 5 mm; AMNH 44207.

Ptyctothyris? *daghaniensis* Muir-Wood, 1935 Figure 18

Heimia jurciliensis (Haas, 1890): Muir-Wood, 1925: 187, pl. 15, fig. 6a-c.

LOCALITY: Subunit 82, Gebel El-Maghara, Sinai.

GEOLOGICAL OCCURRENCE: Upper Callovian.

DESCRIPTION: Shells medium to large (table 12), evenly biconvex, and oval to just subpentagonal in general outline; dorsal valve has well-developed, shallow sulcus originating approximately 15 mm from anterior commissure and bounded on either side by low carinae, or folds, developing anteriorly; some degree of anterolateral constriction present, giving rise to moderately developed paraplicate anterior commissure; ventral umbo short and massive, beak suberect; symphytium obscured and pedicle foramen large and circular; beak ridges not developed.

REMARKS: Muir-Wood (1935: 122, pl. 13, figs. 2a, b) described *Ptyctothyris*? *daghaniensis* from the "Argovian" Daghani section of Somalia. The species was poorly illustrated with one crushed specimen and a series of transverse serial sections (fig. 20). Earlier (1925: 187, pl. 15, figs. 6a-c) she described and figured

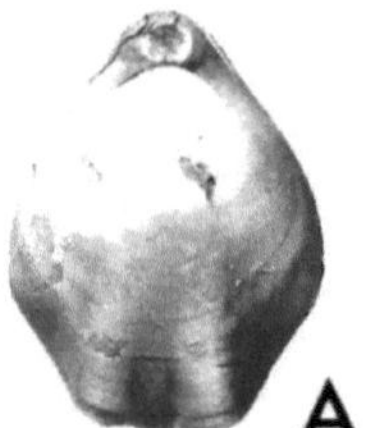

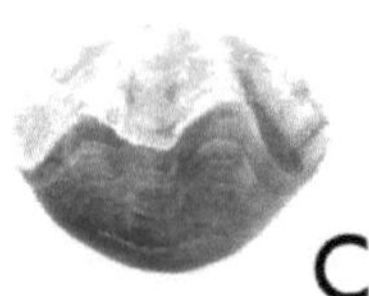

FIGURE 18: *Ptyctothyris daghaniensis* Muir-Wood, 1935. A-C, Dorsal, lateral, anterior views, AMNH 44208, x1.

TABLE 12: Measurements (mm) of *Ptyctothyris*? *daghaniensis*, Muir-Wood, 1935.

Specimen	(L)	(W)	(T)	Subunit
AMNH 44208[a]	34.6	26.0	21.0	82[b]
AMNH 44209	38.4	29.0	23.1	82
AMNH 44210	27.4	23.2	16.8	82
BMNH BB.86940	34.1	28.4	17.9	82
BMNH BB.86941	34.0	28.5	18.7	82
BMNH BB.86942	34.2	28.1	17.5	82

a. Figured specimen.

b. Subunit numbers refer to the section at Gebel El-Maghara unless otherwise noted.

a specimen as *Heimia furciliensis* (Haas) from the Bathonian of the Jordan Valley. The size and morphological features of this specimen agree with numerous examples which she later determined as *Ptyctothyris*? *daghaniensis* in the collections of the Department of Palaeontology of the Natural History Museum, London. The species is not uncommon in the Late Callovian of Gebel El-Maghara where it occurs within the upper part of the Zohar Formation (Subunit 82). As the serial sections of the species given in Muir-Wood's (1935) figure 20 cannot be identified as being congeneric with *Ptyctothyris*, the present taxonomic assignment must be to the genus *Ptyctothyris*? until further investigation of the species can be made.

SUPERFAMILY ZEILLERIOIDEA ALLAN, 1940
FAMILY EUDESIIDAE MUIR-WOOD, 1965
GENUS *SPHRIGANARIA* COOPER, 1989
Sphriganaria cardioides (Douville, 1916)

Figure 19

REMARKS: In his description of *Sphriganaria bramkampi*, Cooper (1989: 105) referred Douvillé's (1916) species *cardioides*, hitherto assigned to the genus *Eudesia*, to his new genus *Sphriganaria* ranging from the Bajocian to

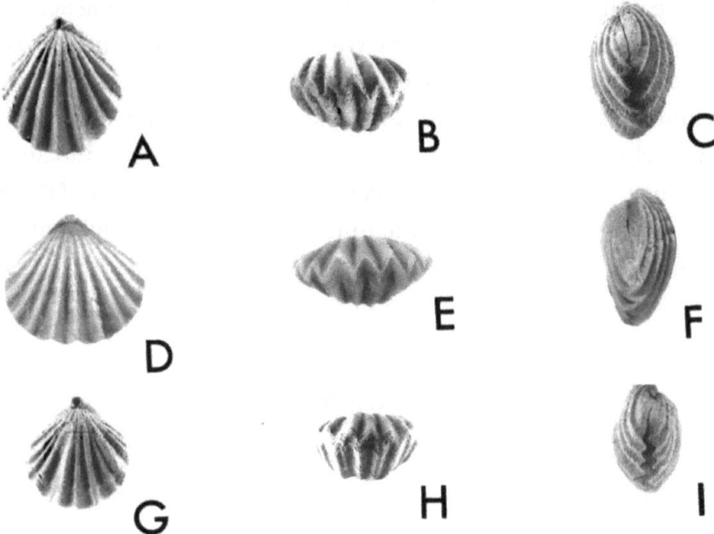

FIGURE 19: *Sphriganaria cardioides* (Douvillé, 1916). A-C, Dorsal, anterior, lateral views, AMNH 44211, x1; D-F, Dorsal, anterior, lateral views, AMNH 44212, x1; G-I, Dorsal, anterior, lateral views, AMNH 44213, x1.

the Kimmeridgian. The three examples figured here (fig. 19) are considered to be of Upper Bathonian age. They compare favorably (table 13) with a series of specimens figured by Cooper (1989: pl. 45, figs. 1-10) which are stated to have come from the Lower Callovian of Engabashi, Gebel El-Maghara, Sinai Peninsula.

TABLE 13: Measurements (mm) of Specimens of *Sphriganaria cardioides* (Douvillé, 1916).

Specimen	(L)	(W)	(T)	Subunit
AMNH 44211[a]	22.8	19.9	13.8	64[b]
AMNH 44212[a]	22.5	23.7	12.8	64
AMNH 44213[a]	18.1	17.2	11.4	64
AMNH 44214	22.5	12.8	12.8	64
BMNH BB.86943	19.7	11.4	11.4	64
BMNH BB.86944	20.4	12.3	12.3	64
BMNH BB.86945	23.1	14.0	14.0	64

a. Figured specimens.
b. Subunit numbers refer to the section at Gebel El-Maghara unless otherwise noted.

ACKNOWLEDGMENTS

This project was funded by grants to Feldman from the National Geographic Society (Nos. 2868-84, 3706-87) and EARTHWATCH and the Center for Field Research, Watertown, MA, during his tenure as Visiting Scientist at the Geological Survey of Israel, Jerusalem. We thank Y. Mimran, Director, Geological Survey of Israel, for providing office space, and laboratory and library facilities. Moshe Arnon, of the same institution, deserves thanks for acting as field assistant and technician on our various excursions to the Negev and Sinai. We acknowledge the critical comments, discussions, and suggestions of the following individuals who have made this study more readable, although we accept full responsibility for all conclusions: D. V. Ager (University College, Swansea), C. H. C. Brunton (BMNH), G. A. Cooper (USNM, retired), and Z. Lewy (GSI).

REFERENCES

Abbate, E., G. Ficcarelii, C. Pirini, Radrizzani, A. Salvietti, D. Torre, and A. Turi. 1974. Jurassic sequences from the Somali coast of the Gulf of Aden. *Revista Italiana di Paleootologiae Stratigraphico* 80: 409-47.

Ager, D. V. 1965. Mesozoic and Cenozoic Rhynchonellacea. *In* R. C. Moore (ed.), Treatise on invertebrate paleontology, Part H, Brachiopoda, H597-625. Lawrence, KS: Geol. Soc. Am., Univ. Kansas Press.

Al Far, D. M. 1966. *Geology and coal deposits of Gabal El Maghara* (N. Sinai). Geological Survey of Egypt, Paper 37: 1-59.

Allan, R. S. 1940. A revision of the classification of the terebratelloid brachiopoda. *Canterbury Museum Records* 4: 267-275.

Arkell, W. J. 1952. Jurassic ammonites from Jebel Tuwayq, central Arabia. *Philosophical Transactions, Royal Society of London* ser. B, 236: 241-313.

------. 1956. *Jurassic Geology of the World.* Edinburgh: Oliver and Boyd.

Bartov, J., and R. Freund. 1968. Columnar section of the Jurassic at Gebel El Minshera (unpublished; deposited at the Hebrew University of Jerusalem and the Geological Survey of Israel).

Buckman, S. S. 1917 [1981]. The Brachiopoda of the Namyau Beds, Northern Shan States, Burma. *Paleontologia Indica* n. ser., 3: 1-299.

Childs, A. 1969. Upper Jurassic rhynchonellid brachiopods from northwestern Europe. *Bulletin of the British Museum of Natural History, Geology* 6.

Cooper, G. A. 1989. Jurassic brachiopods of Saudi Arabia. *Smithsonian Contributions to Paleobiology* 65: 1-213.

Douvillé, H. 1916. Les Terrains Secondaire dans le massif de Moghara à l'Est de l'Isthme Suez. *Academie de Science Paris, Memoire* 54: 1-184.

------. 1925. Le Callovien dans le massif de Moghara: avec description des fossiles par M. Cossmann. *Bulletin de la Société géologique de France* ser. 4, 25: 305-328.

Dubar, G. 1967. Brachiopodes Jurassiques du Sahara Tunisien. *Annales de Paléontology* 53: 1-71.

Farag, I. A. M. 1957. On the occurrence of Lias in Egypt. *Egyptian Journal of Geology* 1: 49-63.

------. 1959. Contribution to the study of the Jurassic formations in the Maghara massif (northern Sinai, Egypt). *Egyptian Journal of Geology* (1961 for 1959) 3: 175-199.

Farag, I. A. M., and W. Gatinaud. 1960a. Un nouveau genre de Terebratulides dans le Bathonien d'Egypte. *Egyptian Journal of Geology* 4: 77-79.

------. 1960b. Six espèces nouvelles du genre *Rhynchonella* dans le roches jurassiques d'Egypte. *Egyptian Journal of Geology* 4: 81-87.

Farag, I. A. M., and S. Omara. 1955. On the occurrence of marine Middle Jurassic deposits at Gebel El Minshera (northern Sinai, Egypt). *Egypt Desert Institute Bulletin* 5: 165-177.

Farag, I. A. M., and A. Shata. 1954. Detailed geological survey of el Minshera area. *Egypt Desert Institute Bulletin* 4: 5-82.

Feldman, H. R. 1987. A new species of the Jurassic (Callovian) brachiopod *Septirhynchia* from northern Sinai. *Journal of Paleontolology* 61: 1156-1172.

Feldman, H. R., and E. F. Owen. 1988. *Goliathyris lewyi,* new species (Brachiopoda, terebratulacea) from the Jurassic of Gebel El-Minshera. American Museum Novitates 2908: 1-12.

Feldman, H. R., F. Hirsch, and E. F. Owen. 1982. A comparison of Jurassic and Devonian brachiopod communities: trophic structure, diversity, substrate relations and niche replacement. *Journal of Paleontology* 56, suppl. 2: 9-10.

Goldberg, M., A. Barzel, P. Cook, and Y. Mimran. 1971. Preliminary columnar section of the Jurassic of Gebel Maghara. *Geological Survey of Israel,* Report MM/1/71.

Gray, J. E. 1840. *Synopsis of the contents of the British Museum,* 44th ed. London: Oxford University.

------. 1848. On the Arrangement of the Brachiopoda. *Annals and Magazine of Natural History* ser. 2, 2: 435-440.

Hirsch, F. 1979. Jurassic bivalves and gastropods from northern Sinai and southern Israel. *Israel Journal of Earth Science* 28: 128-163.

Hoppe, W. von. 1922. Jura und Kreide der Sinai-halbinsel. Z. Dtsch. Palaestina Vereins, 45: 61-79; 97-219.

Huxley, T. H. 1869. *An introduction to the classification of animals.* London: Churchill.

Kuhn, O. 1949. *Lehrbuch der Palaeozoologie.* Stuttgart: E. Schweizerbart.

Lewy, Z. 1981a. A Late Bathonian-Late Callovian unconformity in the Middle East. *Newsletters on Stratigraphy* 10: 27-33.

------. 1981b. Callovian (Mid-Jurassic) stratigraphy of the Middle East. *Geological Survey of Israel, Current Research* 1980: 42-43.

Makridin, V. P. 1955. Nekotorye ûrskie rinhonellidy Evropejskoj časti SSSR. *Zapiski Geologičeskogo Fakulteta Har'kovskogo Universiteta* 12: 81-91. [In Russian]

Moon, F. W., and H. Sadek 1921. Topography and geology of northern Sinai. *Bulletin of Petroleum Research, Cairo* 10.

Moore, R. C. 1965. *Treatise on Invertebrate Paleontology part H, Brachiopoda.* Lawrence, KS: Geological Society of America and University of Kansas Press.

Muir-Wood, H. M. 1925. Jurassic Brachiopoda from the Jordan Valley. *Annals and Magazine of Natural History* ser. 9, 15: 181-192.

------. 1935. *The Mesozoic Palaeontology of British Somaliland: Jurassic Brachiopoda,* 7: 75-147. London: Government of the Somaliland Protectorate.

------. 1965. Mesozoic and Cenozoic Terebratulidina. In R. C. Moore (ed.), *Treatise on invertebrate paleontology, Part H, Brachiopoda,* H762-816. Lawrence, KS: Geological Society of America and University of Kansas Press.

Noetling, F. 1887. *Der Jura am Hermon*. Stuttgart: Ein geognostische Monographie.

Parnes, A. 1974. Biostratigraphic correlation of the Middle Jurassic in Makhtesh Ramon, Gebel Maghara and in Morocco. *Abstract, Geological Society of Israel, Proceedings of Annual Meeting*, Jerusalem.

Picard, L., and F. Hirsch. 1987. *The Jurassic stratigraphy in Israel and the adjacent countries*. Jerusalem: The Israel Academy of Sciences and Humanities.

Range, P. 1920. Die Geologie der Isthmuswueste. *Zeitschrift der Deutschen Gesellschaft für Geowissenschaften* 72: 233-242.

Waagen, W. H. 1882-85. Salt Range fossils, Part 4 (2) Brachiopoda. *Palaeontologia Indica, Memoire*, ser. 13, 1: 329-770.

Weir, J. 1925. Brachiopoda, Lamellibranchiata, Gastropoda and Belemnites. *In* The collection of fossils and rocks from Somaliland made by Mssrs. B. K. N. Wyllie and W. R. Smellie, *Monographs of the Geological Department of the Hunterian Museum, Glasgow University* 1(6): 79-110, pls. 11-14.

------. 1929. Jurassic Fossils from Jubaland, East Africa, Collected by V. G. Glenday. *Monographs of the Geological Department of the Hunterian Museum, Glasgow University* 3: 1-63.

------. 1930. Mesozoic Brachiopoda and Mollusca from Mombasa. *Monographs of the Geological Department of the Hunterian Museum, Glasgow University* 4(4): 77-102.

CHAPTER FIVE

EPI– AND ENDOBIONTIC ORGANISMS ON LATE JURASSIC CRINOID COLUMNS FROM THE NEGEV DESERT, ISRAEL: IMPLICATIONS FOR CO–EVOLUTION

ABSTRACT

Columns of the articulate crinoids *Millericrinus* and *Apiocrinites* from the Upper Jurassic (Upper Callovian) Zohar and Matmor formations of the Negev Desert of Israel display abundant encrusting organisms of about ten species, as well as diverse trace fossils produced by endobionts. Pluricolumnals were colonized by epi- and endobiontic organisms both during life and post-mortem. Skeletonized encrusting organisms include abundant ostreid bivalves (which evidently colonized both live and dead crinoid columnals), two types of serpulid worms, encrusting foraminifera, three species of bryozoans, and small encrusting sclerosponges. Several types of borings are present: *Trypanites* (possibly produced by sipunculids), *Gastrochaenolites* (crypts of boring litholphagid bivalves), elliptical barnacle? borings, and channel-like annelid? borings. In addition, approximately 16% of the pluricolumnals display circular parabolic embedment pits assignable to the ichnogenus *Tremichnus.* They are associated with substantial deformation of the containing columnals and were probably the work of host-specific ectoparasitic organisms. Discovery of *Tremichnus* on Jurassic crinoids extends the range of this trace by almost 100 million years, providing evidence for one of the longest- ranging host-parasite interactions documented thus far (over 200 million years). The relationship of epibionts to the Jurassic crinoids thus ranged from simple utilization of dead hard substrate to probable opportunistic commensalism in forms that colonized the live upright stems, as in some oysters, through host-specific parasitism in the case of *Tremichnus. Crinoid, epibiont, endobiont, Jurassic, parasitism.*

INTRODUCTION

Many fossil crinoid columns, from Ordovician times onward, appear to have been encrusted by a variety of epi- and endobiontic organisms. A wide array of skeletonized encrusting taxa have been reported on ancient crinoid stalks, including sponges, corals, bryozoans, brachiopods, serpulid worms, bivalves, edrioasteroids, and even other crinoids (Lane 1973, 1978; Bell 1975;

McIntosh, 1980; Brett & Eckert 1982; Guensberg 1992; Sandy 1996). In addition to these encrusters, a number of endobiontic organisms have adapted to drilling or embedding their bodies into crinoid skeletal tissue (stereom) to produce swelling sites or to obtain food (Moodie 1918; Franzen 1974; Brett 1981; Warn 1974; Werle et al. 1984). In some instances this drilling was probably parasitic but nonpredatory (Baumiller 1990).

Some of this colonization took place on dead skeletal remains of crinoids, such as pluricolumnals (column segments composed of multiple columnals), although because they are composed of multiple elements it is not entirely clear how long such skeletal elements will remain articulated as large skeletal substrates on the sea floor. Baumiller & Ausich (1992) have recently demonstrated that many crinoid columns disintegrate into segments of relatively similar length, and that these pluricolumnals will remain articulated for considerable intervals of time because of their tight articulations. Hence, post-mortem encrustation and even boring of domichnia (dwelling structures) in pluricolumnals may occur for periods of months to years, as demonstrated here.

In other cases, organisms evidently settled on the stalks of living crinoids. Because these latter organisms colonized living epidermal and stromal tissue they produced reactions on the part of the host, such as gall-like swellings and other deformations that resulted from the production of excess stereom (von Graff 1884; Moodie 1918; Franzen 1974; Welch 1976; Brett 1985). In some cases, the excessive stereomic secretion enveloped or partially enveloped the encrusting organisms, and in such cases it is possible to demonstrate conclusively that the settlement took place on the live, and therefore presumably upright vertical, columns of the crinoids.

Pelmatozoan echinoderms, such as crinoids, provide a potential settling area for epibenthic organisms that are raised above the sea floor to varying distances. Settling epibionts may thus share with their hosts the advantages of being elevated to take advantage of stronger currents above the near-substrate boundary zone, or to escape from stresses of competition from other encrusters, or poorly oxygenated/fluid substrates. Moreover, parasitic epi- or endobionts may potentially tap a certain amount of nutrient from the host.

Nonetheless, the plates of many modern and rapidly buried ancient echinoderms are relatively free of epibionts, as compared, for example, with exoskeletons of arthropods or the shells of mollusks and brachiopods, which are commonly heavily encrusted during life (e.g., Alexander & Brett 1990 and references therein). This is due, in large measure, to the fact that echinoderm plates are covered with an epidermal layer that is apparently capable of inhibiting larval settlement by many organisms (Breimer 1978). The crowns of crinoids, typically but not always, lack encrusting organisms. However, the columns

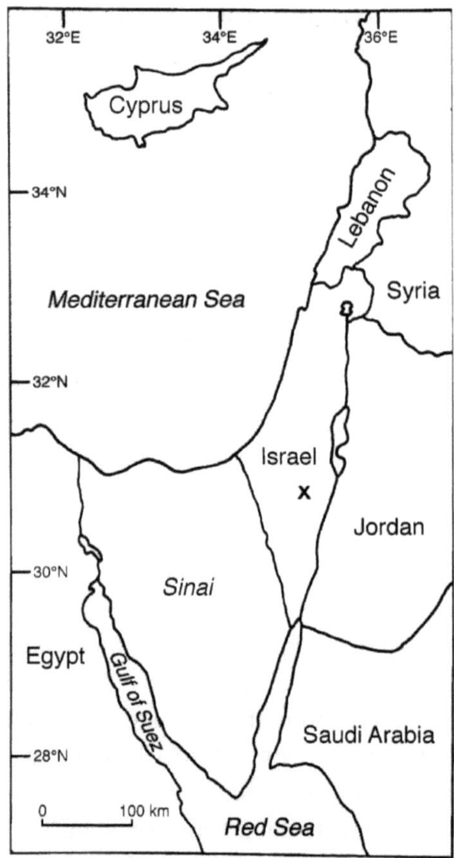

FIGURE 1: Location map showing the collecting area (denoted by an X) in Hamakhtesh Hagadol, Negev Desert, southern Israel (modified from Feldman, 1987).

appear to have been more readily colonized, perhaps because the epidermis is thinner.

Certain organisms successfully elude the crinoid's defenses or even take advantage of the growth potential of a living substrate. Myzostomid annelids, for example, bore into the arms of modern crinoids and/or live within the ambulacral grooves, where they utilize the crinoids' feeding currents or food streams (von raff 1884). Modern pyramidellid gastropods bore into the coelomic cavity of crinoids and obtain their food from the guts of the crinoids (Vaney 1913).

One extinct group of ectoparasites evidently specialized in producing circular-parabolic embedment pits into the stereom of pelmatozoan echinoderm hosts, primarily crinoids. These organisms left distinctive traces, termed *Tremichnus* (Brett 1985), in skeletons of fossil crinoids.

In this chapter we report on epibiontic organisms, including oysters, bryozoans, worm tubes and others, on crinoid remains from the Late Jurassic (Callovian) of Israel. We also report several types of endobiont traces, including *Trypanites, Gastrochaenolites* (lithophagid bivalve crypts), possible barnacle borings, annelid borings, and the youngest known occurrences of *Tremichnus* from the crinoid pluricolumnals. The discovery of these traces in Late Jurassic crinoids greatly extends the known range of this group of ectoparasites.

STRATIGRAPHY

The field area in this study, Hamakhtesh Hagadol, is an elliptical erosional cirque, 15 km long and 6 km wide, with a northeast strike and southeast-directed asymmetry, found in the northern Negev (Fig. 1). Formerly called Makhtesh

Hatira, Hamakhtesh Hagadol cuts through and exposes beds of Upper Callovian-Lower Oxfordian age with a total thickness of about 200 m. A primarily lithostratigraphic section was made in 1962-1963 by Goldberg & Raab of the Geological Survey of Israel and was published by Goldberg (1963) as "A Reference Section of Jurassic Sequence in Hamakhtesh Hagadol" comprising 205 m of beds divided into 75 subunits, with about 30 m of drilled samples from the Kurnub No.2 wildcat (Picard & Hirsch 1987). Hudson (1958) reviewed the paleontology of the area based on collections made by the Petroleum Development Company, a subsidiary of the Iraq Petroleum Company, beginning in the 1930s and continuing through the 1940s. He introduced biozones from the Middle Callovian to the Sequanian (=Upper Oxfordian). Other research that has added to our biostratigraphic knowledge of the study area includes work by Maync (1966) (micropaleontology), Parnes (1961) (echinoids), Reiner (1968), and Hirsch (1979) (bivalves and gastropods), Gill & Tintant (1975) and Lewy (1983) (ammonites).

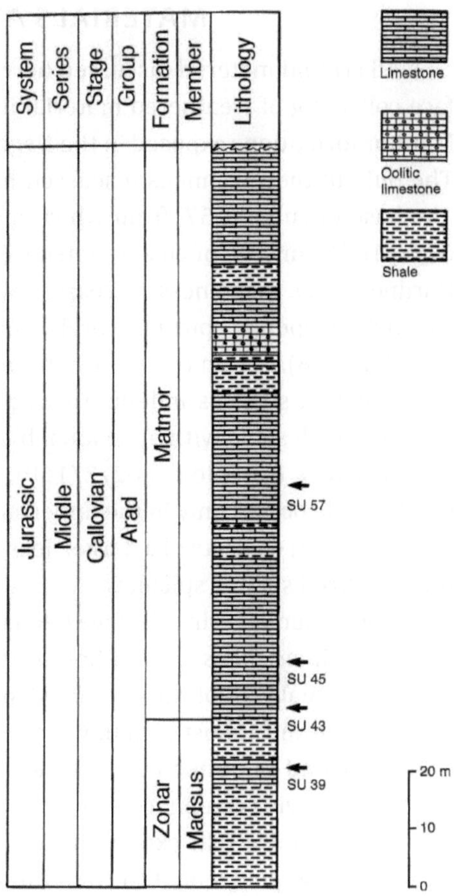

FIGURE 2: Generalized columnar section of the Zohar and Matmor formations at Hamakhtesh Hagadol, Negev Desert, southern Israel. Arrows denote subunit (SU) numbers (after Goldberg 1963 in Picard & Hirsch 1987) from which the crinoids were collected (see Appendix for stratigraphic details).

The crinoids discussed here were collected from Goldberg's (1963) subunits 39, 43, 45, 57, and stratigraphic information is based on Goldberg's (1963) reference section of the Jurassic sequence exposed in Hamakhtesh Hagadol (=Kurnub Anticline), Picard & Hirsch's (1987) stratigraphy, and Feldman's field observations (Fig. 2). Correlative strata (Lohar, Kidod, Beersheba formations; see Feldman 1987; Feldman et al 1991) in northern Sinai were examined for crinoidal material, but none has been found to date. For more detailed stratigraphic information, see Appendix.

MATERIALS AND METHODS

All crinoid material discussed herein was obtained (by HRF) through surface collecting of weathered material in Upper Jurassic beds of the Zohar and Matmor formations exposed in the Negev desert of south-central Israel (Fig. 1). The bulk of the specimens discussed herein are from the Matmor Formation. The base of subunit 57, from which most material was obtained (sample no. 62/2/167), consists of chalky, detrital (mostly bioclastic) brownish limestone. Hardness and coarseness increase toward the upper part, and common fossils include sponge spicules, corals, and stromatoporoids. Just above (sample no. 62/2/168), the limestone is light brown and chalky with regular limonitic stains. Sponge spicules and some ostracodes are present. There is some light yellow, brownish shale with occasional black crusts of gypsum crystals and veins. Slightly above (sample no. 62/2/169), the limestone is rich in fine branching corals. Shale occurs only in the uppermost part. The top of the subunit (sample no. 62/2/170) consists of a darker, less chalky limestone with numerous partly recrystallized sponge spicules.

Two genera of crinoids have been identified in the collections from Israel: calyces of *Apiocrinites* sp. and hold-fasts tentatively attributed to *Millericrinus* (Fig. 3). Virtually all of the encrusted and bored pluricolumnals are of a single morphotype and almost certainly represent a single species. Pluricolumnals are cylindrical and smooth on the exterior. Columnals are low (1 mm high) and all equal in size with no epifacets or cirral junctions. Articular facets of columnals bear fine crenulae that extend inward to a small (<1/4 diameter of columnal facet) roundly pentagonal to round lumen. Diameters of pluricolumnals vary from about 5 to 10 mm.

The morphology of these pluricolumnals (Fig. 3) does not match that seen in *Apiocrinites* columns, which are short and bear fewer, tall, smooth columnals. The encrusted pluricolumnals may belong to *Millericrinus.* Fossils were weathered free of matrix and required only minor cleaning in an ultrasonic cleaner. Surficial weathering and corrosion have altered the surfaces of many crinoid columns, in some cases quite severely. Moreover, most specimens from Subunit 57 have undergone weathering-related chertification. The alteration occurs as a rind of buff chert. Beekite rings obscure surficial details, especially borings, on many of the specimens, occasionally making measurement of features imprecise. The best-preserved specimens were used to assemble data on the sizes and shapes of borings that were measured with digital calipers. Selected specimens exhibiting well-preserved traces of deformations were sawed longitudinally and polished to reveal details of boring cross-sections. A few of the smallest epibionts were also photographed using scanning electron microscopy. Specimens are in repository at the American Museum of Natural History (AMNH) in New York.

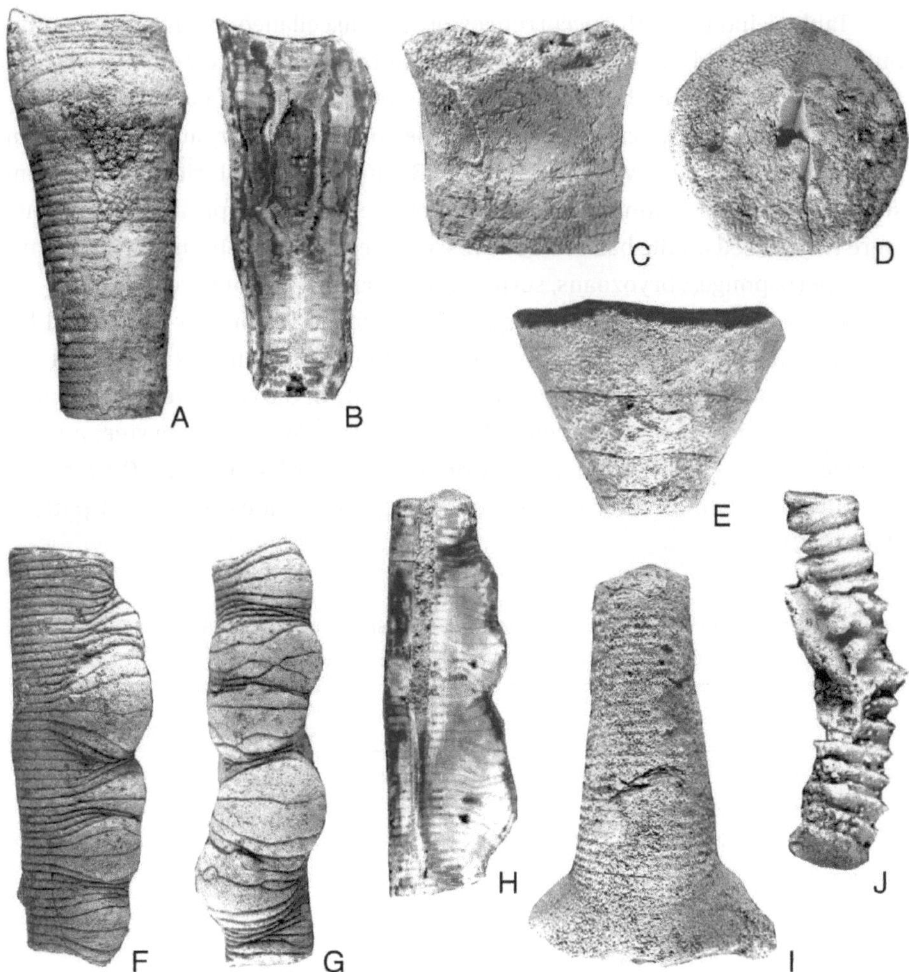

FIGURE 3: A. Pluricolumnal displaying proximal expansion of columnals; x1.4; loc. HHSU57; AMNH 45417. B. Section of pluricolumnal shown in A; note zones of alteration by weathering-related chertification around exterior and lumen; x1.3. C. Proximal columnals and base of dorsal cup of *Apiocrinites*; note encrustation by ostreid bivalves; loc. HHSU57; x2; AMNH 45418. D-E. Two views of well-preserved cup of *Apiocrinites*; loc. HHSU39; x1.4, x1.8; AMNH 45419. F-G. Two views of pluricolumnal showing rounded, gall-like deformations, not associated with boring; loc. HHSU 57; x1; AMNH 45420. H. Polished section of the pluricolumnal shown in F and G; note unidentified black object (bivalve?) that is embedded within the stereom of a welt on the column and exceptionally expanded wedge-like columnal associated with th is object; x l. I. Expanded column and discoidal cemented hold fast of *Apiocrinites;* note small pit of *Trypanites*; loc. HHSU43; x1.7; AMNH 45422. J. Stoloniferous (runner-like) holdfast of *Millericrinus*; note pseudocirral outgrowths of stereom that aided in attachment to the substrate; loc. HHSU45; x1.4; AMNH 45421.

EPIZOANS ON JURASSIC CRINOID STEMS

Table 1 indicates the occurrences of the miscellaneous encrusting epibionts on the crinoid columns. Relative abundance was assessed both on the basis of overall counts of epibionts and in terms of frequency of occurrence (presence of at least one specimen of each species on a particular pluricolumnal), in a sample of 126 roughly similar sized (2-3 cm in length) pluricolumnals from subunit 57 of the Matmor Formation. Eleven species of epi- and endobionts were recognized on the basis of body fossils; these include encrusting foraminifers, sclerosponges, bryozoans, serpulid polychaetes, and bivalves.

The most frequent encrusters (164 specimens) belong to the ostreid bivalve species *Exogyra nana*?. About 44% of the pluricolumnals possessed encrusting oysters. Nearly all specimens consist only of the cemented lower valve, the upper having been disarticulated and removed. These small oysters appear to have nonselectively colonized a wide array of hard substrates. Oysters are typically clustered on the columns, indicating a gregarious settlement pattern (Fig. 4A-C).

TABLE 1: Distribution of epi- and endobiontic organisms on Upper Jurassic crinoids from the Upper Jurassic Matmor Formation, Negev Desert, Israel.

Taxon	Total	Pluricolumnals (Total *N*=126)	% Frequency
Foraminifera			
small agglutinated foraminifer	1	1	0.8
conical foraminifer	18	15	
Porifera			
unidentified sclerosponge	6	5.5	
Bryozoa			
Stomatopora cf. *dichotomoa*	2*	2*	1.6
Small encrusting cyclostome	8	8	6.3
Ctenostome boring	1	1	0.8
Annelida			
Serpula sp. A	17	9	6.3
Serpula sp. B	8	5	4.0
Spirorbis	2	2	1.6
Anthropoda			
Rogerella (barnacle borings)	69	35	27.89
Mollusca-Bivalvia			
Exogyra cf. *nana*	164	56	44.4
lithophagids	48	25	19.8
Incertae Sedis			
Tremichnus cf. *paraboloides*	60	20	16.0

Evidently, some of the oysters encrusted on the columns of live hosts, as the crinoids reacted by producing slight swellings of stereom around the periphery of the oyster shells (Fig. 4D). Larvae were capable of colonizing the crinoid column, and presumably the oysters may have benefited slightly from being elevated above the sea bottom for improved filter feeding.

Normally, the bivalves' accretionary growth outpaced secretion of stereom by the crinoids. In one case, however, an articulated oyster was completely immurated (Fig. 4E); this may represent an individual that died of other causes and was gradually entombed by growth of the crinoid's stereom.

However, in some instances, colonization of pluricolumnals was clearly post-mortem. For example, in several instances, the *Exogyra* valves encrusted over an articular facet of a pluricolumnal (Fig. 4F). One cluster of oysters is also elongated along the column in a double row with a consistently clear (non-colonized) area occupying about a third of the column diameter. This may have been a column fragment that lay horizontally on the sea floor and was colonized post-mortem by the oysters. Such observations confirm the inference of Baumiller & Ausich (1992) that some crinoid pluricolumnals may resist disarticulation for considerable periods of time.

FIGURE 4: Ostreid bivalves on Jurassic crinoids. A, B. Cluster of *Exogyra nana*? on pluricolumnal; note close packing of adjacent bivalves; loc. HHSU45; x 1.3; AMNH 45423. C. Pluricolumnal displaying tightly clustered *Exogyra*; note conformation of valve edges to one another; loc. HHSU57; x1.7; AMNH 45424. D. Lateral view of single *Exogyra* valve; note slight expansion of columnal on opposite side of pluricolumnal that indicates crinoid was alive when oyster encrusted the column; loc. HHSU57; x1.7; AMNH 45425. E. Polished section of pluricolumnal showing immurated small ostreid (black lines-arrow); note thickening of columnals above the embedded bivalves; loc. HHSU57; x1.8; AMNH 45426. F. End (articular facet) view of pluricolumnal showing encrusting ostreid bivalves; in contrast to last examples this specimen was encrusted post-mortem; loc. HHSU45; x2; AMNH 45427.

Serpulid polychaete tubes of at least two kinds are relatively common encrusters on the Israel crinoid columns. A large form with a central carina (Fig. 5A, B) is less common (8 specimens, 4% of pluricolumnals) and generally occurs as isolated individuals, whereas smaller sinuous tubes commonly occur in small clusters on about 6% of the pluricolumnals (Fig. 5D).

Serpulids, which first appear in the Silurian (Robison 1987), are among the most ubiquitous of modern epibiontic organisms. Presently, they occur over a broad range of depths and appear to be relatively non-selective early successional colonizers of hard substrates (Jackson 1977), although Rasmussen & Brett (1985) observed persistent associations of clusters of one species of serpulid with encrusting demosponges. Jurassic serpulids appear to have been facultative encrusters on the crinoid columns. That they commonly grew on live, upright columns is indicated by the fact that the longer tubes spiral around the column. In several specimens the upper portion of the serpulid tube has been eroded away, and in two instances a borehole *(Trypanites)* penetrates through the exposed floor of the tube into the crinoid column beneath. Such evidence indicates that, as in modern epibiontic assemblages, serpulids colonized relatively early and, in some instances, had already died and had their skeletons partially abraded or corroded away before burial of the crinoid columns.

Worm tubes of various types have commonly encrusted crinoid columns throughout geologic history. Small spiral forms *(Spirorbis)* and the problematic *Cornulites* are common epibionts on middle Paleozoic crinoid columns as well as on many with other skeletons. Again, they appear to have been relatively non-selective of host substrates, but Bordeaux & Brett (1990) did note some

FIGURE 5: Encrusting worm tubes. A, B. *Serpula (Dorsiserpula)* sp.; two views of serpulids with dorsal carina; note tube curves around the column; loc. HHSU45; x1.6; AMNH 45428; HHSU45, x1.9; AMNH 45429. C. Serpulid tube that has been breached open, indicating that both pluricolumnal and attached serpulid underwent some corrasion prior to burial; loc. HHSU57; x1.4; AMNH 45430. D. Cluster of small vermiform serpulid tubes on large column; loc. HHSU57; x3; AMNH 45431.

degree of selection by *Spirorbis* for certain species of brachiopods. Although *Spirorbis* are common in modern environments, they are very rare on the Jurassic crinoids from Israel; a single specimen tentatively assigned to *Spirorbis* is illustrated (Fig. 7C).

A third rare type of epibiont on the Jurassic crinoid columns (1.6%) is the vine-like cyclostomate bryozoan *Stomatopora* sp. (Fig. 6C, D). *Stomatopora* is one of the longest-ranging invertebrate genera, with a span from Ordovician to Recent. Taylor & Furness (1978) documented the occurrence and substrate selection of these simple bryozoans from the Jurassic of England. Morphologically, it is similar to other cyclostomes, such as *Corynotrypa* and *Hederella,* that are common encrusters of mid-Paleozoic crinoid columns. Hence, this represents an ecologically conservative encrusting guild common through much of the Phanerozoic.

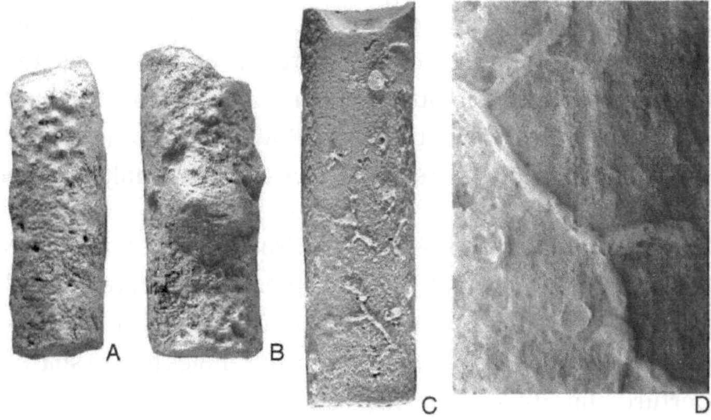

FIGURE 6: Bryozoans and sclerosponges on pluricolumnals. A. Unidentified sclerosponge; note mammelons and calices; loc HHSU57; X1.5; AMNH 45432. B. Small mound-like cyclostome bryozoan; loc. HHSC57; x2; AMNH 45433. C, D. *Stomatopora* cf. *dichotoma,* a dichotomously branching runner-like cyclostome bryozoan; loc. HHSU57; x2; x11; AMNH 45434.

FIGURE 7: Foraminifera(?) and *Spirorbis*. A. SEM photograph of volcano-like conical encrusting foraminifer(?); loc. HHSU57; x32; AMNH 45435. B. SEM photograph of foraminifer(?); loc. HHSU57; x32; AMNH 45436. C. *Spirorbis* sp. SEM photograph of partially corroded specimen; loc. HHSU57; x39; AMNH 45437.

A few small patches of an unidentified cyclostome bryozoan are present on the crinoid columns, and three specimens of a crustose mammelon-bearing sclerosponge (Fig.6A, B) were also noted. It is likely in these cases that the epizoans encrusted live crinoids because they are partially wrapped around the columns.

A final type of epibiont that is relatively common on the crinoid columns is a minute conical mound, approximately 2 mm in diameter at the base (Fig. 7 A). This is apparently a small encrusting foraminifer. A single specimen of small foraminifer(?) was also noted (Fig. 7B).

TRACE FOSSILS ON JURASSIC CRINOID COLUMNS

The crinoid columns from the Jurassic of the Negev Desert revealed a large number of trace fossils, including cylindrical, channel-like, and dendritic borings, probably produced by acanthroracican barnacles, *Gastrochaenolites,* the flask-like crypts of endolithic bivalves, and circular-parabolic pits assignable to the ichnogenus *Tremichnus* (Table 2). The latter were probably all produced on live crinoid hosts and represent embedment structures. However, the other borings are less clearly life associations and, in some cases, were demonstrably postmortem.

Trypanites.—Small circular borings, ranging from 0.3 to 3.5 mm and averaging 1.43 mm in diameter (SD 0.67 mm), are the most abundant traces and are present on 57% of the crinoid columns examined; they are also present on associated oysters, bryozoans and other fossils. The boreholes have a sharply defined, circular aperture and are slender and conical in cross section. These simple

TABLE 2: Frequencies of occurrence and sizes of endobiontic traces in Upper Jurassic crinoid columns from Israel. N = total number of measured specimens; 0-1 = diameter (circles) or length of long axis (ellipses); D-2 = length of short axis (ellipses); SD = standard deviation; frequency refers to the number percentage of pluricolumnals in a total sample that display at least one example of the trace.

Trace/Description	N	Mean D-l	SD D-1	Mean D-2	SD D-2	Frequency (number/ %pluricolumnals)
Trypanites (circular-conical borings)	154	1.4	0.7	--	--	72/57
Branching channel	3	1.5	--	--	--	3/2
Channel-like borings	5	1.5	--	--	--	5/4
Barnacle borings	69	1.6	0.5	0.8	0.2	35/28
Gastrochaenolites (flask-like lithophagid crypts	48	4.4	2.4	2.6	1.3	25/20
Tremichnus (circular-parabolic embedment pits)	60	2.4	1.4	2.3	1.2	20/16

borings are tentatively assigned to the ichnogenus *Trypanites* (Fig. 8A-C). They are not associated with columnal deformation and were most likely produced on partially disarticulated pluricolumnals on the sea floor. That they represent true borings (rather than embedment structures, see below) is evident in samples that penetrate exoskeletons of mollusks. Several such holes were observed to enter the columnals along articular facets and thus were obviously formed after death and disarticulation of the crinoid column. These represent dwelling borings (domichnia), possibly of a sipunculid worm (Pemberton et al. 1980). In some instances, *Trypanites* may have been confused with small embedment structures (*Tremichnus*) that lacked associated deformation.

Channel-like borings.—A small number of crinoid pluricolumnals (4%) display narrow (1-1.5 mm) elongate channel-like borings (Fig. 8E, F). These grooves resemble the Paleozoic annelid boring *Vermiforichnus*. The channels are unbranched, suggesting production by a solitary organism, such as a solitary polychaete.

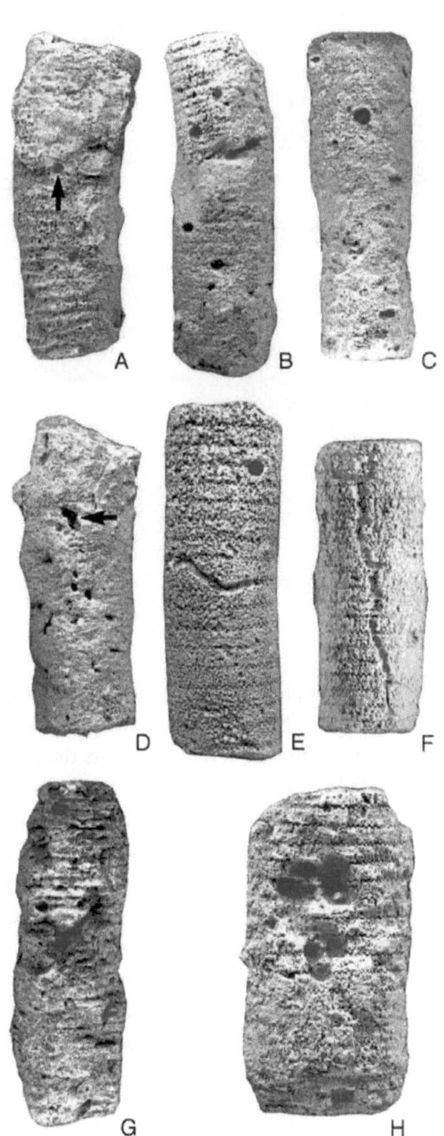

FIGURE 8: *Trypanites* and other borings. A. *Trypanites* on pluricolumnal, including one (arrow) penetrating an ostreid shell, proving that this is a true boring; loc. HHSU57; x1.7; AMNH45438. B, C. Sharply defined circular borings *(Trypanites)* on crinoid pluricolumnals; loc. HHSU57; x1.4; AMNH45427; loc. HHSU57; x1.3; AMNH45439. D. *Trypanites* and *Rogerella*?; elliptical borings probably produced by acrothoracid barnacles; also note figure-8 shaped boreholes of lithophagid bivalves (arrow) and encrustation by sclerosponge in upper right; loc. HHSU57; x 1.3; AMNH45432. E. Single *Trypanites* and channel-like boring; loc. HHSU57; x1.5; AMNH 45440. F. Branching channel-like boring; loc. HHSU57; x1.7; AMNH 45441. G, H. Heavily corroded and bored pluricolumnals; loc. HHSU57; x1.2; AMNH 45442; loc. HHSU57, x1.9; AMNH 45443.

A second type consists of a dichotomously branching, sharply entrenched groove about 1-2 mm in width. This trace is similar to *Tarrichnium,* a branching groove found previously only on columns of the crinoid *Balanocrinus* from the Miocene of the West Indies (Wanner 1938). This trace has been attributed to a stoloniferous hydrozoan.

A single example of a sediment-filled longitudinal boring subparallel to the column was observed in a polished cross section of a pluricolumnal (Fig. 9I). It is unclear how this boring was formed or even where it extends to the exterior.

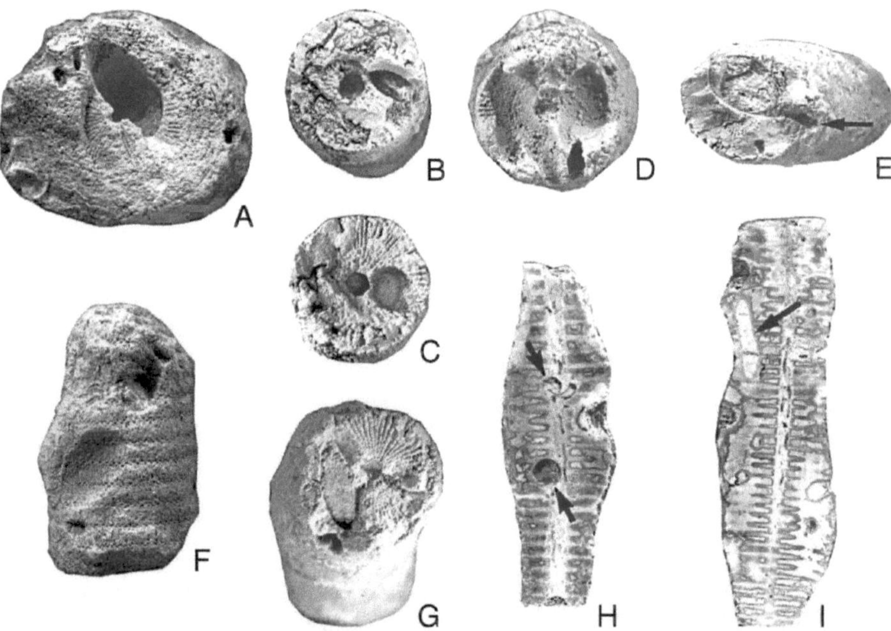

FIGURE 9: Lithophagid bivalve borings *(Gastrochaenolites)* in Jurassic pluricolumnals. A-D. Natural cross sections of bivalve crypts shown on columnal articular facets. A. Large borehole showing shape of bivalves; note that boring cross-cuts the axial canal of the crinoid columnal; loc. HHSU57; x1.9; AMNH45444. B, C. Bivalve crypts that approach but do not cross-cut columnal lumens; loc. HHS 57; x1.9; AMNH 45445; loc. HHSU57; x2; AMNH 45446. D. Pluricolumnal facet showing three bivalve crypts; loc. HHSU57; x2.3; AMNH 45447. E. Complete outline of bivalve crypt; note narrow neck (siphonal canal; arrow) leading inward to expanded chamber; articulated bivalves are enclosed within the matrix filling the boring; loc. HHSU57; x1.8; AMNH45432. F. Corroded pluricolumnal showing remnants of breached open bivalve crypts; loc. HHSU57; x2.1; AMNH45448. G. Pluricolumnal facet showing a sediment-infilled bivalve shell still *in situ* within crypt; loc. HHSU57; x2; AMNH45427. H. Bivalve borings on polished section; note spar-filled crypts cutting sediment fill of lumen (top arrow) and boring cutting lumen without spar filling (bottom arrow); loc. HHSU57; x2; AMNH45449. I. Polished section showing bivalve crypts; note remains of shells in sediment fill. Also note longitudinal, sediment-filled boring that cross-cuts several columnals (arrow); loc. HHSU57; x1.3; AMNH 45450.

Barnacle borings.—A number of Jurassic crinoid columns (28%) possess small elliptical borings averaging 1.56 mm in long dimension (SD 0.5) and about 0.84 mm (SD 0.2) in width. These pits rarely overlap one another. They do not possess elevated rims or special swellings. These small pits are relatively simple in profile and not particularly deep (1.5-2.0 mm). Their simple elliptical form suggests that these borings may be the work of acrothoracid barnacles (Hantzschel 1975). The absence of special swellings and deformations around most of the borings indicates that many were not produced on live hosts. A few of the traces do occur on swellings, but it is not clear whether these were produced earlier by other causes, or whether they were caused by the barnacle borings.

Gastrochaenolites: Lithophagid borings.—Many of the crinoid columns (20%) display relatively large, oval to flask-shaped borings, assignable to the ichnogenus *Gastrochaenolites.* These are best observed on column articular facets where natural cross sections have been exposed by disarticulation of the columnals (Fig. 9B-D). The borings expand inward from relatively small (0.5 to 2.0 mm) oval-shaped apertures. The apertures are similar in size to the elliptical barnacle(?) borings noted above, but have a more rounded outline; best-preserved examples display a slight figure-eight configuration to the apertures, as if composed of two overlapping holes of equal diameter. These two holes may represent the positions of incurrent and excurrent siphons of the bivalve. The crypts are 3-10 mm deep (mean 4.43 mm; SD 2.45) and 3-5 mm wide (mean 2.56 mm; SD 1.34) at the broadest area, which is about two-thirds of the depth of the borehole. A narrow "neck" (siphon canal) leads inward from the aperture to the expanded crypt, which is filled with fine micritic sediment or, rarely, with sparry calcite. In weathered natural cross sections, on column articular facets, the sediment fill may be removed leaving the crypt walls exposed (Fig. 9B-D). Four sediment-filled crypts observed in this study preserve cross sections of the trace makers themselves: articulated thin-shelled lithophagid bivalves (Fig. 9E).

These bivalve borings were apparently produced after the death of the crinoids in many cases. There is no associated deformation of the columnals around the borings. Three crypts observed in cross section penetrate the lumen of the column. In two cases this results in continuity between fine-grained sediment inside the borehole and that of the lumen. However, in one instance, the crypt is spar-filled and cross-cuts the sediment filling the lumen (Fig. 9H). This indicates that the boring was produced in the pluricolumnal after the crinoid had died and disarticulated. The tissue that originally occupied the lumen had been removed by decay, and the canal filled with sediment. Furthermore, the sediment infill had apparently been indurated such that the base of the boring is sharply defined. Such specimens indicate that pluricolumnals must have resisted disarticulation for a period of at least months. The frequent "natural cross sections" of the borings indicate disarticulation prior to burial and after colonization by the bivalves.

Tremichnus.—In addition to the various borings discussed above, several of the crinoid columns (16%) recovered from the Matmor Formation displayed circular to slightly oval pits with smoothly parabolic cross sections (Fig. 10). These traces, which range in diameter from 0.5 mm to 5.5 mm (mean 2.28 mm; SD 1.18 mm for minimum dimension), are assignable to the ichnogenus *Tremichnus.* In most cases, these pits are located near the crests of low swellings produced by asymmetrical growth and distortion of the crinoid columnals. Three pluricolumnals display clusters of pits of varying diameter on gall-like protuberances. The ichnogenus *Tremichnus* was established by Brett (1985) to encompass circular-parabolic pits in the stereom of echinoderms, with or without associated deformation on the part of the host.

Some larger examples of *Tremichnus* (3 mm) deviate strongly from circular in outline (mean minimum diameter 2.28 mm; mean maximum diameter 2.41 mm). The pit is invariably somewhat broader in a transverse direction (parallel to the column facets) (Fig. 11B). Ratios of width to length average about 1.4.

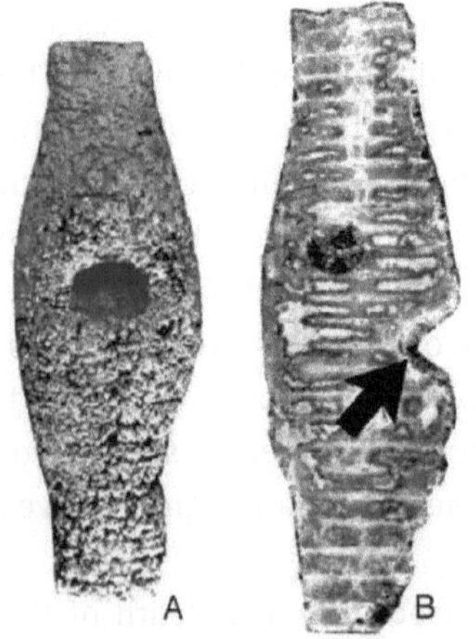

FIGURE 10: *Tremichnus* cf. *paraboloides* on Jurassic pluricolumnals. A. Single large and slightly elliptical pit centered on swelling of column; loc. HHSU57; x2; AMNH 45449. B. Cross section of *Tremichnus* (arrow) shown in A; note simple, parabolic profile; borings near the column lumen are *Gastrochaenolites* (bivalve borings; see also Fig. 9).

Several *Tremichnus* examples exhibit a blurring of the pit perimeter and, in one case, a second concentric rim about 0.1 mm wide parallels the original pit outline (Fig. 11E). These features were evidently produced by growth of new stereom and partial infilling of unoccupied pits. All such evidence shows clearly that these pits were formed on live crinoid hosts.

Brett (1985) recognized that these structures were distinct from cyst-like swelling on crinoid arms, previously termed *Myzostomites,* as well as several other types of deformations in fossil crinoids. Franzen (1974) illustrated various types of pits in Silurian crinoid columns and calyces and suggested they were the work of epizoans that settled on living crinoids.

As in Paleozoic examples of *Tremichnus,* it is evident that these circular-parabolic pits were produced by the process of embedment, in which the parasite settled on the epidermis of the crinoid and inhibited the growth of stereom in its near vicinity. Normal growth of the crinoids' endoskeletons resulted in the development of an embedment pit, a region of "non-growth" occupied by the epibiont itself. Moreover, the consistent orientation of distorted (slightly elliptical) larger pits with respect to the columnals is best explained by differential growth of the columnar stereom. In some cases the epibiont may have done some excavating of the stereom, as in rare instances in which the borehole penetrates through the plates of the crinoid (Brett 1978, 1985). Hypertrophy of the crinoid stereom produced general swelling around the embedment pits. Clearly the pits are not simply coincident with bumps of stereom, but rather the epibiont that produced the embedment pits induced the crinoid to produce extra skeletal secretions. However, in most cases the Jurassic *Tremichnus* do not occur on an elevated mound or gall, but rather are surrounded by simple swellings associated with asymmetrical radial accretion of the columnals. This contrasts with Paleozoic occurrences of *Tremichnus cysticus,* which occur on localized mounds or galls of stereom that were added over normal diameter columnals.

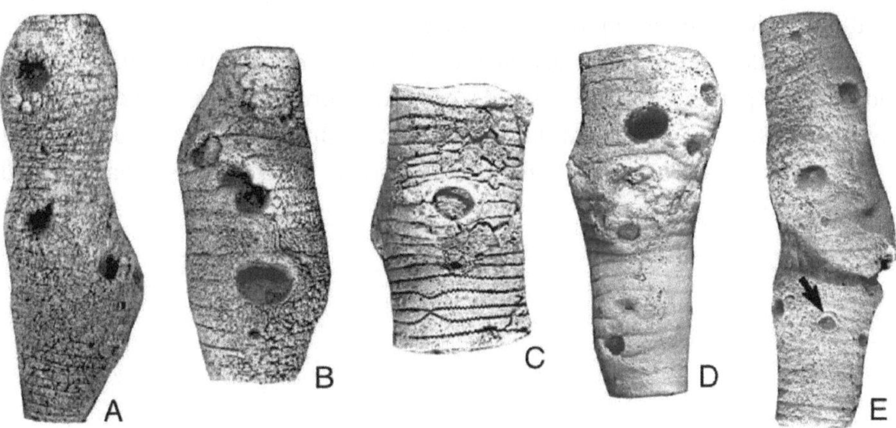

FIGURE 11: Deformed pluricolumnals with *Tremichnus.* A. Specimen with three subequal-sized *Tremichnus;* note that pits correspond in position to swelling of the columnals; also note that pit outlines are rounded; for cross sections of the pits in center see Fig. 10; loc. HHSU57; x1.2; AMNH 45450. B. Series of subequal-sized pits aligned obliquely and distinctly elliptical large hole with long axis perpendicular to column axis; loc. HHSU57; x1.9; AMNH 45451. C. Single elliptical pit centered on suture between symmetrically deformed columnals; loc. HHSU57; x1.9; AMNH 45452. D. Strongly deformed column; note warping of columnals in vicinity of pits and elliptical (broadened) shape of largest pit; loc. HHSU45; x1.2; AMNH 45453. E. Distorted column; note smoothing of some pit outlines and concentric rim on one small pit (arrow); loc. HHSU45; x1.1; AMNH 45454.

It is noteworthy that many of the Jurassic crinoid columns display local swellings or galls (Figs. 3F, G; 12C, D) that are not obviously associated with epibionts. Such a condition has not been observed commonly amongst collections of Paleozoic crinoids. Furthermore, it is clear that some of the columns that possess attached epibionts, such as *Exogyra,* also display swelling in the vicinity of these encrusters (Fig. 4D). The epibionts responsible for producing *Tremichnus* may have taken advantage of a pre-existing condition within crinoid hosts that enabled them to produce excessive secretions, thereby rapidly embedding the epibiont within the walls of skeletal stereom and producing dwelling structures (domichnia) that helped to support and protect the epibiont. The epibiont may have further enhanced the development of its protective cavity by some boring, perhaps by chemical secretions.

It is not clear whether the producers of *Tremichnus* harmed their hosts any more so than do various epibiontic oysters or bryozoans, despite their production of swellings in the columns. Did they actually obtain food from the stromal tissues of the crinoids? This possibility has been suggested previously for Paleozoic occurrences, partly because of the high degree of host selectivity (Brett 1978). However, it is unclear that a large amount of nutrient could be obtained from the rather dense stereom of the crinoid columns, and it is perhaps more likely that suspension-feeding epibionts simply used the crinoids' unique mode of skeletal secretion to produce swelling sites.

The oldest known occurrence of *Tremichnus* is represented by pitted calyx plates of the cladid crinoid *Carabocrinus,* illustrated by Lewis (1982) from the Middle Ordovician Bromide Formation. Eckert (1988) illustrated rare occurrences of *T. cysticus* from Early Silurian crinoid columns.

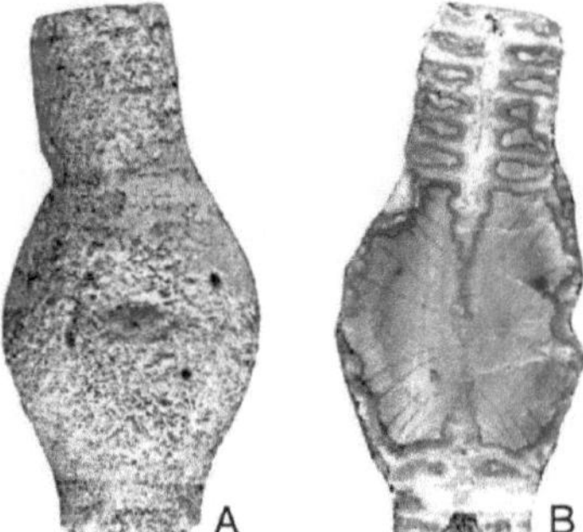

FIGURE 12: Gall -like swelling on pluricolumnals. A, B. External and cross sectional views of swollen pluricolumnal; note vaguely defined ellipsoidal pit (*Tremichnus*?) in center of A and extreme degree of di stortion of outer edges of columnals in B; loc. HHSU57; x1.8; AMNH 45455. C, D. Lateral and frontal view of cup-like swelling on side of pluricolumnal. D shows circular depressions that appear to be remnants of *Tremichnus* pits; loc. HHSU57; x1.5; AMNH 45456.

Tremichnus is particularly common in Late Silurian crinoids, both in North America and Gotland. Brett (1978) demonstrated that the circular-parabolic pits, later termed *Tremichnus,* displayed a considerable degree of host-selectivity among large collections of crinoids and blastozoans from the Silurian (Wenlockian) of New York. Flexible crinoids, especially *Ichthyocrinus,* dendrocrinid inadunates and a few species of camerates were common hosts. Moreover, there is evidence that *Tremichnus* of distinctive morphologies (e.g., *T. paraboloides, T. puteolus*) were consistently associated with certain lineages of crinoids for prolonged periods of time.

Tremichnus is not abundant in most Devonian fossil assemblages, but pitted calyces of the flexible crinoid *Synaptocrinus* are known. These suggest the persistence of host-parasite relations in the ichthyocrinid lineage. Rare pitted and deformed crinoid columns have been observed from the Middle Devonian of North America and Morocco (Franzen 1974).

Crinoid columns with *Tremichnus* are abundant in Carboniferous collections, and a few taxa commonly display pits on the calyces. For example, survey of large collections of crinoids from the Mississippian Burlington, Lake Valley, and Fort Payne formations revealed a high frequency of small *Tremichnus* in platycrinitid and dichocrinid carnerates. Large parabolic pits are reported from the Pennsylvanian flexible crinoid *Amphicrinus* (see Brett 1978). Until the present study, these specimens represented the youngest known occurrence of *Tremichnus,* and Brett (1985) suggested that the ichnogenus *Tremichnus* became extinct by the end of the Paleozoic along with crinoid hosts.

The occurrence of parabolic embedment structures, identifiable as *Tremichnus,* in crinoid columns from the Upper Jurassic (Callovian) of Israel permits a range extension of over 100 million years for this ichnogenus. These trace fossils demonstrate the persistence of a crinoid-parasite interaction from Middle Ordovician to the Late Jurassic, a total span of nearly 200 million years and through the greatest crisis in the history of marine life: the Permo-Triassic extinctions.

There are too few data on various types of Jurassic crinoids to test carefully for host specificity. At present, the pits are known only on columns of a single morphotype, possibly assignable to *Millericrinus.* A number of cups, plates, and columns of the genus *Apiocrinites* are known, but none have proven to have *Tremichnus* on them. Possibly, there was host selection for *Millericrinus.*

Nonetheless, as all known occurrences of *Tremichnus* are in pelmatozoans (mainly crinoids), and because Paleozoic forms display a high degree of host selectivity, we can infer that the association between the *Tremichnus*-formers and the pelmatozoan echinoderm host was an obligate one. Hence, we may also assume that the epibionts managed to survive the Late Paleozoic and Triassic extinctions on the rare host species of crinoids that persisted through these times of crisis. Not all inadunates became extinct at the end of the Permian; a handful of species obviously survived this crisis and subsequently gave rise to the Subclass Articulata in the Early Triassic. Surprisingly, the pit-forming epibionts, despite their apparent restrictions to relatively few Paleozoic crinoid hosts, persisted on at least one species of surviving crinoid. At least one species capable of sustaining the *Tremichnus*-producing epibionts also must have survived through the major Late Triassic crisis. The epibionts evidently proliferated again in the Jurassic, for the pits are abundant in columns of *Millericrinus* from Israel.

The *Tremichnus*-crinoid association is one of the most long-ranging of all known parasitic co-actions. It is, of course, impossible to determine whether the producers of the embedment pits belonged to the same species or even the same genus as did those of the later Paleozoic time. Clearly, the crinoid hosts have evolved substantially. The Jurassic crinoids belong in the Subclass Articulata, whereas the youngest previously known occurrences are in flexible crinoids and cladid inadunates. Nonetheless, the size and morphology of the embedment pits, together with their associated effects on the crinoid host stereom, indicate that the biology of the producers was probably fundamentally similar to that of *Tremichnus* producers in much earlier times.

This evidence provides one of the strongest cases known in the geologic record for the stability of host-parasite interactions. Such co-evolutionary links must have been extremely resilient to change. As emphasized by Boucot (1990), organism interactions develop over geologically brief time spans and may persist for very extensive interludes reflecting prolonged co-evolution among symbiotic co-actors. In extreme instances, co-action pairs may persist in the face of major biotic crises such as mass extinctions. Whereas some other types of epibionts on the Jurassic crinoids were completely new (e.g., oysters and various types of newly evolved serpulid worms), this coaction appears to have been truly ancient and to have persisted with little change for almost a quarter of a billion years. One must wonder at what point this tight association did become extinct. Further study of crinoid material throughout the geologic column will be required to

Ichnogenus *Tremichnus,* Brett 1985
Ichnospecies *Tremichnus* cf. *cysticus*

Description: Pits circular to elliptical in largest samples, regularly parabolic in cross section, about 2/3 as deep as wide; ranging diameter from approximately 0.1 to over 6.5 mm in minimum dimension (mean 2.28 mm; SD 1.18 mm;

N = 60). Pits occurring entirely on crinoid columns. Some pits are surrounded by swellings produced by differential growth of columnals in the area of the pit. A few pits display slight concentric rims.

Remarks: All the *Tremichnus* pits in the Jurassic crinoids occur on columns, and nearly all are associated with swellings of the column, as in *T. cysticus* from the Middle Paleozoic. Gall-like protuberances are rare, in contrast to Paleozoic examples. The mean size of these embedment pits is slightly larger than that observed for *Tremichnus* on Middle Paleozoic crinoids (see Brett 1985), and the maximum diameters (6.5 mm) are substantially larger than the largest Paleozoic *Tremichnus* (see Brett 1978). Cases of overlap between adjacent pits are very rare, in part because clustering of pits is less dense on the Jurassic crinoids. *Tremichnus* on Jurassic crinoids from Israel do not appear aligned along sutures, although none are known from cups of crinoids; so only columnal sutures can be assessed here.

At present, the *Tremichnus* from the Late Jurassic of Israel are the geologically youngest known occurrences of this ichnogenus. No examples of *Tremichnus* are known from the columns of Cretaceous or Cenozoic crinoids. However, additional fossil evidence of this type of interaction should be sought in collections of crinoid material throughout earth history, including disarticulated columns and calyx plates.

SUMMARY

Crinoid pluricolumnals discussed in this paper from the Jurassic (Upper Callovian-Lower Oxfordian) beds of south-central Israel were heavily encrusted and bored by a variety of organisms. Eleven species of epi- and endobionts were recognized, including encrusting foraminifers, sclerosponges, bryozoans, serpulid polychaetes, and bivalves. A variety of trace fossils are also present on 57% of pluricolumnals; these include *Trypanites* on over 50%, acrothoracid barnacle borings, channel-like borings and bivalve crypts *(Gastrochaenolites)* on 20% of columns.

The identification of embedment pits of *Tremichnus* in crinoid columns from the Upper Jurassic (Callovian) of Israel permits a range extension of over 100 million years for the ichnogenus. A crinoid-parasite interaction persisted from the Middle Ordovician to the Late Jurassic (nearly 200 million years) and through the Permo-Triassic extinction, the greatest crisis in the history of marine life.

These observations provide insights into the evolution of echinoderm encrusting associations, as well as the longevity of host-parasite interactions.

Certain of the Jurassic epi- and endobionts, such as the bryozoan *Stomatopora* and *Tremichnus*, represent archaic lineages, associated with crinoids since at least the Middle Ordovician. Others, including ostreid and lithophagid bivalves, are representative of the "Modern fauna" (see Sepkoski 1981). Hard-substrate communities tend to be evolutionarily conservative (Brett 1988). Jurassic echinoderm substrates retain most of the guilds and many of the higher level taxa of epi- and endobionts that occur on Paleozoic crinoids; however, the addition of new taxa resulted in an overall higher diversity of colonizers and a larger number of guilds of organisms on these Mesozoic than on Paleozoic crinoid columns.

ACKNOWLEDGMENTS

Feldman acknowledges support from the National Geographic Society (grant nos. 4739-92 and 5666-96) and wishes to thank Dr. Yaacov Mimran, former Director, Geological Survey of Israel, for providing office space, laboratory, and library facilities during his tenure as visiting scientist for two field seasons. We are grateful to Dr. Francis Hirsch for his assistance in the field and for many hours of discussion on the stratigraphy of Israel. We also thank Susan Klofak and Andrew Modell, both of the American Museum of Natural History (AMNH), New York, for preparing the specimens and taking the photographs, respectively. Peting Melville (AMNH) helped with the scanning electron micrographs and Aliza Eisdorfer assisted in laboratory preparation and statistical treatment of the specimens. Richard Alexander and Christina Franzen reviewed the manuscript and provided numerous suggestions for its improvement.

REFERENCES

Alexander, R, and C. E. Brett. 1990. Symposium on Paleozoic epibionts: Introduction. *Historical Biology* 4: 151-153.

Baumiller, T. K. 1990. Non-predatory drilling of Mississippian crinoids by platyceratid gastropods. *Palaeontology* 33: 743-748.

Baumiller, T. K., and W. I. Ausich. 1992. The broken-stick model as a null hypothesis for crinoid stalk taphonomy and as a guide to the distribution of connective tissue in fossils. *Paleobiology* 18: 288-298.

Bell, B. M. 1975. A study of North American Edrioasteroidea. *Memoirs New York State Museum and Science Service* 21: 1-447.

Bordeaux, Y. L. and C. E. Brett. 1990. Substrate specificity of epizoans on Middle Devonian brachiopods. *Historical Biology* 4.3: 203-220.

Boucot, A. J. 1990. *Evolutionary Paleobiology of Behavior and Coevolution.* Amsterdam: Elsevier.

Brett, C. E. 1978. Host-specific pit-forming epizoans on Silurian crinoids. *Lethaia* 11: 217-232.

------. 1981. Hold fast adaptations in pelmatozoan echinoderms. *Lethaia* 14: 343-370.

------. 1985. *Tremichnus:* A new ichnogenus of circular-parabolic pits in fossil echinoderms. *Journal of Paleontology* 59: 625-635.

------. 1988. Paleoecology and evolution of marine hard substrate communities: An overview. *Palaios* 3: 373-378.

Brett, C. E., and J. D. Eckert. 1982. Palaeoecology of a well preserved crinoid colony from the Silurian Rochester Shale in Ontario. *Royal Ontario Museum Contributions to Life Sciences* 131: 1-20.

Breimer, A. 1978. General morphology, Recent crinoids. In R. C. Moore and C. Teichert (eds.), *Treatise on Invertebrate Paleontology, Pt. T, Echinodermata,* T9-T58. Lawrence, KS: Geological Society of America and University of Kansas Press.

Eckert, J. D. 1988. The ichnogenus *Tremichnus* in the Lower Silurian of western New York. *Lethaia* 21: 281-283.

Feldman, H. R. 1987. A new species of the Jurassic (Callovian) brachiopod *Septirhynchia* from northern Sinai. *Journal of Paleontology* 61: 1156-1172.

Feldman, H. R., E. F. Owen, and F. Hirsch. 1991. Brachiopods from the Jurassic of Gebel El-Maghara, northern Sinai. *American Museum Novitates* 3006: 1-28.

Franzen, C. 1974. Epizoans on Siluro-Devonian crinoids. *Lethaia* 7: 287-301.

Gill, G. A., and H. Tintant. 1975. Les ammonites Calloviennes du sud d'lsrael. *Israel Journal of Earth Sciences* 14: 122-138.

Goldberg, M. 1963. Reference section of Jurassic sequence exposed in Hamakhtesh Hagadol (Kurnub Anticline). Detailed binocular sample description, including field observations. Unpublished Internal Report, Oil Division, *Geological Survey of Israel.*

Graff, L. von. 1884. Report on the Myzostomida collected during the voyage of the *H.M.S. Challenger* 1873-1876. *Report on the Scientific Results of the Exploring Voyage of H.M.S. Challenger, Zoology* 10: 1-126.

Guensberg, T. 1992. Paleoecology of hardground encrusting and commensal crinoids, Middle Ordovician, Tennessee. *Journal of Paleontology* 66: 129-148.

Hantzschel, W. 1975. Trace Fossils and Problematica (1st supplement). In C. Teichert (ed.), *Treatise on Invertebrate Paleontology, Pt.* W *(2nd Ed.), Miscellanea,* 1-269. Lawrence, KS: Geological Society of America and University of Kansas Press.

Hirsch, F. 1979. Jurassic bivalves and gastropods from northern Sinai and southern Israel. *Israel Journal of Earth Sciences* 28: 128-163.

Hudson, R. G. S. 1958. The Upper Jurassic faunas of southern Israel. *Geological Magazine* 95: 415-425.

Jackson, J. B. C. 1977. Competition on marine hard substrata: The adaptive significance of solitary and colonial strategies. *American Naturalist* 111: 743-769.

Lane, N. G. 1973. Paleontology and paleoecology of the Crawfordsville fossil site (Upper Osagian, Indiana). *University of California Publications in Geological Sciences* 99: 1-141.

Lane, N. G. 1978. Mutualistic relationships of fossil crinoids. In R. C. Moore and C. Teichert (eds.), *Treatise on Invertebrate Paleontology, Pt. T, Echinodermata,* T345-T347. Lawrence, KS: Geological Society of America and University of Kansas Press.

Lewis, R. 1982. Holdfasts. In J. Sprinkle (ed.), *Echinoderm faunas from the Bromide Formation (Middle Ordovician) of Oklahoma. The University of Kansas Paleontological Contributions, Monograph* 1: 1-369.

Lewy, Z. 1983. Upper Callovian ammonites and Middle Jurassic geological history of the Middle East. *Geological Survey of Israel Bulletin* 76: 1-56.

Mayne, W. 1966. Microbiostratigraphy of the Jurassic of Israel. *Geological Survey of Israel Bulletin* 40: 1-56.

Mcintosh, G. C. 1980. Dependent commensal associations between Devonian crinoids and two species of tabulate corals. *Geological Society of America Abstracts with Programs* 12(7): 481.

Moodie, R. L. 1918. On parasitism of Carboniferous crinoids. *Journal of Parasitology* 4: 174-176.

Parnes, A. 1961. On the occurrence of *Pseudopygurus* Lambert in southern Israel. *Israel Research Council Bulletin* 10G: 216-222.

Pemberton, S. G., D. R. Kobluk, and M. J. Risk. 1980. The boring *Trypanites* at the Silurian-Devonian disconformity in southern Ontario. *Journal of Paleontology* 54: 1258-1266.

Picard, L., and F. Hirsch. 1987. *The Jurassic Stratigraphy in Israel and the adjacent Countries.* Jerusalem: The Israel Academy of Sciences and Humanities.

Rasmussen, K., and C. E. Brett. 1985. Taphonomy of Holocene cryptic biotas from St. Croix, Virgin Islands: information loss and preservational biases. *Geology* 13: 551-553.

Reiner, W. 1968. Callovian gastropods from Hamakhtesh Hagdol (southern Israel). *Israel Journal* of *Earth Sciences* 17: 171-198.

Robison, R. A. 1987. Annelida. In R. S. Boardman, A. H. Cheetham, and A. J. Rowell (eds.), *Fossil Invertebrates,* 194-204. Palo Alto, CA: Blackwell.

Sandy, M. R. 1996. Oldest record of peduncular attachment of brachiopods to crinoid stems. Upper Ordovician, Ohio, U.S.A. (Brachiopoda; Atrypida; Echinodermsta; Crinoidea). *Journal of Paleontology* 70: 532-534.

Sepkoski, J. J., Jr. 1981. A factor analytic description of the Phanerozoic marine fossil record. *Paleobiology* 7: 36-53.

Taylor, P. O., and R. W. Furness. 1978. Astogenetic and environmental variation of zooid size within colonies of Jurassic *Stomatopora* (Bryozoa, Cyclostomata). *Journal of Paleontology* 52: 1093-1102.

Vaney, C. 1913. L'adaptation des gastropodes au parasitisme. *Bulletin Scientifique de la France et de la Belgique* 47: 1-87.

Wanner, J. 1938. Beitrage zur Paläontologie des Ostindischen Archipels. XV. *Balanocrinus sunldaicus* n. sp. und sein Epoke aus dem Altmiocän der Insel Madurs. *Neues jahrbuch Mineralogie, Geologie, Paläontologie, (Beilage-Band)* 79: 385-402.

Warn, J. M. 1974. Presumed myzostomid infestation of an Ordovician crinoid. *Journal of Paleontology* 48: 506-513.

Welch, J. R. 1976. *Phosphannulus* on Paleozoic crinoid stems. *Journal of Paleontology* 50: 218-225.

Werle, N. G., T. I. Frest, and R. H. Mapes. 1984. The epizoan *Phosphannulus* on a Pennsylvanian crinoid stem from Texas. *Journal of Paleontology* 58: 1163-1166.

Appendix

Detailed stratigraphic information for subunits 57, 45, 43, and 39, Zohar and Matmor formations (Callovian), Hamakhtesh Hagadol, Negev Desert, Israel.

Subunit 57

Lithology.—The base of the subunit (sample no. 62/2/167) consists of chalky, detrital (mostly bioclastic) limestone, brownish and light yellow-brownish, with limonitic stains. Hardness and coarseness increase toward the upper part with sponge spicules, corals, stromatoporoids, and sclerosponges common. Just above (sample no. 62/2/168), the limestone is light brown, chalky with regular limonitic stains. Sponge spicules and some ostracodes are present. There is some light yellow, brownish shale with occasional black crusts of gypsum crystals and veins. Slightly above (sample 62/2/169), the limestone is rich in fine branching corals. Shale occurs only in the uppermost part. The top of the subunit (sample no. 62/2/170) consists of a darker, less chalky limestone with numerous partly recrystallized sponge spicules and, in places, large, poorly preserved pelecypods.

Formation.—Matmor Formation.

Age.—Callovian.

Number of meters above base.—146-149.

Subunit 45.

Lithology.—Subunit 45 contains Goldberg's (1963) "second" reef complex. The base of the subunit (sample no. 62/2/113) consists of detrital limestone, partly pseudoolitic, light yellow-brown in color with limonitic stains. Sponge spicules are common, as are corals. The rock is bouldery at weathered surfaces. Just above (sample no. 62/2/114), the limestone becomes soft, highly chalky, and argillaceous with cherry corals. Sample 62/2/115, just above, is rich in corals and "fillings" of argillaceous material responsible for the boulder-like appearance. Many nautiloids, brachiopods, and echinoid spines are present. Sample no. 62/2/116 is identical to sample no. 62/2/114, whereas sample no. 62/2/117 is similar to sample no. 62/2/115, with the exception that here the rock is an almost pure coral-stromatoporoid limestone. The top of the subunit (sample no. 62/2/118) appears as an irregular coral-stromatoporoid limestone, softer at the upper part.

Formation.—Matmor Formation.

Age.—Callovian.

Number of meters above base.—114-118.

Subunit 43

Lithology.—Subunit 43 contains Goldberg's (1963) "first" reef complex. The base of the subunit (sample no. 62/2/104) consists of light yellow-brown, partly pseudoolitic, detrital limestone with some limonitic staining. Numerous sponge spicules, foraminifera, corals, stromatoporoids, brachiopods, gastropods, and ammonites are present. The rock is jointed, with limonitic brown crusts and hematitic concretions. Above (sample no. 62/2/105), the rock is similar, but harder. In parts the limestone is coral-rich. partly replaced by drusy calcite. Gastropods, echinoids, and ammonites are present, along with some sponge spicules. Laterally, a small branching coral reef develops (up to 1 m thick). Sample 62/2/106 is similar but lighter in color, softer, and partly chalky. The top section (sample 62/2/107) is also similar but somewhat argillaceous.
Formation.—Matmor Formation.
Age.—Callovian.
Numbers of meters above base.—106-110.

Subunit 39

Lithology.—This subunit is equivalent to Goldberg's (1963) "Big Calcarenite." The base (sample no. 62/2/86) consists of light yellow-brown-gray, partly pseudoolitic, slightly argillaceous limestone with limonitic staining. There is a rich megafauna: sponge spicules, foraminifera, corals, stromatoporoids, echinoids, brachiopods, gastropods, pelecypods, and ammonites. The lower part is harder and the upper part softer, chalky, and slightly more argillaceous. Sample 62/2/87, just above, is similar but darker, whereas sample 62/2/88 is harder and lighter in color. The next higher sample (62/2/89) consists of calcareous shales. The same rich megafauna noted in sample 62/2/86 occurs throughout the entire subunit.
Formation.—Zohar Formation, Madsus Member.
Age.—Callovian.
Number of meters above base.—93-96.

Chapter Six

The Jurassic of the Southern Levant. Biostratigraphy, Palaeogeography and Cyclic Events

ABSTRACT

During the Jurassic, the Levant was part of the Gondwanian Tethys platform-shelf. The palaeotectonic setting of the Levant consisted in relative lows and highs, controlled by differential rates of subsidence: the shallow Negev high in southern Israel (900-2000 m), the Maghara basin in northern Sinai (3000 m), the central Israel Judean embayment (3000 m), extending to south Antilebanon (Hermon), the northern platform of Galilee (1500-2000 m), extending to Lebanon and northern Syria. Clastics are the result of the wearing down of the Arabian Massif in the South and South East. Yet, thickening of the early Oxfordian shales in the offshore wells, points to derivation from some land-mass in the West, now hidden below the Mediterranean. The Gevar'Am trough, alongside the present Levant-coast, formed during the Tithonian. Volcanics mark the beginning and end of the Jurassic (Asher at 197 my and Tayasir from 140 my onward). The stratigraphic sequence consists at its base of lateritic palaeosols (Mishhor Formation). Marine deposition set during Pliensbachian times. In the North, monotonous platform carbonates (Haifa Formation) are interrupted by the Oxfordian volcanics of Devorah. To the South, the fluviatile clastic intervals of the Toarcian-Aalenian (Inmar Formation) and Bathonian (Sherif Formation) characterize the Negev facies. Shales (Karmon Formation) straddle the Bathonian-Callovian limit. A middle-late Callovian hiatus occurs in the Coastal plain. The deposition of Oxfordian shales (Majdal Shams Formation) is restricted to the basins of North Sinai, central Israel to south Antilebanon (Judean embayment). Late Jurassic bioherms and reefs (Nir' Am Formation) form a belt along the coastal plain area. In the Kimmeridgian, oolitic facies, shales and occasional sands recur (Nahar Sa'ar Formation). The Tayasir volcanics and Gevar'Am turbidites cover a truncation surface.

RÉSUMÉ

Le Jurassique du sud Levant.
Biostratigraphie, paléogéographie et cycles sédimentaires.

Durant le Jurassique, la région du Levant faisait partie de la plate-forme nord gondwanienne de la Téthys. La subsidence varie selon les régions considérées.

On distingue dans le Sud d'Israël, le haut fond du Negev (900-2000 m); dans le Sinaï le bassin de Maghara (3000 m); au centre d'Israël le bassin de la baie de Judée (3000 m), prolongé jusqu'a l'Antiliban (Hermon); la plate-forme septentrionale de la Galilée (1500-2000 m) qui s'étend au Liban et au Nord de la Syrie. L'essentiel des apports clastiques provient de l'érosion du Massif arabe au Sud et au Sud-Est. Cependant, l'épaississement des shales oxfordiens dans les sondages offshore traduisent vraisemblablement l'existence d'une terre émergée située en Méditerranée. Au Tithonien, la fosse de Gevar'Am s'individualise parallèlement a la côte actuelle du Levant. Le début et la fin du Jurassique sont marqués par des épisodes volcaniques (Asher à 194 Ma et Tayasir à partir de 140 Ma). La succession lithologique est la suivante. Succédant à des dépots latéritiques (Formation de Mishhor), à la base du Jurassique, le régime marin s'établit au Pliensbachien. Au Nord, se déposent des calcaires monotones de plates-formes (Formation de Haifa), interrompus à l'Oxfordien par l'épisode volcanique de Devorah. Vers le sud, des grès fluviatiles s'intercalent dans les intervalles Toarcien-Aalénien (Formation de Inmar) et Bathonien (Formation de Sherif). La limite Bathonien-Callovien est marquée par le dépôt de shales (Formation de Karmon). Dans la plaine côtiére, on constate une lacune du Callovien moyen et supérieur. Les shales oxfordiens (Formation de Majdal Shams) se sont déposés dans les bassins de Maghara, du centre d'Israël et dans la partie méridionale de l'Antiliban (baie de Judée). Au Jurassique supérieur, la Formation corallienne de Nir' Am s'étire le long de la plaine côtiére. Au Kimméridgien, des faciès oolitiques, des shales et parfois des sables caractérisent la Formation de Nahar Sa'ar. L'épisode volcanique de Tayasir, et les turbidites de Gevar'Am, d'âge tithonien, reposent sur une troncature d'érosion.

INTRODUCTION

Since the monograph on the Jurassic of Israel and adjacent countries (Picard & Hirsch, 1987; Hirsch & Picard, 1988), contributions have been added, in the field of palaeontology and biostratigraphy: palynomorphs (Conway, 1990), brachiopods (Feldman, 1987; Feldman et al., 1991), ostracods (Rosenfeld et al., 1987a, 1987b, 1988, 1991), ammonites (Cariou, this paper), foraminifera and algae (Bassoullet & Grossowicz, this paper). In the more general field of palaeogeography, interpretation of seismic profiles in the east Mediterranean Levantine basin and the drilling activity offshore the Sinai coast (Hirsch et al., 1995) have shed new light on the existence of a thick sedimentary succession, including the Jurassic.

The biostratigraphy of the Jurassic in the Levant often exhibits an endemic character, related to that of the southern Tethys shelf. The Jurassic succession of the Levant is well known from the boreholes in Israel and from the outcrops of

Gebel Maghara (Sinai, Egypt), Makhtesh Ramon, Hamakhtesh Hagadol (Negev, Israel) and Mt. Hermon (Antilebanon). The Jurassic facies of the Levant shares its particularities with African, Apulian and Arabian shelf-platforms. Boreholes supply a good insight in the development of the Jurassic in the Levant, linking the over 2000 m of late Liassic to Kimmeridgian strata of Sinai, similar to the Negev series, to the 2500 m of Middle to Upper Jurassic strata of Mount Hermon, which give a good insight to the succession buried under central Israel. The Jurassic of Lebanon resembles the succession of northern Israel. The Lower and Middle Jurassic series in Makhtesh Ramon (central Negev) provides the stratotypes of the Mishhor, Ardon and Inmar formations. The Middle-Upper Jurassic part of the Zohar Formation and the late Callovian Matmor formation are exposed in Hamakhtesh Hagadol (northern Negev). The latter represents a unique facies of the late Callovian in the Middle East. The almost complete Middle to Upper Jurassic of Mount Hermon (Antilebanon) exposes the Hermon, Majdal Shams and Nahar Sa'ar formations. The latter, also exposed in eastern Samaria, is unconformably covered by the Tayasir volcanics.

The following sections were selected in the present study: the northern Israel Devorah borehole (I), the Hermon exposure and the southern coastal plain Helez-Deep borehole (II); the northern Negev Kurnub borehole and Hamakhtesh Hagadol exposure and western Negev Qeren borehole (III) and the northern Sinai Hallal borhole and Maghara exposure (Figs 1 and 2). In the stratigraphic subdivision of Israel, the Jurassic is comprised within the Arad group. The present review attempts to reduce the number of formations to units, well defined in exposed stratotypes, though in a few cases subsurface units, exposed nowhere, had to be retained (marked by *):

> Mishhor, Asher*, Ardon, Inmar, Qeren, Haifa*, Daya*, Sherif*, Hermon, Zohar, Matmor, Niram*, Majdal Shams, Nahar Sa'ar and Gevar'Am*.

STRATIGRAPHIC OVERVIEW

LIASSIC

The Mishhor Formation (Goldberg, 1969) consists of a bauxitic-lateritic palaeosol and flint clays, resting on an eroded Triassic landsurface. Its average thickness does not surpass 50 m. In the North of the country it is replaced by the Asher volcanics. In the northern Sinai (Egypt) borehole of Gebel Ralal, 500 m of red marls, some dolomites and sands of Norian (*Granuloperculatipollis rudis* Van Erve, 1977) through early Liassic (*Araucariasites australis* Van Erve, 1977) underlay the typical Mishhor pisolites and red clays. Normally the Mishhor Formation overlies a karstic surface leaching the Carnian Mohilla or Noro-Rhaetian Shefayim Formation (Ga'ash borehole).

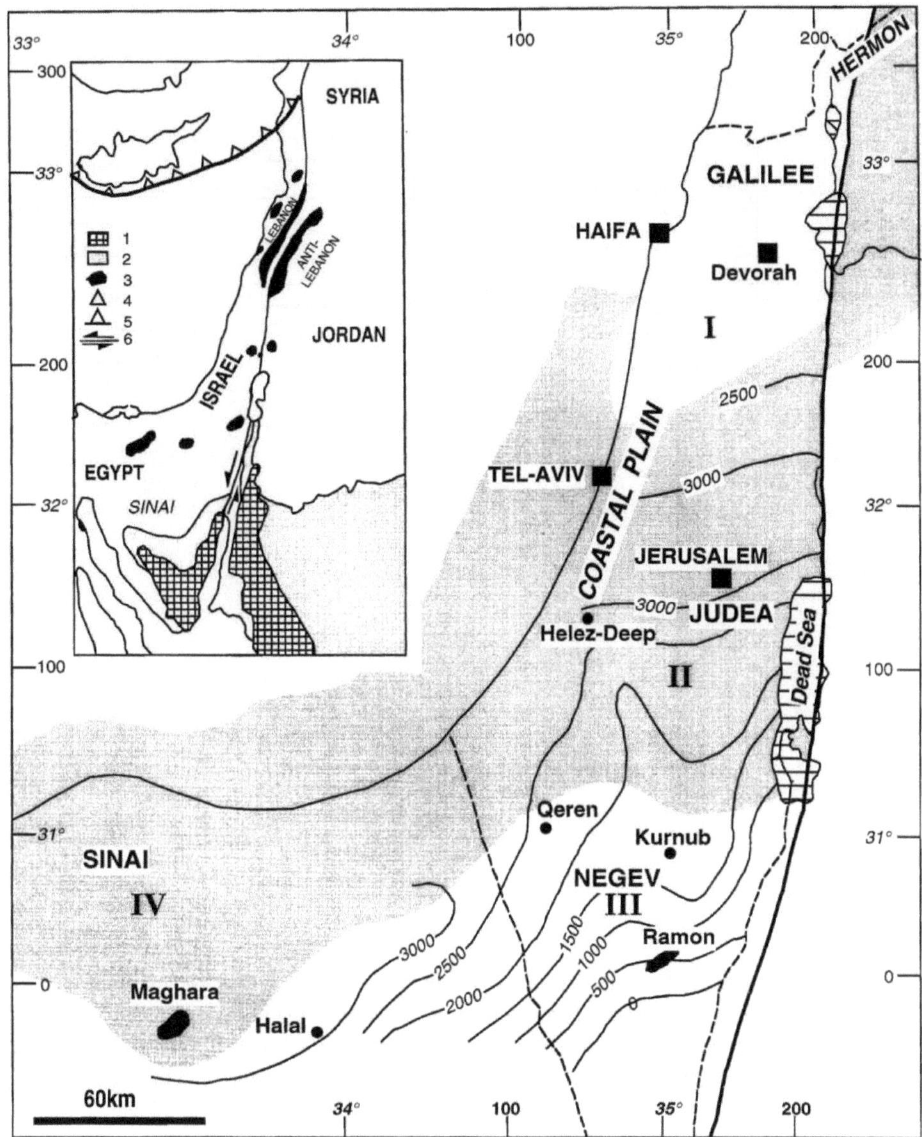

FIGURE 1: Location map and reference map to the Levantine countries. 1, Precambrian; 2, Palaeozoic-Triassic; 3, Jurassic; 4, Mount Hermon; 5, Alpine thrust-front; 6, Neogene sinistral transform. (I) Galilee High; (II) Judean embayment; (III) Negev High; (IV) Sinai Deep, with extension of lower Oxfordian Majdal Shams shales (shaded area) and total Jurassic isopachs.

FIGURE 1: Carte de situation et de référence de la région levantine. 1. Précambrien; 2, Paléozoïque-Trias; 3, Jurassique; 4, Mont Hermon; 5, chevauchement alpin; 6, transformante sénestre Néogène. (I) Haut fond de Galilée; (II) Baie de Judée; (III) Haut fond du Neguev; (IV) Bassin du Sinaï; Isopaques du Jurassique; (en grisé) les shales oxfordiens de Majdal Shams.

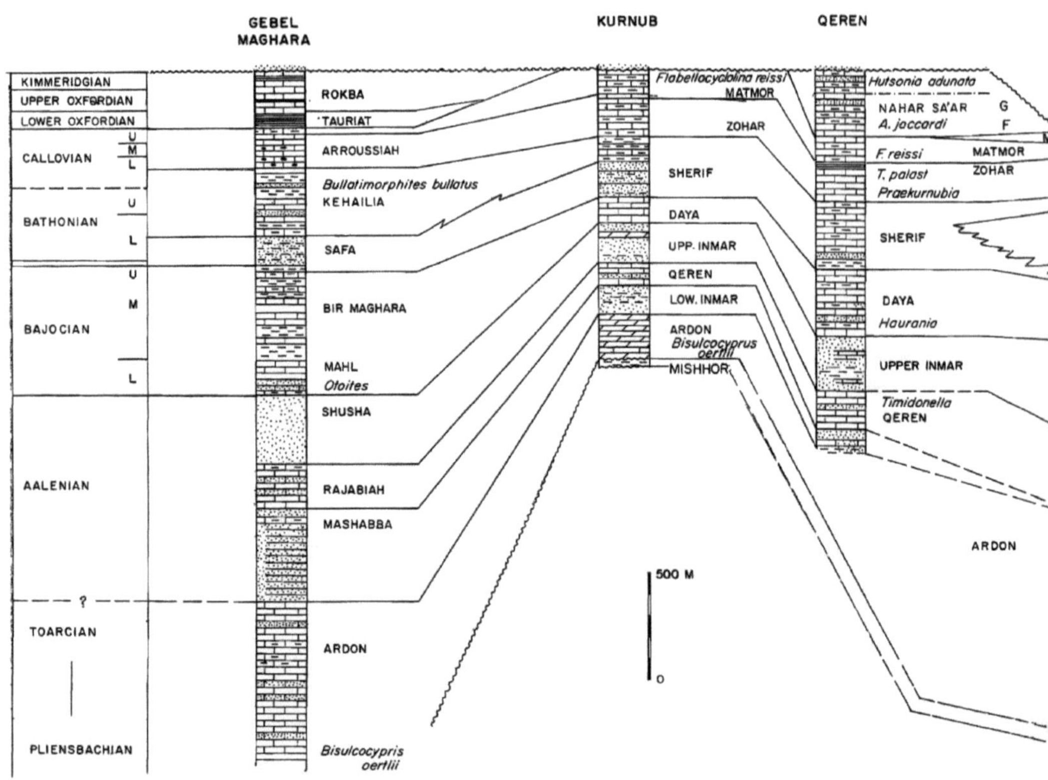

Figure 2: Stratigraphic sections. (I) Galilee High Devorah: the Liassic Asher volcanics are covered by the Middle Jurassic carbonatic Haifa Formation, the volcanic Devorah tuffs, interstratified within the Oxfordian, followed by the Oxfordian-Kimmeridgan Nahar Sa'ar Formation. (II) Judean embayment Helez-Hermon: in the Helez borehole, the Liassic Mishhor laterites are followed by Pliensbachian-Toarcian carbonates (Ardon). In the Ramallah borehole Aalenian carbonates (Qeren) directly overlie the Mishhor laterites. Middle Jurassic is primarily represented by carbonates of Haifa-Hermon type. The lower Oxfordian is characterized by the development of the Oxfordian Majdal Shams shales, which extend to the NW Negev and Sinai (Egypt). (III) Negev High (Kurnub-Qeren): The Negev section is characterized by numerous clastics in the Kurnub borehole, becoming more carbonatic toward the Qeren borehole. The Liassic consists of the Mishhor and Ardon formations. The Aalenian Inmar Formation is split by the Qeren carbonates. The Middle Jurassic is diversified, comprising the Daya, Sherif, Zohar and Matmor formations. The Upper Jurassic is truncated in Kurnub, its development being well represented at Qeren by the Nahar Sa'ar Formation. (IV) Maghara Deep: Gebel Maghara exposure and Halal borehole: In the more clastic succession of the Maghara Deep, the Negev units are still well represented, thought the total thickness has more than doubled. A-G: palynozones (after Conway, 1990).

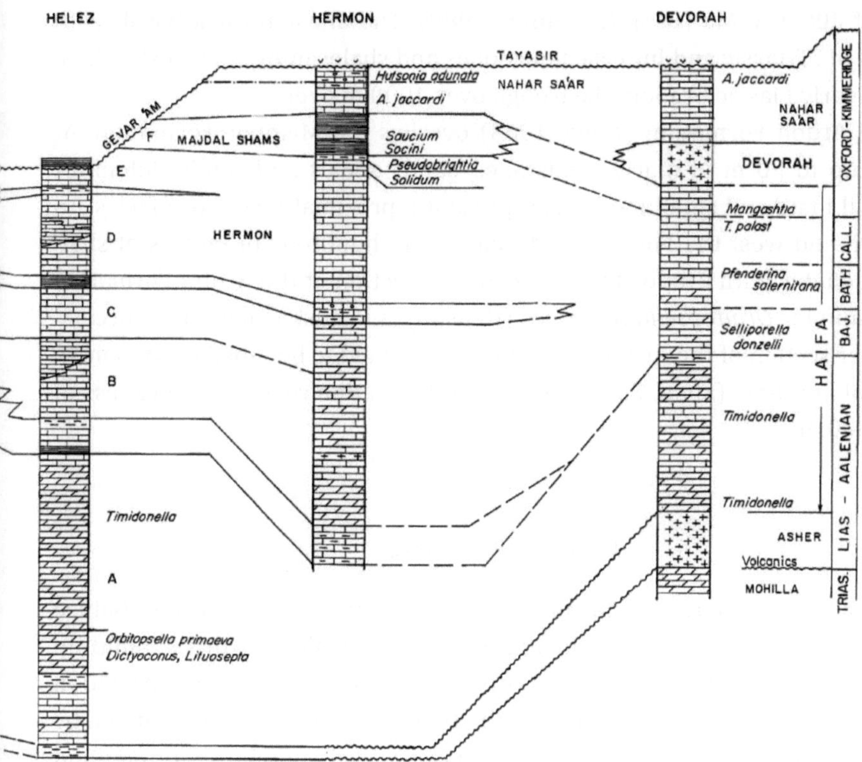

FIGURE 2: Coupes stratigraphiques. (I) Plate-forme de Galilée: les roches volcaniques de Asher sont recouvertes par la formation carbonatée de Haifa, les tuf volcaniques de Devorah interstratifiés à l'Oxfordien et la formation oxfordo-kimmeridgienne de Nahar Sa'ar. (II) La Baie de Judée de Helez à Hermon: dans le forage de Helez, les latérites liasiques de Mishhor sont suivies des carbonates du Pliensbachien-Toarcien (Ardon). Dans le forage de Ramallah, les carbonates de l'Aalénien (Qeren) reposent directement sur les latérites du Mishhor. Le Jurassique moyen est représenté par les carbonates du type Haifa-Hermon. L'Oxfordien se caractérise par le développement des marnes oxfordiennes de Majdal Shams, qui s'étendent au nord-ouest de Neguev et au Sinaï (Egypte). (III) Haut-fond du Neguev (Kurnub-Qeren): la coupe du Neguev se distingue par son abondance en roches clastiques dans le sondage de Kurnub, devenant plus carbonatéen direction de Qeren. A l'Aalénien, les carbonates de Qeren s'imbriquent dans la Formation de Inmar. Le Jurassique moyen est diversifié et comprend les Formations de Daya, Sherif, Zohar et Matmor. Le Jurassique supérieur est tronqué à Kurnub tandis que dans le sondage de Qeren la Formation de Nahar Sa'ar est bien développé. (IV) Bassin de Maghara: dans la coupe du Gebel Maghara et du sondage de Hallal (Sinaï), les unités du Neguev sont représentées, tandis que leur épaisseur totale est doublée. A-G: paynozones (d'après Conway, 1990).

The Asher volcanics in northern Israel (Atlit, Haifa, Asher and Devora) consist of basalt, agglomerates and tuffs (with few thin dolomite and shale interbeds), of 200-180 Ma (Lang & Steinitz, 1989). The unusual thick basalt with enclaves of ?Triassic and Jurassic limestones and shales in Atlit suggests a Late Triassic-early Liassic Graben-like trough over 4000 m deep.

The Ardon Formation (Nevo, 1963) overlies the Mishhor Formation. At Ramon, up to 50 m of marly sandstones, sandy shales and sandy dolomites with algal mats and small molluscs represent a proximal nearshore facies. To the north and west the Formation develops into hundreds of meters of shallow carbonates with evaporites. In the lowest section of the Ardon Formation the foraminifer *Orbitopsella primaeva* (Henson), a Pliensbachian taxon occurs. The upper section of the Ardon Formation is assumed to be mostly Toarcian. In the Hallal borehole (Sinai, Egypt) the Ardon Formation reaches a thickness of almost 800 m.

MIDDLE JURASSIC

From the northern coastal plain (Ga'ash) to the north (Carmel, Galilee and Lebanon) over 1000 m of dolomicrites, occasionaly anhydritic, build up an undifferentiated Aalenian to Oxfordian succession of the Haifa Formation (Derin, 1974). The Aalenian interval of the Haifa Formation is characterized by the foraminifer *Timidonella* sp. The Bajocian interval contains *Haurania* and the Bathonian interval *Paleopfenderina salernitana* Sartoni & Crescenti. The Callovian is represented by a lower interval characterized by an assemblage of *Kilianina* sp., and an upper interval with *Trocholina palasteniensis* (Henson, 1948). The uppermost part of the uniform sedimentary regime of the Haifa Formation comprises micrites yielding the algae *Pseudoclypeina* and the foraminifer *Levantinella egyptiensis* (Fourcade et al., 1983) (ex *Manghastia)* of questionable late Callovian-early Oxfordian age.

Aalenian

The Inmar Formation (Nevo, 1963), consisting of cross-bedded sandstones, interbedded with kaolinitic clays, coally flint-clays and alunite horizons (Goldberry, 1982) was compared by Lorch (1967) with the Aalenian Yorkshire lower estuarine beds. A thin marine intercalation at Ramon yields Toarcian brachiopods (Parnes, 1980) and separates further North and West the Inmar clastics into a lower and upper part. This marine intercalation, known as the Qeren Formation (Goldberg, 1964), consists of carbonates that in the boreholes west of Ramon, interwedge the fluvio aeolian-deltaic Inmar clastics, replacing gradually the clastic lower part of the Inmar Formation until it merges with the underlying Ardon carbonates. These pelbiomicrites, oosparites and dolomicrites yield the Aalenian-lower Bajocian foraminifer *Timidonella* sp. (Bassoullet) accompanied in the oolitic shoals at Helez by *Dictyoconus cayeuxi* (Lucas) of the same age. At Maghara, an ammonite fragment was attributed to *Grammoceras* or *Sonninia,* confering a Toarcian-Aalenian age to the Qeren Formation (Arkell, 1956). In the Negev boreholes (Qeren, Makhtesh Qatan, Kurnub, Daya, Barbur, Boqer, Rekhme) and Sinai (Halal) the upper part of the Inmar Formation consists of sandstones and shales. The Inmar Formation passes from predominantly continental in Ramon to shallow marine to the west and north, consisting of shaly marls (Rosh Pinna) with ostracods and calcareous nannoplancton (Moshkovitz & Ehrlich, 1976). Further west and north the Inmar facies wedges entirely out, replaced by the carbonate facies of the Haifa Formation.

Bajocian

Nearly 100 m of marine cross-bedded sands with ripple marks, ferruginous oolitic dolomites and plantiferous shales yield late middle and late Bajocian ammonites at Ramon (Parnes, 1981). The arenaceous facies of the central Negev develops into the foremostly carbonatic facies of the Daya Formation (Eliezri, in Coates et al., 1963) of the Negev boreholes, which is defined by the foraminifer *Haurania.* It reaches in the Hermon section over 900 m of carbonates with volcanic intercalations.

In central Israel, the coastal plain and Judea, oosparites (Sederot facies) and micrites replete with poriferan spiculae (Barnea facies) may reach up to 400 m. Differing from the latter, the marly 500 m of the Mahl and Bir Maghara formations (Al Far, 1966) in North Sinai, yield a plantiferous clastic succession of shaly limestone, shales, shaly sandstone and coal streaks, followed by massive oncolitic coralline limestones, interstratified by thick marls and shales, rich in ammonoids (Douvillé, 1916; Arkell, 1952, 1956; Parnes, 1981), topped by black shales, from which the ammonite *Thambites planus* Arkell 1956 was determined (Parnes, 1981).

On the Syrian side of the Hermon, the 170 m medium-to fine-bedded oolitic organo-clastic, ochreous-stained limestone (J2bt) yield brachiopods and bivalves (Razvalyaev, 1966). These “calcaires ocres” (Vautrin, 1934) may correspond to a shaly horizon, interwedging the oosparites and spiculites of the Sederot and Barnea facies, close to the Bajocian-Bathonian boundary in the southern coastal plain, around Helez (Derin, 1974).

Bathonian

In most of the Negev boreholes, the Sherif Formation (Eliezri, in Coates et al., 1963) consists of alternating sandstone, shale, pelmicritic and some sparitic limestone. The more shaly upper part of the Sherif Formation corresponds in the coastal plain and northernmost Negev to shales, rich in calcareous nannoplankton, ostracods and foraminifers of late Bathonian-early Callovian age. The lower part of the Sherif Formation consists at Gebel Maghara of cross-bedded hematitic stained sandstones alternating with limonitic shales with abundant plant remains and several coal seams (Safa Formation). The upper part of the Sherif Formation encompasses thin oolitic limestone with frequent shale and occasional thin sandstone beds (Kehailia Formation). The ammonite *Micromphalites pustuliferus* (Douvillé) in its lower part belongs to the early Bathonian. The shaley-carbonatic upper part yields *Bullatimorphites bullatus* (d'Orbigny), indicating the upper Bathonian or lower Callovian.

The bright bluish-grey limestones and dolomites of the Hermon Formation (Dubertret, 1960) with corals and brachiopods, that attain 700 m, are according to Razvalyaev (1966) and Kutznetsova & Dobrova (1995) entirely of Callovian age (J2Cl). According to Derin (in Goldberg, 1969), the lower 600 m of the Hermon Formation is equivalent to the Bathonian interval of the subsurface Haifa Formation.

Callovian

The alternating limestone, marl and shale series that was identified in the northern Negev boreholes as the Zohar Formation (Coates et al., 1963) belongs entirely to the Callovian. The upper half of the Zohar Formation is well exposed at Hamakhtesh Hagadol (Kurnub). It consists in its lower part of stromatoporoid and coral limestone, shale and sandstone, marl and limestone, placed in the middle Callovian *coronatum* Zone (Gill & Tintant, 1975; Gill et al., 1985). Its upper part of alternating marl, limestone and shale, is placed in the late Callovian *athleta* Zone (Gill & Tintant, 1975 and Lewy, 1983). The section at Hamakhtesh Hagadol, above the Zohar Formation, consists of the alternating limestone and marl of the Matmor Formation (Goldberg, 1963; emend. Hirsch & Roded, 1996). Its lower part yields ammonites of the late Callovian *athleta* Zone (Gill & Tintant, 1975; Lewy, 1983), with a bivalve and gastropod fauna still similar to that found in the Zohar Formation. The upper part of the Matmor Formation consists of an oolite near base, alternating yellow shales, white reefoidal and lagunar limestones. In its fauna of large bivalves and gastropods, the genera *Purpuroidea* and *Eunerinea* (Hirsch, 1979) show affinities with European Oxfordian taxa. The ammonite *Peltoceras solidum* Spath still belongs into the late Callovian *athleta* Zone, whereas *Pseudobrightia* sp. puts the uppermost interval to the late Callovian *lamberti* Zone. This interval also yields the foraminifera *Flabellocyclolina reissi* Hottinger and *Kurnubia* gr. *palastiniensis* Henson. The Matmor Formation occurs in the boreholes of the NW Negev and NE Sinai, among which Hallal (Sinai) and Qeren (Negev).

The upper 100 m of the Hermon Formation also consist of massive limestone and yield foraminifera of Callovian age. Along the S-E slopes of Mount Hermon, the uppermost part of the Hermon Formation consists of alternating limestone and marl beds that yield abundant late Callovian belemnites, bivalves and ammonites. Only a reduced, truncated Callovian interval is found in the coastal plain of Israel.

UPPER JURASSIC

A reefoidal facies of biolithites of stromatoporoidea, corals, poriferans and bryozoans occurs under the coastal plain around Caesarea and in the Kokhav field is identified as the Nir Am Formation (Derin, 1974). In the Beer Yaakov, Gerar, Kfar Darom, Nirim, and Kissufim wells it overlies thin Majdal Shams shales. The position of the Nir Am Formation, between Hermon and Gevar 'Am Formations confers probably a mostly Oxfordian age to this Formation, that reaches a thickness of 400 m in the Carmel borehole.

Lower-Middle Oxfordian

The Majdal Shams Formation (Saltzman et al., 1968) (200 m) consists of fossiliferous shales at base, platy limestones and siliciclastics at top. The shales, with limonitic concretions, are interbedded by marls and occasional thin limestone or dolomitic beds. The basal beds are extremely rich in small pyritized ammonites, identified since Noetling (1887) as the early Oxfordian *Brightia socini* zone, put by Arkell (1956) to the "*Mariae*" Zone. In his monograph, Haas (1955) studied 7600 ammonites from Majdal Shams, the exact level of which, within the 43 m thick lower marls, was retrieved by the bed by bed collection reported by Razvalyaev (1966). The nuculide bivalve infaunal mobile detritus feeders confirm the statement of Haas that the paramount darkish shale facies with pyritic dwarf forms intimate to less aerated depositional environment—remarkably also occurring in the Renggeri Zone with equal lithofacies in Switzerland, England and southern Russia. Vautrin's overlying "Lusitanian" consist of argillaceous marls and yellow limestone intercalated by marls holding fossils that Arkell (1956) puts to the *Transversarium* Zone. Razvalyaev (1966) assigns the upper pelitomorphic limestones and alternating dark marls "most probably to the *Bimammatum* Zone." The shale interval of the Majdal Shams Formation is present in the Israeli subsurface and in the nearly 70 m Tauriat shale of Gebel Maghara (Sinai, Egypt). Thin bedded dark platy limestones with marl interbeds and a "tripoli"-like yellow siliciclastic ledge of spiculitic limestone close the succession of the Majadal Shams Formation in the Hermon section. The dark highly pyritic shale with nodules and pyritised fossils, interbedded with few limestones or dolomites (Goldberg & Friedman, 1974) that overlie the Zohar carbonates in the Zohar-Kidod gas-field and in the coastal plain Helez-Kokhav oil-field, was termed "Kidod" by Coates et al. (1963), a term largely followed by exploration geologists. The unequivocal early Oxfordian age (Mariae Zone) of these shales on the base of ammonite-biostratigraphic data (Lewy, 1983; Razvalyaev, 1966) in the exposures of Gebel Maghara (Sinai, Egypt) and Majdal Shams (Mt. Hermon) was emphasized by Picard & Hirsch (1987) and Hirsch (1996). These shales are absent in the area of Hamakhtesh Hagadol and boreholes of Qeren, Rehkme, Boqer and Gebel Hallal.

In many coastal plain wells, early Oxfordian shales rest directly on the thin calcarenite of the lower Zohar Formation (Callovian), showing karstification (Goldberg & Bogosh, 1978; Buchbinder, 1981).

The Majdal Shams Formation reaches over 200 m in Central Israel (from the Helez Area eastwards to Judea and Jordan Valley). In most of the Northern Negev and Northern Sinai, it is reduced to an average of 100 m. In the offshore borehole of Delta-1, NW of Caesarea, a 200 m serie is still present, comprising a carbonitic upper and shaley lower section. The 80 m lower section consists foremostly of black and green shales rich in dinoflagellates, spores and coccolithes (Moshkovitz & Ehrlich, 1980; Cousminer & Conway, 1990), of early Oxfordian age. Fragments of volcanics and of bio-intrasparites were interpreted by Friedman et al. (1971) as cobbles or pebbles deriving from the shelf edge and transported by turbitity currents into the depositional area of mudstone called "Delta facies."

No trace of Oxfordian has been left in the larger and deep erosion funnel around the Helez wells (Barnea, Gevar Am, Helez, Talme Yafe).

Upper Oxfordian—Kimmeridgian

The uppermost part of the Hermon section exposes along its SE flank the succession, totaling up to 180 m, of the Nahar Sa'ar Formation (Saltzman et al., 1968). The echinoid limestone at base consists of ca. 40 m of massive limestones, yielding abundant spines as well as a few complete specimens ot the echinoid *Balanocidaris glandifera* Muenster, for which a late Oxfordian-early Kimmeridgian age is attributed by the foraminifer *Alveosepta jaccardi* (Schrodt, 1894; Derin, 1974). Follow 140 m of well bedded yellow oolites, fossiliferous marls, with a massive limestone and more shale in its upper part at Wadi E Shatr, yielding the indicative ostracod *Hutsonia adunata* Bischoff, 1990 (Rosenfeld & Honigstein, this paper). Razvalyaev (1966) reports from dark-grey limestone and clayey limestone the brachiopods *Septaliphoria jordanica* (Noetling), *Somalirhynchia africana* Weir, and *S. somalica* Daque.

The lower subdivision correlates with 90 m of massive limestone rich in echinoid radioles in Lebanon, the flinty limestone, interstratified with yellow marls at Gebel Rokba at Gebel Maghara (Sinai, Egypt). The upper subdivision is equivalent to a much thicker succession in Lebanon (Dubertret, 1975). There the 80 m of brown to yellow oolitic limestones and marls of the "couches jaunes inférieures" or "couches d'Azour," still with *Alveosepta jaccardi* (Schrodt, 1894), represent a lower unit. Interwedged at this level, occurs the 180 m thick Bhannes volcanic complex, consisting of basaltic sheets and weathered pillows, volcanic tuffs interfingered by lignitiferous and marls alternating with thin limestones. Follows the middle unit of the limestone ledge of the "falaise de Bikfaya" and the upper unit of the "couches jaunes supérieures" or "couches de Salima" in which Bischoff (1964) mentions *Berriasella richteri* (Opp.), an upper Tithonian form for Arkell (1956). The subsurface succession of the Nahar Sa'ar Formation is subdivided in the lower massive limestone of the Beersheba Formation (Coates et al., 1963), followed by the oolite, shale and sand succession of the Haluza Formation (Coates et al., 1963). The foraminifer *Alveosepta jaccardi* (Schrodt, 1894) confers a late Oxfordian-early Kimmeridgian age to the lower 2 thirds of this section. In the Devora borehole of Galilee interformational volcanics occur. A younger unit, consisting of the upper massive carbonate ledge, the so-called Haifa Bay Formation (Derin, 1974), apparently equivalent to the Lebanon "falaise de Bikfaya," was identified in the northernmost Israel boreholes (Haifa Bay, Hazon, Hula). This ledge, though reduced in thickness, is also found in the NW Negev boreholes (Qeren, Haluza), and consists of carbonates associated with *Campbeliella striata* Carozzi and *Alveosepta jaccardi personata* Tobler, 1928, confering a Kimmeridgian age to this part of the Formation. Such an age can now also be attributed on the base of the ostracod *Hutsonia adunata* Bischoff, 1990, found in Lebanon in the Salima Formation and topmost part of the Nahar Sa'ar Formation at Ein Qinia (Hermon).

Tithonian

Generally regarded as a Lower Cretaceous unit (Beriassian-Hauterivian), the gray dark sandy shales Gevar 'Am Formation (Derin & Reiss, 1966), in borehole Negba-1 (from a core) yielded ammonites of Late Jurassic (Tithonian) affinity (Raab, 1962). This assumption is strenghtened by the finding by Moshkovitz &

Ehrlich (1980) of *Conusphaera mexicana* Trejo, 1969 in the Gevar 'Am Formation of offshore well Delta-1. Recently De Haan (1997) found Late Jurassic palynofloras in the Gevar 'Am shales of a number of wells of the coastal plain.

BIOSTRATIGRAPHY

The biostratigraphy of the Jurassic of Israel has been reviewed by Picard & Hirsch (1987). It is originally based on ammonites (Parnes, 1981; Lewy, 1983; Gill et al., 1985), calcareous nannoplankton (Moshkovitz & Ehrlich, 1976), foraminifers, algae and ostracods (Maync, 1966; Derin, 1974), bivalvia and gastropods (Hirsch, 1979), and palynomorphs (pollen, spores, dinocysts and acritarchs) (Conway, 1990). In the present paper, substantial additions are presented on ammonites, palynomorphs, ostracods, brachiopods and foraminifers.

AMMONITES

The zonal subdivision proposed by Parnes (1981) was revised by Enay & Mangold (1994). The Jurassic of Israel and of the Hermon (south Antilebanon) belong palaeogeographically to the Arabian platform (Hirsch, 1976; Dercourt et al., 1993).

Provincialism

Ecologic constraints particular to such a large and shallow area as the Arabian Platform, gave ammonites a remarkable endemic imprint (Gill et al., 1985; Enay et al., 1987). This provincialism is the cause of difficulties in the establishment of a precise stratigraphic correlation with European reference sequences.

The proposed biostratigraphy uses species of the Arabian biota (Enay and Mangold, 1994) at the level of sub-zones. The establishment of a stratigraphic correlation with the European standard is based on taxa common to both provinces. These are in general rather rare, or even absent, due to ecological or sedimentary (hiatus) reasons. Such conditions prevail generally in the Jurassic of Israel and the Hermon, with the exception of the transgressive pyritic ammonite bearing marls of the lower Oxfordian. This short time span of very relative opening of the Arabian platform edge to the pelagic Tethys-platforms eased the coming in of Mediterranean Tethyan taxa, enabling stratigraphic correlation with Europe. However, even during the Lower Oxfordian, the confinement of the ammonite-population remained strong, characterized by the dominance of Hecticoceratinae, with forms specific to the region. As a result, a sample of 340 specimens from the basal level of the marls, was composed of 96% Hecticoceratinae, 1% Aspidoceratinae, 1% Perisphinctinae and less than 1% Phylloceratinae. Due to this provincialism the proposed correlations (Table 1) remains necessarily approximate.

Biostratigraphy

The zonal subdivision for the Hermon, Negev and northern Sinai proposed by Parnes (1981), Lewy (1983) and Gill et al. (1985), is now revised on the base of the subdivision for Arabia proposed by Enay and Mangold (1994).

The late Bajocian age of *Thambites planus* Arkell, 1956 and the early Bathonian age of *Micromphalites* were proposed by Enay and Mangold (1994) and differ from Parnes's original assumption. The presence of genuine middle Callovian (*Coronatum* zone) and late Callovian (*Athleta* and *Lamberti* Zones) is now clearly established at Hamakhtesh Hagadol. The *Lamberti* Zone is also well represented at Majdal Shams, encompassing numerous ammonites. New biostratigraphic results were obtained in the Callovian and Oxfordian stages.

LATE CALLOVIAN.—*Athleta* Zone:

May be subdivided into two subzones, using indo-Malagasy and/or Arabian rather than the submediterranean index-species (Gill et al., 1985):

—*Pachyerymnoceras* sp. Subzone: The association comprises *Kurnubiella compressa* Gill, Thierry & Tintant, *Peltoceras* sp. juv., *Collotia* sp. and *Reineckeia nodosa* Till. The living chamber of *Pachyerymnoceras* sp. has a diameter of 190 mm. This relatively involute form belongs to the group of *Pachyerymnoceras kmerense* Mang., which together with other species characterize the Trezeense Subzone in Algeria (Mangold, 1988). It differs from the Algerian species by a larger number of undivided ribs, becoming strong toward growth-end (10 on the last half whorl). This fauna is found between subunits 33 and 41 of the Hamakhtesh Hagadol section of Goldberg (1963).

—*Solidum* Subzone: The Indian species *Peltoceras solidum* Spath was selected as index-species by Enay and Mangold (1994) for Arabia. In Arabia, it occurs in the upper part (T3) of the Tuwaiq Limestone Formation. An almost complete specimen was collected by Lewy in subunit 48 (Goldberg, 1963) at Hamakhtesh Hagadol. The species is frequent toward the top of the Hermon Limestone.

"*Lamberti*" Zone:

—*Pseudobrightia* sp. horizon

The equivalent of the *Lamberti* Zone in the southern Antilebanon section of Majdal Shams is found at the top of the Hermon Formation. A thin horizon (0.30 m thick) of condensed platy marly limestone with phosphatic grains is rich in ammonites, mostly flattened calcareous molds of Hecticoceratinae: *Pseudobrightia* sp., *Hecticoceras (Putealiceras) intermedium* (Spath), *H. (Kheraites) ferrugineus*

(Spath), *H. (Brightia)* aff. *salvadorii* (P. & B.), *H. (Putealiceras) douvillei* (Jean.), *Pachyerymnoceras levantinense* Lewy. *Pseudobrightia s. st.* is recognizable at the base of the *Lamberti* Zone in Europe (Cariou, 1984). The association found at the Hermon is of the same age, corresponding *pro parte* to the *Poculum* Subzone of the submediterranean province (Cariou, 1973; Cariou et al., 1985). The specimen of *Pseudobrightia* (coll. F. Hirsch) in the scree of subunit 74 (Goldberg, 1963), top of Matmor Formation, at Hamaktesh Hagadol, is figurated as *Brightia* sp. (Lewy, 1983, Pl. 1, Figs 17-18). Consequently the top of the Matmor Formation is still late Callovian and not early Oxfordian as assumed previously. It is interesting to note that the end-Callovian condensed outer platform limestones at Mt. Hermon express clearly a transgressive system tract, like the deposits of the same age in western Europe (Rioult et al., 1991).

LOWER OXFORDIAN.—The lower Oxfordian at the Hermon is represented by tens of meters of marl with small pyritized ammonites (Majdal Shams Formation). This facies resembles the "*Creniceras renggeri* marls" of the Jura Range, with which it shares several species (Enay, 1966), among which *Creniceras renggeri* (Opp.). The Hermon fauna, first described by Noetling (1887) and globaly attributed to the *Socini* Zone, was described in greater detail by Haas (1955), who correctly interprets the *Socini* Zone as the southern Tethyan equivalent of the *Mariae* Zone. Razvalyaev (1966) gave a precise stratigraphic ammonite species distribution. Frebold (1928) has proposed a subdivision into 2 zones: the *Hecticoceras socini* Zone at the base and the *Oecotraustes scaphitoides* Zone at top. The outcrop conditions do not permit continuous bed by bed collection. Two levels have delivered hundreds of *in situ* pyritized ammonites, one 2.50 m thick, at the very base of the sequence, the other, less abundant, ca. 40 m higher. The two assemblages are clearly distinct:

—*Socini* horizon: abundant *Hecticoceras socini* (Noetling), associated with *H. kautzschi* (Noetling, in Haas), *H. separandum* Haas, *H. solare* Haas, *H. schumacheri* (Noetling), *H. chatillonense* de Lor., *H.* cf. *guthei* (Noetling), *H. caelalum* Coquand, *Taramelliceras* cf. *langi* (de Loriol), *Euaspidoceras subbabeanum* (Sintzow), *E. douvillei* (Collot), *E. subcostatum* (in Haas, non Spath), *Perisphinctes bernensis* de Lor., *Mirosphinctes* aff. *robyi* (de Loriol), *Sowerbyceras helios* (Noetling).

—*Socium* horizon: the very characteristic *Hecticoceras socium* Haas is associated with *H. syriacum* Haas, *H. bonarelli* de Lor., *H. kautzschi* (Noetling), *Creniceras renggeri* (Oppel), *Euaspidoceras* sp., *Properisphinctes* sp., *Mirosphinctes* sp.

Correlation of both horizons with Europe can be made on the base of comparison with figurated European forms, the stratigraphic distribution of which in the fold of the *Mariae* Zone is precisely known (Schirardin, 1958; Gygi, 1990; Vidier et al., 1993; Fortwengler et al., 1997.

TABLE 1: Ammonite levels in the Levantine series and their probable equivalences with submediterranean Europe zonal scheme. The units with asterisks are new.

TABLEAU 1: Niveaux à ammonites des séries du Levant et leurs équivalences probables avec le schéma de zonation d'Europe sub-méditerranéenne. Les unités avec un astérisque sont nouvelles.

Europe (Submediterranean Province)				Levant
Stage	Substage	Zones	Subzones	(Characteristic ammonites)
Tithonian				?*Virgatosphinctes*
Oxfordian	Middle	*Plicatilis*		*Euaspidoceras* gr. *Perarmatum*
	Lower	*Cordatum*		?*Peltoceratoides*
		Mariae	*Praecordatum* *Scarburgense*	*Hecticoceras socium** *Hecticoceras socini*
Callovian	Upper	*Lamberti*	*Lamberti* *Poculum*	*Pseudobrightia* sp.*
		Athleta	*Collotiformis* *Trezeense*	*Peltoceras solidum* *Pachyermnoceras* sp.*
	Middle	*Coronatum*		*Kurnubiella ogivalis*
Early Callovian-Late Bathonian?				*Bullatimorphites* cf. *bullatus*
Bathonian	Lower	*Zigzag*		*Micromphalites*
Bajocian	Upper	*Parkinsoni*		*Thambites planus*
		Garantiana		*Ermoceras mogharense*
		Niortense		*Ermoceras runcinatum*
		Laeviuscula		*Dorsetensia Otiotes*

The *Socini* horizon is equivalent to the *Scarburgense* Subzone. The *Socium* horizon would represent the top of the *Scarburgense* Subzone and the *Praecordatum* Subzone.

PALYNOMORPHS

The zonation was established on the base of the first appearance of the Jurassic palynomorphs in Israel and their correlation with their occurrence in other parts of the world.

LATE PLIENSBACHIAN TO MIDDLE BAJOCIAN APPLANOPSIS TURBATUS ZONE.—Terrestrial palynomorphs belong to a floristic province extending across north Africa, Arabia and into Iran and Afghanistan. Boreal floras are dominated by coniferalean disaccate pollen, whereas, northern Gondwanean floras are subtropical, characterized by non-disaccate coniferalean pollen (*Corollina,*

Applanopsis); as well as, Filiciales *incertae sedis* and Bennettitalean floral elements (Lorch, 1967; Conway, 1990), disaccate pollen are absent to rare. Poor floristic diversity and limited evolution (at least until late Oxfordian times) indicates strong climatic restrictions in the Levant Province.

Strata from the Mishhor, Ardon, Inmar and Qeren formations, although some are marginally marine, contain dinocysts of the *Applanopsis turbatus* Zone (including *Nannoceratopsis pellucida* Deflandre, 1938 and *Dapsilidinium? deflandrei* (Valensi, 1947)) that correlate well with coeval boreal suites.

LATE BAJOCIAN KORYSTOCYSTA KETTONENSE ZONE (LATE BAJOCIAN) DICHADOGONYAULAX SELLWOODII ZONE (LATEST BAJOCIAN).—In Israel, Mid-Jurassic assemblages from shallow shelf depocentres indicate normal marine salinities, they are dominated by proximate dinocysts with epitractal archeopyles (*Dichadogonyaulax* and *Korystocysta*), together with small proximate cysts with apical archeopyles (*Ellipsoidictyum, Sentusidinium*). These elements typify Tethyan warm water faunas elsewhere. Despite this close affinity and not withstanding their limited diversity the assemblages correlate well with boreal assemblages. Differences on the species level are probably due to ecological factors. In the Negev, strata from the upper part of the Inmar and Daya formations contain the taxa defining this Zone. In northern and central Israel it defines a portion of the Haifa Formation.

BATHONIAN: ENERGLYNIA ACCOLARIS ZONE.—There is a continuity of ecological conditions in the Levant. No major evolutional changes occur, ctenidodinoid elements remain dominant, as in Europe, and are reinforced by the entry of *Energlynia.* The skolochorate cyst *Systematophora* appears, whose acme is younger.

Low diversity characterizes Bathonian assemblages of Israel, similar assemblages occur in contemporaneous intervals from NE Spain, Portugal, and in NE Libya (Sargeant, 1976; Fenton et al., 1980; Thusu & Vigran, 1985). Although boreal assemblages appear more diverse, provincialism is difficult to define; many taxa are widespread, even with global distribution. This zone defines the Sherif Formation and coeval portions of the Haifa and Hermon formations.

CALLOVIAN: POLYSTEPHANEPHORUS CALATHUS ZONE.—In Israel; low diversity again characterizes Callovian assemblages, and elsewhere in the Tethyan Realm. The most distinct change in warm neritic Late Jurassic dinocyst assemblages is the acme of skolochorate cysts (*Adnatosphaeridium, Compositosphaeridium, Emmetrocysta, Polystephanephorus, Surculosphaeridium* and *Systematophora*). There is a decline of ctenidodinoid elements and small proximate cysts with apical archeopyles (*Ellipsoidictyum, Sentusidinium*). Large gonyaulaccid cysts with precinglar archeotypes are scarce. The Zone defines the Zohar and Matmor formations as well as the corresponding interval of the Haifa Formation.

OXFORDIAN: MILLIOUDODINIUM NUCIFORMIS ZONE (EARLY TO MIDDLE OXFORDIAN) AND EPIPLOSPHAERA RETICULOSPINOSA ZONE (MIDDLE TO LATE OXFORDIAN).—In Israel, Oxfordian assemblages indicate normal marine salinities and continue to be characterized by skolochorate cysts morphologically related to *Systematophora,* which are more numerous in the Tethyan Realm. Cavate dinocysts are rarer than in boreal assemblages. These features indicate some degree of isolationism or provincialism, although, there are broad similarities with western Neotethyan suites in Europe, differences being largely at species level. There is a dramatic entry of large gonyaulaccid dinocysts (*Hystrichogonyaulax, Millioudodinium*) that rise to dominate Oxfordian assemblages of Israel. These zones define the Majdal Shams and Nahar Sa'ar formations. The lower zone also defines an age-equivalent interval of the Haifa Formation.

OSTRACODS

The data on Jurassic Ostracods that were described from Gebel Maghara (Sinai, Egypt), Makhtesh Ramon, Hamakhtesh Hagadol (Negev) and Majdal Shams (Hermon), are summarized herein (Rosenfeld et al., 1987a, 1987b, 1988, 1991).

LOWER-MIDDLE LIASSIC *BISULCOCYPRIS OERTLII* ASSEMBLAGE ZONE.—At Makhtesh Ramon (south Israel) shales at the base of the Ardon Formation yield the non-marine ostracods *Bisulcocypris oertlii* Gerry, *Fabanella ramonensis* Rosenfeld & Honigstein and *Laevicythere* sp. In drillings, the marine *Ektyphocythere* cf. *vitilis* (Apotolescu, Magne & Malmoustier) may set in, but not in the same layers as the nonmarine taxa, pointing to alternating environments of fresh-brackish water and shallow sea. The nonmarine fauna is endemic at the specific level, whereas the marine form shows European affinity.

TOARCIAN-AALENIAN *EKTYPHOCYTHERE BUCKI* ASSEMBLAGE ZONE.—At Gebel Maghara (Sinai, Egypt) the clastic upper Inmar Formation yields frequent *Cytherella*? *toarcensis* (Bizon), rare to common *Ektyphocythere bucki* (Bizon), *Isobythocypris ovalis* (Bate & Coleman) and *Praeschuleridea inmarensis* Rosenfeld & Gerry and *Kinkelina kadeshensis* Rosenfeld & Gerry. The environment of deposition was warm shallow marine. The assemblage shows affinities with European forms.

BAJOCIAN *GLYPTOGATOCYTHERE MAGHARAENSIS* ASSEMBLAGE ZONE.—In the shales, that alternate with the carbonates of the Daya Formation at Gebel Maghara, the frequent *Glyptogatocythere magharaensis* Rosenfeld & Gerry and the common to rare *Cytherella bashai* Rosenfeld & Gerry, *Monoceratina striata* Triebel & Bartenstein, *Rutlandella transversiplicata* Bate & Coleman and *Ektyphocythere zerqaensis* Basha are characteristic and restricted to this assemblage zone, whereas *Glyptocythere huniensis* Basha and *Bairdia* aff. B Jones proceed higher. The environment of deposition was warm shallow marine, though *Monoceratina* indicates somewhat deepening of the facies. The assemblage still shows affinities with European forms, though more taxa and the dominant taxon are endemic (Israel, Jordan).

BATHONIAN *PROGONOCYTHERE HONIGSTEINI-FASTIGATOCYTHERE BAKERI ASSEMBLAGE* ZONE.—In the Sherif Formation at Gebel Maghara, this assemblage zone is dominated by *Progonocythere honigsteini* Rosenfeld & Gerry, *Fastigatocythere bakeri* (Basha), *Praeschuleridea hornei* Rosenfeld & Gerry, and *Zerqacythere subiehensis* Basha. Further occur for the first time: *Glyptogatocythere malzi* Basha, *Ektyphocythere shulamitae* Rosenfeld & Gerry, *E. aardaensis* Basha and *Terquemula goldbergi* Rosenfeld & Gerry. The environment of deposition was warm shallow marine and faunal affinity remains dominantly provincial (Israel, Jordan).

CALLOVIAN *EKTYPHOCYTHERE ZOHARENSIS* ASSEMBLAGE ZONE.—In the rather rare shale intercalations of the Zohar Formation at Gebel Maghara *Ektyphocythere zoharensis* Rosenfeld & Gerry and *Terquemula gublerae* (Bizon) are found. Throughout the Zohar and lower Matmor formations at Hamakhtesh Hagadol (Negev, Israel) an assemblage is found that comprises *E. zoharensis* Rosenfeld & Gerry, *Bairdia* aff. *hilda* Jones, *Afrocytheridea faveolata* Bate, *Progonocythere* aff. *parastilla* Whatley and *Micropneumatocythere laevireticulata* Rosenfeld & Honigstein. In the upper part of the Matmor Formation, the assemblage of *E. zoharensis* Rosenfeld & Gerry has a different composition. Here, along with the taxa of the assemblage found below, now also appear *Exophthalmocythere? kidodensis* Rosenfeld & Gerry, *Mandelstamia hirschi* Rosenfeld & Honigstein, *Cytherella index* Oertli and *Oligocythereis* aff. *fullonica* (Jones & Sherborn). The environment of deposition of the Zohar and Matmor Formations is shallow warm marine and the faunal affinity is widely endemic (Israel, Saudi Arabia).

EARLY OXFORDIAN *EXOPHTHALMOCYTHERE? KIDODENSIS* ASSEMBLAGE ZONE.—The Tauriat shales (Majdal Shams Formation) at Gebel Maghara yield

Exophthalmocythere? kidodensis Rosenfeld & Gerry, *Progonocythere* aff. *parastilla* Whatley and *Terquemula gublerae* (Bizon). At the Hermon, the Majdal Shams Formation yields an abundant assemblage, comprising next to *E.? kidodensis* Rosenfeld & Gerry and *T. gublerae* (Bizon), also *T.* cf. *martini* (Bizon), *Cytherella* cf. *umbilica* Bate, *Monoceratina stimulea* (Schwager) and *M.* cf. sp. B Bate, as well as *Cytherelloidea atlantolevantina* Rosenfeld & Honigstein, *Eucytherura oxfordiana* Rosenfeld & Honigstein, *Acrocythere dubertreti* Rosenfeld & Honigstein, *Homerocythere hermonensis* Rosenfeld & Honigstein and *Oligocythereis irregularis* Rosenfeld & Honigstein. The environment of deposition of the Majdal Shams Formation at the Hermon is definitely of deeper water with euxinic bottom conditions. The faunal affinity of ostracods remains endemic, notwithstanding a slight relation to western Europe, the Tethyan oceanic barrier held this province apart from its boreal Eurasiatic counterpart north of the Tethys.

LATE OXFORDIAN-KIMMERIDGIAN *HUTSONIA ADUNATA* ASSEMBLAGE ZONE.—In the Nahar Sa'ar Formation at Ein Quniya (Hermon) as well as in boreholes (Qeren, Haluza, Boger, Kohal, Beersheva and Hazon) an assemblage identical to the one described from the couches d'Aazour and calcaires de Salima of Lebanon occurs (Bischoff, 1990). There stratigraphically important representatives of the genera *Schuleridea* and *Hutsonia* have a range from late Oxfordian-Kimmeridgian and possibly Tithonian.

BRACHIOPODS

Brachiopods of the region have been described by Douvillé (1916), Muir-Wood (in Hudson, 1958), Feldman (1987) and by Feldman et al. (1982, 1991). The affinity of these faunas is generally endemic to the Ethiopian Province, the geographical distribution of which is wide, including a Gondwanian shelf crescent from east to north Africa (Somalia, Kenya, Tanzania, Arabia, the Levant, Egypt, Tunisia and the Maghreb). Middle to Upper Jurassic brachiopods described from Saudi Arabia, Israel, Egypt and Tunisia, while containing many genera and species common to all four subregions, also contain elements that are characteristic of the area of origin, an observation also made for bivalves, e.g. *Eligmus* (Hirsch, 1979).

The Ethiopian Province is suspected to have been invaded in the Early Jurassic from the north (Europe) and thereupon isolated for the remainder of the Jurassic, developing special morphological features which distinguish them from their original stock, a phenomenon also common in Callovian nerineids. Though Jurassic brachiopods from Cutch (India) were broadly correlated with lower Bathonian to Oxfordian faunas of Europe (Kitchin, 1900; Spath,

1927-1933; Arkell, 1956), it is suggested that a number of genera and species from Israel, Saudi Arabia and Cutch are very closely related. The rhynchonellid *Pycnoria* described by Cooper (1989) from the upper Bathonian-Callovian of Saudi Arabia is almost certainly congeneric with *Rhynchonella fornix* Kitchin, 1900 and *R. nobilis* Kitchin of Cutch. The genus *Schizoria* described by Cooper (1989) from the upper Bajocian of Saudi Arabia appears to be represented by *R. assymetrica* Kitchin and the specimens of *Globirhynchia crassa* Cooper from late Bajocian beds, are similar to specimens described as *R. versabilis* Kitchin. Among Terebratulida, species of *Kutchithyris,* encountered in Israel and Saudi Arabia, are very similar in external morphology to those of Cutch. The works of Weir (1925, 1929) and Muir-Wood (1925) give the impression of a rhynchonellid-dominated fauna of comparatively little diversity in the middle Upper Jurassic of Somalia. A similar impression is obtained from Tunisia (Dubar, 1967), leading to the conclusion of a closer faunal and ecological relationship of these areas than actually exist. Likewise, Bathonian to Oxfordian beds in south Israel draw a closer comparison with the Somalian fauna than with the Saudi Arabia fauna monographed by Cooper (1989). Yet, a number of terebratulid species that occur in both South Arabia and Israel have not been recognized from Somalia and Tunisia. Rhynchonellida common to all four areas include ubiquitous *Somalirhynchia* and *Daghanirhynchia,* but the genera *Amydroptychus, Conarosia, Echyrosia, Colpotoria, Eurysites, Lirellarina, Nastosia* and *Strongyloria* of Cooper (1989) do not occur in southern Israel, Somalia or Tunisia. However the genus *Pycnoria* occurs in Bathonian to Callovian beds in Gebel El-Maghara (Feldman, 1987) and Saudi Arabia (Cooper, 1989) and may also occur with forms such as *Schizoria* from the Bajocian, *Burmirhynchia* Bajocian-Callovian and *Globirhynchia* in both Gebel El-Maghara (Sinai, Egypt) as well as Cutch (India).

FORAMINIFERA

The extensive studies of Maync (1966), Derin & Reiss (1966), Derin & Gerry (1972) and Derin (1974) have laid the basis for a comprehensive microfacies distribution, mainly based on foraminifera and algae. Jurassic foraminifera, in need of a revision in the field of taxonomy and biostratigraphy, were given special attention in the Peri-Tethys project. Lately, Kuznetsova & Dobrova (1995) and Kuznetsova et al. (1996) have added information on Jurassic foraminifera from Syria.

LIASSIC.—From the few outcrops of the Ardon Formation in Maktesh Ramon (Negev), the exact age of the marine Jurassic basal transgression in that part of the Levant is still uncertain. The Formation only yielded *Glomospira* sp.

with alga *Thaumatoporella parvovesiculijera* Raineri, in restricted lagoonal facies, taxons without chronostratigraphical value.

A revision of the subsurface Liassic confirmed the presence of *Orbitopsella primaeva* (Hottinger) in the lowest portion of the Ardon Formation in boreholes of the coastal plain (Derin, 1974), indicating a Pliensbachian age for the Liassic transgression. Forms previously identified as *Orbitopsella* aff. *praecursor* (Gümbel) and considered as indicative of the "Orbitopsella praecursor Zone" (Pliensbachian) are referable to *Timidonella* sp. In the same interval occur *Gutnicella* gr. *cayeuxi* (Lucas), *Spiraloconulus* cf. *perconigi* (Allemann & Schroeder) and at the top, *Limognella* sp. All these forms are elsewhere known in the Tethys from Aaleno-Bajocian time span.

MIDDLE JURASSIC.—To the Bajocian are related *Amijiella amiji* (Henson) and dasyclad *Selliporella donzellii* Sartoni & Crescenti, and to the Bathonian *Paleopfenderina salemitana* Sartoni & Crescenti.

However, the previously quoted *Meyendorffina bathonica* Aurouze & Bizon and *Orbitammina elliptica* (d'Archiac), well known late Bathonian markers in western Europe, are actually a new species of flattened *Kilianina.* Its range is to be placed in early and/or middle Callovian and it is often associated with *Praekurnubia* and small primitive *Kurnubia (K) variabilis* Redmond. The presence of *Satorina apuliensis* Fourcade & Chorowicz was not confirmed. In the *Coronatum* Zone (middle Callovian) *Praekurnubia crusei* Redmond and small *Kurnubia* (*K variabilis* Redmond, *K bramkampi* Redmond) are widespread. At Hamakhtesh Hagadol, the larger *Kurnubia* (*K palastiniensis* (Henson), *K.* cf. *wellingsi* (Henson)), together with small species, *Flabellocyclolina reissi* Hottinger and *Steinekella* cf. *steinekei* Redmond appear. *F. reissi* seems only present in the upper Callovian.

LATE JURASSIC.—Lower Oxfordian can be characterized by "*Mangashtia*" *egyptiensis* Fourcade et al. (Derin, 1974). The widespread *Alveosepta jaccardi* biozone is well represented (upper Oxfordian-Kimmeridgian). At the top of the range zone of this taxon *Anchispirocyclina praelusitanica* Maync also occurs (Maync, 1966). Algae include *Pseudoclypeina* and *Clypeina jurassica* Favre.

PALAEOGEOGRAPHY

For a correct understanding of the Jurassic palaeogeographic setting of Israel prior to the infracretaceous truncation, one has to move the Golan-Jordan side about 100 km back to the south. In the south, the Negev High consists of the Avdat and Massada-Jordan "blocks," separated by the Kurnub "basin" (Goldberg & Friedman, 1974). Plunging toward the NW, the "nose" of the Negev High separates the depotcenters of the North Sinai (Maghara-Halal) from the Central Israel trough (Helez-Ramallah-Hermon) of the Judean embayment (Hirsch, 1985). To the North, the overall thinner Jurassic column represents the Galilee High that extends to the adjacent Lebanon and northern Antilebanon (Hirsch, 1985).

LOWER JURASSIC (Fig. 3).—Major uplift took place at the Rhaetian-Liassic boundary, resulting in denudation, karst and truncation processes. In northern Israel, a transversal trough was filled by pre-Toarcian Asher volcanics, the wearing down of which produced the materials found in the laterites of the Mishhor Formation, covering most of Israel and adjacent areas (Picard & Hirsch, 1987).

During the middle-late Liassic transgressions, true marine platform carbonate regime developed fast in the depotcenters of north Sinai and of the Judean embayment. Clastics, the result of the wearing down of the Arabian massif to the South and South East, and interspersed oolites and evaporites mark the proximal sides of the basins. Shales replace the sandstones of the upper Inmar Formation and the carbonate bodies of Ardon and Qeren merge toward the distal part of the basin. In northern Israel, Lebanon and north Antilebanon

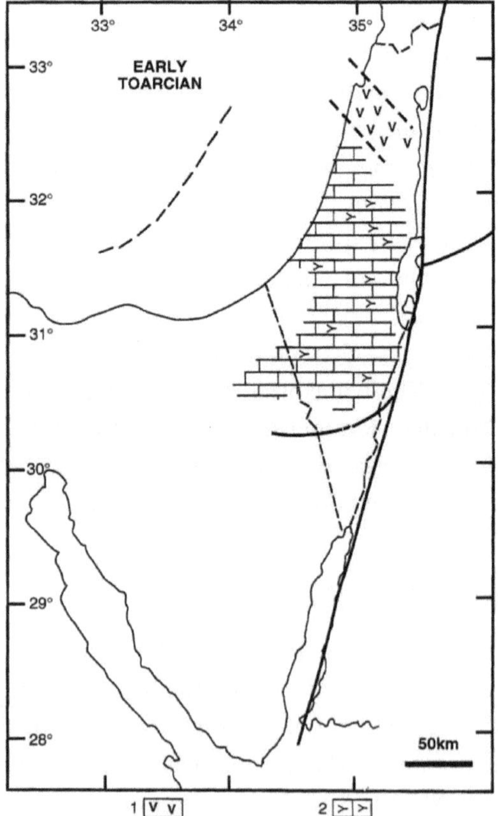

FIGURE 3: Early Toarcian facies map. The early Toarcian is represented by an interval within the huge mass of platform carbonates with evaporitic layers that build up the Ardon Formation. Its existence toward Lebanon is not clear and the interval may onlap directly the Asher Basalts (VV) in the northern part of Israel. It is estimated to consist on the slope toward the Judean embayment of over 200 m, thickening towards the central axe of the embayment to the order of 500 m. 1, basalts (193-194); 2, carbonates and evaporites.

FIGURE 3: Carte des faciès au Toarcien inférieur. Le Toarcien inférieur représentée un intervalle dans la masse immense des carbonates de plate-forme et évaporites qui constituent la Formation de Ardon. Son extension vers le Liban n'est pas claire, intervalle pouvant couvrir les basaltes de Asher (VV) dans le nord d'Israël. En bordure de la baie de Judée l'intervalle peut atteindre 200 m et s'épaissit vers l'axe du bassin (de l'ordre de 500 m). 1, basaltes (193-194); 2, carbonates et évaporites.

platform all clastics have vanished and carbonates persist without interruption from the Liassic to the Upper Jurassic (Haifa Formation).

MIDDLE JURASSIC (Fig. 4).—From the northern Israel (Haifa Formation) to the central Israel and Israel coastal plain and Hermon (Hermon Formation), calcarenites are interwedged with micrites.

In southern Israel (Negev) the interplay of continental (Arabian massif) arenaceous with littoral-tidal shales and carbonates characterize the Zohar Formation. The micrites are interbedded with shale and a few thin sandstones of subordonate non fluviatile origin.

Contrasted with the apparent continuous platform sedimentation regime of north Israel, the emersion of part of the coastal plain belt, witnessed by karst phenomena in several boreholes (Buchbinder et al., 1984), caused erosion and truncation of the Callovian carbonates. This infra-Oxfordian unconformity (POX) resulted in the filling of a karstic landscape by the transgressive early Oxfordian Majdal Shams shales.

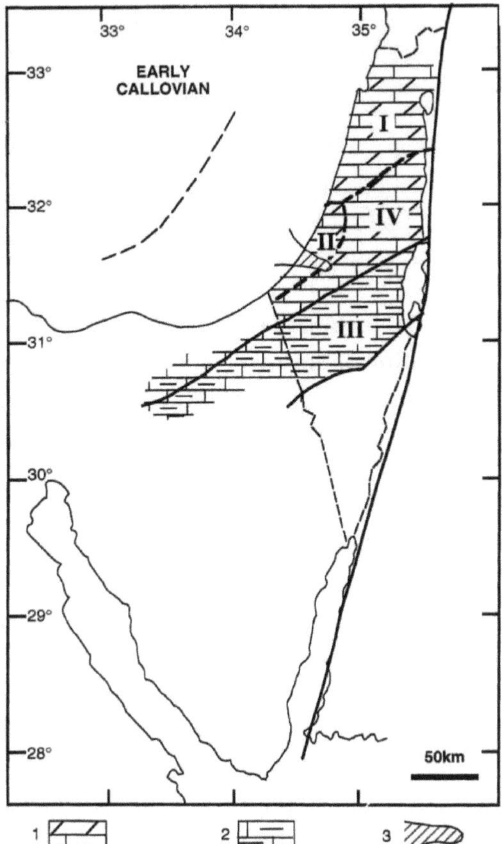

FIGURE 4: Early Callovian facies map. This ca. 100 m thick interval subdivides in tectono-sedimentary zones: I. calcarenites (Haifa); II. calcarenites (Brur) of the "unstable zone," truncated by a karstic event; III. shales (Upper Sherif-Karmon), calcarenites and marls (lower Zohar); IV. massive calcarenites (Hermon). 1, carbonates; 2, limestones and marls; 3, eroded.

FIGURE 4: Carte des faciès du Callovien inférieur. L'intervalle d›environ 100 m d'épaisseur se subdivise en zones tecto-no-sédimentaires: I, calcarénites (Haifa); II, les calcarénites (Brur) de fa "zone instable" sont tronquées par un événement karstique; III, shales (Sherif supérieur-Karmon), calcarénites et marnes (Zohar); IV, calcarénites massives (Hermon). 1, carbonates; 2, calcaires et marnes; 3, érodé.

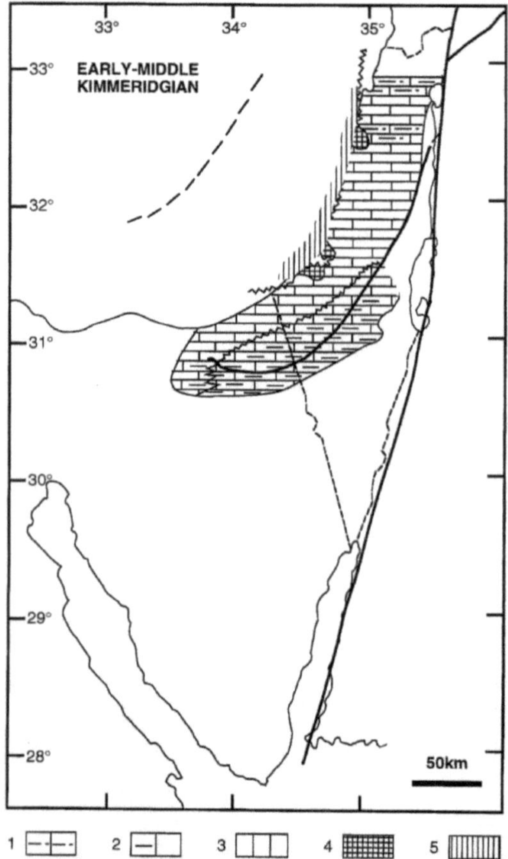

FIGURE 5: Early-middle Kimmeridgian facies map: consists of reefs, oolites, shales and sands (Nahar Sa'ar). Eroded by the infra-Cretaceous uplift along the present Levant coast and in the SE Negev. 1, oolithes; 2, marls; 3, limestones; 4, reef; 5, eroded.

FIGURE 5: Carte des faciès du Kimméridgien inférieur et moyen: se compose de récifs, oolites, shales et sables (Nahar Sa'ar). Tronqué lors du soulèvement infra-Crétacé le long de la côte actuelle du Levant et dans le Neguev du SE. 1, oolithe; 2, marnes; 3, calcaires; 4, récif; 5, érodé.

LATE JURASSIC.—The Oxfordian palaeogeography is well delimited by the distribution of the lower-middle Oxfordian shale-facies (Majdal Shams). Absent from the northern High of Galilee (I) and southern High of the Negev (III), the shales reach 200 m thickness along the axis of the Judean embayment (II) and less than 100 m in the Sinai deep (IV). Volcanic tuffs occur in northern Israel (Devorah). In northern and central Israel, including the NW Negev, the lower Kimmeridgian (Fig. 5) is developped in the shallow marine facies of oolites, marls and shales of the Nahar Sa'ar Formation.

A different palaeogeography is initiated by the Tithonian (Fig. 6) regional uplift, generating the Infracretaceous unconformity. In the wells drilled into the canyons of the "Helez erosion embayment" nearly all Callovian and Upper Jurassic formations are missing, as they were removed by submarine Tithonian-Hauterivian chanelling of the turbiditic Gevar 'Am shale. Subaereal truncation reached down to the Dogger in the Central Negev and to Triassic-Palaeozoic levels in the southern Negev.

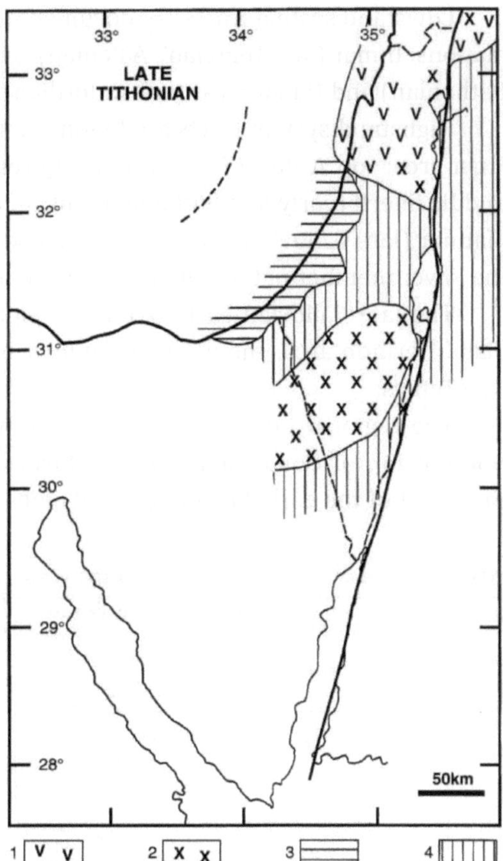

FIGURE 6: Late Tithonian facies map. This interval belongs virtually to the infra-Cretaceous tectono-environmental phase. It is marked by wide denudation. Tithonian-Hauterivian volcanics (Tayasir) and turbidites (Gevar'Am canyon). 1, alkalin basalts (140-115); 2, basic intrusion; 3, shales; 4, eroded.

FIGURE 6: Carte des faciès du Tithonien supérieur. Cet intervalle appartient virtuellement à la phase tectono-environnementale infra-Crétacé. Il est marqué par une dénudation étendue, un volcanisme tithonique-hautérivien (Tayasir) et par des turbidites (Canyon de Gevar'Am). 1, basaltes alcalins; 2, intrusions basiques; 3, shales; 4, érodé.

TECTONO-EUSTATIC CYCLIC EVENTS

Several transgressive systems tracks or highstands were individualized, characterized by condensed facies and ferruginous oolites. Ammonites enable precise dating at Bajocian, early Bathonian, early late Bathonian, middle-late Callovian and latest Callovian-lower Oxfordian intervals. Their relatively good synchronism with system tracts of same signification in Europe shows the importance of tectonoeustatism as factor of sedimentary control, including 3rd order scale phenomena in the sense of Haq et al. (1988). In the longer term, the Jurassic transgression is here neatly expressed by a sedimentary drift to a more marine pole, up to the Oxfordian. During this retrogradation, abundant and thick sandy intercalations up to the Bathonian, become scarcer and thin in younger levels until their disappearance. From the Kimmeridgian onwards, the sedimentary dynamic is reversed with typically prograding shallow bioclastic and corallian limestone deposits, as in Europe.

Lowstand system tracts are documented by the regressive intervals of formations: Inmar (late Toarcian?-Aalenian), Safa (early Bathonian), Karmon (late Bathonian) and Nahar Sa'ar (late Oxfordian).

Highstand system tracts are found primarily in the more proximal Negev-Sinai area: top Ardon (Toarcian), top Qeren (Aalenian), top Daya (Bajocian), middle Sherif (early late Bathonian), intra-Zohar (early late Callovian) and top Matmor (late Callovian). Such conditions occur over the entire region at top of the lower member of the Nahar Sa'ar Formation (late Oxfordian).

The late Callovian-early Oxfordian highstand system tracts, in the more distal Hermon area, shows evidence for connection with the pelagic western Neo-Tethys.

The regional Jurassic cyclic evolution versus global events in the levantine portion of Gondwana (Fig. 7), starts with post-Triassic period of karst, the result of the early Liassic regional emersion of Haq et al. (1988) long term

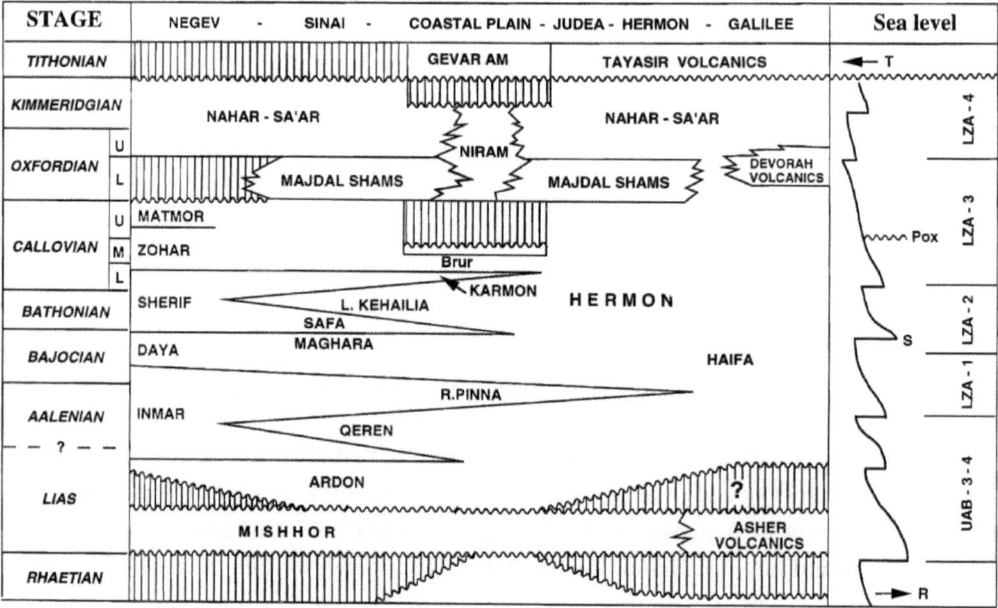

FIGURE 7: Schematic diagram of Jurassic formations with tectono-eustatic curve. Hiatus (vertical lines), unconformities (undulated), lateral facies-change (zigzag lines), mainly clastic facies (dots), shales (horizontal lines), mostly carbonatic (white). Tectono-eustatic curve and second order super-cycles (Haq et al., 1988).

FIGURE 7: Diagramme schématique des formations jurassiques et courbe téctono-eustatique. Hiatus (lignes verticales), discontinuités (ondulée), variations latérales de faciès (ligne en zigzag), faciès clastique dominant (points), shales (lignes horizontales), prédominance carbonaté (blanc). Courbe tectono-eustatique et supercycles de second ordre d'apres Haq et al. (1988).

sea level-drop upper Absaroka B (UAB 1-2), filled with Mishhor laterites. ?Pliensbachian-Toarcian Ardon evaporites with shallow lagoon and reefoid carbonates, alternating with deltaic and paralic lower Inmar sands, express the long term rise with minor short term oscillations, characterizing the end of UAB. Aalenian upper Inmar (Rosh Pinna) clastics abruptly set in with the onset of lower Zuni-A -1 sea-level drop and subsequent rise, reaching a significant highstand in the Bajocian Daya transgression (*Laeviuscuia-Otoites*) and following characteristic *Ermoceras* and *Eligmus* assemblages, ending in *Parkinsoni-Thambites* lowstand, persisting into the early Bathonian. The early Bathonian lower Sherif/Safa sands and paralic coals seem to be either delineating a regression, which in terms of Haq et al. (1988) global cycles is contradictory. This tectono-eustatic event appears to be characteristic of the entire Levantine-Arabian platform, witnessing a local epirogenic pulse (S) (Hirsch, 1987, 1988).

Next in the sequence is the LZA-2.2 initial Bathonian phase of drowning (lower Kehailia) with ammonites (*Micromphalites*) in Israel and Arabia and the late Bathonian-early Callovian rise (LZA-3) with *Bullatimorphites* (upper Kehailia). According to Haq et al. (1988), the late Bathonian is a lowstand, followed by an overall transgressive early Callovian.

The middle and upper Callovian reach a regional high, inundating most of the Arabian platform with typical "*Zohar*/Matmor" type carbonates, that yield Nerineacea, foraminifers (*Kurnubia*-lineage), brachiopods and abundant bivalves (*Eligmus*), mostly related to Ethiopian-Somalian taxa.

The emersion along the coastal plain of Israel, causing karst-phenomena (Buchbinder et al., 1984) and the hiatus of a significant part of the Callovian (Picard & Hirsch, 1987) is, in the present authors' interpretation, apparently the result of pre-Oxfordian (POX) epirogenic movements.

The early Oxfordian (*Mariae* Zone) Majdal Shams shale-onlap covered the foregoing hiatus in the Judean embayment, extending from Maghara (Sinai) to the Hermon, possibly interfingering laterally with the Galilee carbonatic type facies.

The late Oxfordian-Kimmeridgian shallowing and progressive regression (*Alveosepta jaccardi*) occured in contrast to the global trend of a general sea level rise (LZA-4), generating the emersion of the Syrian (Mardin) promontory, the development of thick evaporites in Central Arabia and the deposition of sands and shales from the Negev to north Egypt. Such progressive regression may occur in contrast of the global trend of a general sea level rise, and conform to the general evolution found in Europe. A generalized sea level rise, resulting in the extension of marine facies, may rapidly fill the available space, that due to tenfold sediment-production (biogenic in the present case), produces a shallow platform, until a regressive facies is reached as the filling up is progressing.

Epirogenic movements obscured the Late Jurassic global regression by juxtaposing "Tithonian" graben-fillings with emersions, continental deposits and basaltic flows.

Changing ecological and tectono-environmental conditions in the Middle Eastern Levantine Jurassic are echoed by the composition of floral and faunal assemblages, endemism and cosmopolitism. Palaegeographical and environmental evolution to a wide extent match the cycles of Haq et al. (1988), although it suggests that Levantine "African-Arabian-Apulian" sea level anomalies, due to "noises" generated by the Proto-Atlantic opening and southern Neo-Tethys rifting, are perturbing the global sea level record.

ACKNOWLEDGMENTS

We express our thanks to R. Enay (Lyon) for reading an early draft of the manuscript. We are indebted to J. Thierry (Dijon) and M. Mouty (Damascus) for reviewing the paper. We very much appreciated their constructive remarks, which contributed to the improvement of the paper. We thank P. Grossman for drafting of the figures.

Part of the research was performed within the frame of the Peri-Tethys Program. Partial funding was provided to H-R Feldman by the National Geographic Society, grants 4739-92 and 5666-96.

REFERENCES

Al Far, D. M. 1966. *Geology and coal deposits of Gebel El Maghara* (N. Sinai). Geological Survey of Egypt, Paper 37: 1-59.

Arkell, W. J. 1956. *Jurassic Geology of the World.* London: Oliver and Boyd.

------. 1952. Jurassic Ammonites from Jebel Tuwaiq, Central Arabia. *Philosophical Transactions of the Royal Society of London* Ser. B, 236: 241-314.

Bischoff, G. 1964. Die Gattung Cytherelloidea im Oberen Jura und in der Unterkreide. *Senckenbergiana Lethaea* 45: 1-27.

Bischoff, G. 1990. The Genus *Hutsonia* in the Late Jurassic. *Senckenbergiana Lethaea* 70: 397-429.

Buchbinder, L. G. 1981. Dolomitization, porosity development and late mineralization in the Jurassic Zohar (Brur Calcarenite) and Sederot formations in Ashdod-Gan Yavne. *Israel Journal of Earth-Sciences* 30: 64-80.

Buchbinder, L. G., M. Margaritz, and M. Goldberg. 1984. Stable isotope study of karstic related dolomitization: Jurassic rocks from the Coastal Plain, Israel. *Journal of Sedimentary Petrolology* 54: 236-256.

Cariou, É. 1973. Ammonites of the Callovian and Oxfordian. In A. Hallam (ed.), *Atlas of Palaeobiogeography*, 287-295. Amsterdam: Elsevier.

------. 1984. Biostratigraphic subdivisions of the Callovian stage in the Subtethyan province of ammonites, correlations with the Subboreal zonal scheme. In O. Michelsen and A. Zeiss (eds.), International Symposium on Jurassic Stratigraphy, Erlangen. *Geological Survey of Denmark,* 2: 315-326.

Cariou, É., D. Contini, J.-L. Dommergues, R. Enay, J. Geyssant, Ch. Mangold, and J. Thierry. 1985. Biogéographie des ammonites et évolution structurale de la Téthys au cours du Jurassique. *Bulletin de la Société géologique de France* 1(5): 679-697.

Coates, J., E. Gottesman, M. Jacobs, and E. Rosenberg. 1963. Gas discoveries in the western Dead Sea Region. *Sixth World Petroleum Congress, Francfort* 1: 21-36.

Conway, B. H. 1990. Paleozoic-Mesozoic Palynology of Israel. II: Palynostratigraphy of the Jurassic Succession in the Subsurface of Israel. *Geological Survey of Israel Bulletin* 82: 1-39, 18 pls.

Cooper, G. A. 1989. Jurassic brachiopods of Saudi Arabia. *Smithsonian Contributions to Paleobiology* 65: 1-213.

Cousminer, H. L., and B. Conway. 1981. Initial results and potential of Israeli palynology program. *Geological Survey of Israel, Current Research* 1980: 57-61.

Dercourt, J., L. E. Ricou, and B. Vrielynck (eds.). 1993. *Atlas Tethys Palaeoenvironmental Maps.* Paris: Gauthier-Villars.

Derin, B. 1974. The Jurassic of Central and Northern Israel. Ph. D. Thesis, Hebrew University, Jerusalem, Israel. [in Hebrew, English summary].

Derin, B., and Z. Reiss. 1966. *Jurassic Microfacies of Israel.* Israel Institut of Petroleum, Special Publication, Tel Aviv: 1-43.

Derin, B., and E. Gerry. 1972. Jurassic Biostratigraphy and Environments of Deposition in Israel. 5th African Micropaleontological Colloquium, Addis Abeba. *Israel Institut of Petroleum, Micropaleontological laboratory, Tel Aviv, report* 2/12: 1-20.

Douvillé, H. 1916. Les terrains secondaires dans le Massif du Moghara à l'est de l'Isthme de Suez d'après les explorations de M. Couyat Barthoux. *Mémoires de l'Académie des Sciences, Paris, Paléontologie* 54(2): 1-184.

Dubar, G. 1967. Brachiopodes Jurassiques du Sahara Tunisien. *Annales de Paléontologie* 53: 1-71.

Dubertret, L. 1960. *Feuille Hermon, Notice explicative.* Beyrouth: République Libanaise, Ministère des Travaux Publics.

------. 1975. Introduction à la carte géologique à 1/50 000 du Liban. *Notes et Mémoires sur le Moyen-Orient* 13: 345-403.

Enay, R. 1966. L'Oxfordien dans la moitié sud du Jura français: étude stratigraphique. *Nouvelles Archives du Muséum d'Histoire naturelle de Lyon* 1(8): 1-323.

Enay, R., Y.-M. Lenindre, Ch. Mangold, J. Manivit, and D. Vaslet. 1987. Le Jurassique d'Arabie Saoudite centrale: nouvelles données sur la lithostratigraphie, les paléoenvironnements, les faunes d'ammonites, les âges et les corrélations. *Geobios,* M.S. 9: 13-65.

Enay, R., and Ch. Mangold. 1994. Première zonation par ammonites du Jurassique d'Arabie Saoudite, une référence pour la province arabique. In É. Cariou and P. Hantzpergue (eds.), 3rd International Symposium on Jurassic Stratigraphy, Poitiers 1991. *Geobios,* M.S.: 161-174.

Feldman, H. R., F. Hirsch, and E. Owen. 1982. *A Comparison of Jurassic and Devonian Brachiopod Communities: Trophic Structure, Diversity, Substrate Relations and Niche Replacement. Third North American Paleontological Convention, Proceeding.* Vol. 1. B., 169-174. Montreal: Business and Economic Services.

Feldman, H. R. 1987. A new species of the Jurassic (Callovian) Brachiopod *Septirynchia* from Northern Sinai. *Journal of Paleontology* 61(6): 1156-1172.

Feldman, H. R., E. Owen, and F. Hirsch. 1991. Brachiopods from the Jurassic of Gebel El-Maghara, Northern Sinai. *American Museum Novitates* 3006: 1-28.

Fenton, J. P. G., R. Neves, and K. M. Piel. 1980. Dinoflagellate cysts and acrytarchs from upper Bajocian to Middle Bathonian strata of central and southern England. *Palaeontology* 23: 151-170.

Fourcade, E., A. A. Arafa, and J. Sigal. 1983. Description of a new species of foraminifera in the Malm of the Middle East: *Mangashlia? egyptiensis* n.sp. *Revue de Micropaléontologie* 27: 21-29.

Fortwengler, D., D. Marchand, and A. Bonnot. 1997. Les coupes de Thuoux et de Savournon (se de la France) et la limite callovien-oxfordien. *Geobios* 30(4): 519-40.

Frebold, H. Von. 1928. Die stratigraphische Stellung der Grenzschichten des syrischen Callovien und Oxford. *Zentralblatt for Mineralogie, Geologie und Paläontologie* Abt. B, 3: 183-201.

Friedman, G., A. Barzel, and B. Derin. 1971. Paleoenvironments of the Jurassic in the central coastal belt of northern and central Israel and their significance in the search for petroleum reservoirs. *Geological Survey of Israel, Report,* OD/1/71: 1-26.

Gill, G. A., and H. Tintant. 1975. Les Ammonites Calloviennes du sud d'Israël. Stratigraphie et relations paléogeographiques. *Comptes Rendus de la Société géologique de France* 4: 103-106.

Gill, G. A., J. Thierry, and H. Tintant. 1985. Ammonites calloviennes du Sud d'Israël: systématique, biostratigraphie et paléobiogéographie. *Geobios* 18(6): 705-751.

Goldberg, M. 1963. Reference Section of Jurassic Sequence in Hamakhtesh Hagadol (Kurnub Anticline). Detailed binocular sample description, including field observations. Unpublished Internal Report, Oil Division. *Geological Survey of Israel*: 1-50.

Goldberg, M. 1964. Problems in the Jurassic stratigraphy of Southern Israel. *Israel Journal of Earth-Sciences* 13: 169-171.

------. 1969. The Jurassic of Majdal Shams area, Mt. Hermon. Measured and sampled by M. Goldberg, M. Brown, M. Raab, B. Derin. In Remarks on the Jurassic lithostratigraphy of Mt. Hermon. *Israel Geological Society, the Golan Heights meeting proceedings, Jerusalem*: 7-8.

Goldberg, M., and R. Bogosh. 1978. Dolomitization and hydrothermal mineralization in the Brur Calcarenite (Jurassic), southern coastal plain. *Israel Journal of Earth-Sciences* 27: 36-41.

Goldberg, M., G. Friedman. 1974. Paleoenvironments and paleogeographic evolution of the Jurassic system in southern Israel. *Geological Survey of Israel Bulletin* 61: 1-44.

Goldberry, R. 1982. Paleosols of the Lower Jurassic Mishhor and Ardon formations ("Laterite Derivative Facies"), Makhtesh Ramon, Israel. *Sedimentology* 29: 669-690.

Gygi, R. 1990. The Oxfordian ammonite succession near Liesberg B. E. and Pery BE, northern Switzerland. *Eclogae geologicae Helvetiae* 83(1): 177-199.

Haan, P.J. (DE). 1997. *Early Cretaceaous Palynomorphs from the Southern Coastal Plain of Israel and a study of the Fern Family Schizaeaceae.* Ph. D. Thesis, University of California.

Haas, O. 1955. Revision of the Jurassic ammonite fauna of Mount Hermon, Syria. *American Museum of Natural History Bulletin* 108: 1-210.

Haq, B., J. Hardenbol, and P. Vail. 1988. Mesozoic and Cenozoic chronostratigraphy and cycles of sea-level change. *SEPM Special Publication* 42: 71-108.

Hirsch, F. 1976. Sur l'origine des particularismes de la faune du Trias et du Jurassique de la plate-forme africano-arabe. *Bulletin de la Société géologique de France* part 7, vol. 18(2): 543-552.

------. 1979. Jurassic bivalves and gastropods from northern Sinai and southern Israel. *Israel Journal of Earth-Sciences* 28(4): 128-163.

------. 1985. The Arabian sub-plate during the Mesozoic. In *The geological evolution of the eastern Mediterranean. Special Publication: Geological Society of London* 17: 217-223.

------. 1987. The Gondwanian Triassic and Jurassic Tethys shelf: Sephardic and Ethiopian faunal realms. *Shallow Tethys* 2, *Balkema*: 215-232.

------. 1988. Jurassic biofacies versus sea level changes in the Middle eastern Levant (Ethiopian province). *2nd International Symposium Jurassic Stratigraphy, Lisbonne*: 963-981.

------. 1996. Geology of the Southeastern slopes of Mount Hermon. *Geological Survey of Israel, Current Research* 10: 22-27.

Hirsch, F., and L. Picard. 1988. The Jurassic facies in the Levant. *Journal of Petroleum Geology* 11(3): 277-308.

Hirsch, F., A. Flexer, A. Rosenfeld, and A. Yellin-Dror. 1995. Palinspastic and crustal setting of the eastern Mediterranean. *Journal of Petroleum Geology* 18(2): 149-170.

Hirsch, F., and R. Roded. 1996. The Jurassic stratigraphical nomenclature in Hamakhtesh Hagadol. *Geological Survey of Israel, Current Research* 10: 10-14.

Hudson, R. G. S. 1958. The Upper Jurassic faunas of Southern Israel. *Geological Magazine* 95: 415-425.

Kitchin, F. L. 1900. Jurassic Fauna of Cutch. pt. I: The Brachiopoda. *Paleontologica Indica* ser. 9, 3: 1-87, pl. 1-15.

Kuznetsova, K. I., A. A. Grigelis, J. Adjamian, E. Jarmakani, and L. Hallaq. 1996. *Zonal Stratigraphy and Foraminifera of the Tethyan Jurassic (Eastern Mediterranean).* Amsterdam: Gordon and Breach.

Kuznetsova, K. I., and M. R. Dobrova. 1995. Endemic and cosmopolitan assemblages of Foraminifers and Ostracods from the Middle Jurassic basins of Syria. *Stratigraphy and Geological Correlations* 3(2): 134-146.

Lang, B., and G. Steinitz. 1989. K-Ar dating of Mesozoic magmatic rocks in Israel: A review. *Israel Journal of Earth Sciences* 38(2-4): 89-104.

Lewy, Z. 1983. Upper Callovian ammonites and Middle Jurassic geological history of the Middle East. *Geological Survey of Israel Bulletin* 76: 1-56.

Lorch, J. 1967. A Jurassic flora of Makhtesh Ramon, Israel. *Israel Journal of Botany* 16: 177-188.

Mangold, Ch. 1988. Les *Pachyerymnoceras* (Pachyceratidés, Périsphinctacés, ammonites) du Callovien moyen et supérieur de la région de Saida (Algérie occidentale). Origine phylétique et biogéographie des Pachyceratidés. *Geobios* 21(5): 567-609.

Maync, W. 1966. Microbiostratigraphy of the Jurassic of Israel. *Geological Survey of Israel Bulletin* 40; *The Institute for Petroleum Research and Geophysics Report* 1007: 1-56.

Moshkovitz, S., and A. Ehrlich. 1976. *Schizosphaerella punctulata* Deflandre et Dangeard and *Crepidolithus crassus* (Deflandre, Noel), Upper Liassic calcareous nannofossils from Israel and Northern Sinai. *Israel Journal of Earth Sciences* 25: 51-57.

------. 1980. Late Jurassic calcareous nannofossils in Israel's offshore and onland areas. *Geological Survey of Israel, Current Research*: 65-72.

Muir-Wood, M. 1925. Jurassic Brachiopods from the Jordan Valley. *Annals and Magazine of Natural History* 15: 181-192.

Nevo, A. 1963. The Jurassic strata of Makhtesh Ramon. *Israel Geological Society, Summary of Lectures, Makhtesh Ramon Symposium*: 11-12.

Noetling, F. 1887. *Der Jura am Hermon.* Stuttgart: E. Schweizerbart'sche Verlagsbuchhandlung, 1-46.

Parnes, A. 1980. Lower Jurassic (Liassic) Invertebrates from Makhtesh Ramon (Negev, Southern Israel). *Israel Journal of Earth-Sciences* 29: 107-113.

Parnes, A. 1981. Biostratigraphy of the Mahmal Formation (Middle and Upper Bajocian) in Makhtesh Ramon (Negev, Southern Israel). *Geological Survey of Israel Bulletin* 74: 1-55.

Picard, L., and F. Hirsch. 1987. *The Jurassic Stratigraphy in Israel and the Adjacent Countries.* Jerusalem: The Israel Academy of Sciences and Humanities.

Raab, M. 1962. Jurassic-Early Cretaceous Ammonites from the Southern Coastal Plain, Israel. *Geological Survey of Israel Bulletin* 34: 24-30.

Razvalyaev, A. V. 1966. Explanatory Notes to the Geological Map of Syria, 1: 200 000. *Syrian Arab Report of the Ministry of Industry, Department of Geological and Mineral Research, Damacus*: 1-122.

Rioult, M., O. Dugue, R. Jan Du Chene, C. Ponsot, G. Fily, J. M. Moron, and P. R. Vail. 1991. Outcrop sequence stratigraphy on the Anglo-Paris basin, Middle to Upper Jurassic (Normandy, Maine, Dorset). *Bulletin des Centres de Recherches Exploration-Production Elf Aquitaine* 1: 101-194.

Rosenfeld, A., E. Gerry, and A. Honigstein. 1987a. Jurassic Ostracodes from Gebel Maghara, Sinai, Egypt. *Revista Española de Micropaleontologia* 19: 251-280.

Rosenfeld, A., H. Oertli, A. Honigstein, and E. Gerry. 1987b. Oxfordian Ostracodes from the Kidod Formation of the Majdal Shams Area, Mount Hermon, Golan Heights. *Bulletin des Centres de Recherches Exploration-Production Elf Aquitaine* 11(2): 233-248.

Rosenfeld, A., A. Honigstein, E. Gerry, H. Oertli, and A. Flexer. 1988. Early Jurassic Ostracodes from the Ardon Formation in Israel and Sinai. *Geological Survey of Israel, Current Research* 6: 50-56.

Rosenfeld, A., and A. Honigstein. 1991. Callovian-Oxfordian Ostracodes from the Hamakhtesh Hagadol Section, Southern Israel. *Revista Espanola de Micropaleontologia* 23(3): 133-148.

Saltzman, U., Y. Bartov, G. Gottlib, M. Magharitz, and Y. Mimran. 1968. Composite columnar section of sequences exposed at the SE flank of Mt. Hermon. In U. Saltzman (ed.), *The Geology of the Southern Hermon Slopes: Internal Report.* Tel Aviv: TAHAL, 1-46.

Sargeant, W. A. S. 1976. Dinoflagellate cysts and acritarchs from the Great Oolite Limestone (middle Jurassic: Bathonian) of Lincolnshire, England. *Geobios* 9: 4-45.

Schirardin, J. 1958. Les ammonites de l'Oxfordien du Jura alsacien de la région de Ferrette; *Bulletin du Service de la Carte géologique d'Alsace et de Lorraine* 11(1): 1-50.

Spath, L. F. 1927-1933. Revision of the Jurassic Cephalopod fauna of Kachh (Cutch). *Palaeontologia Indica* 9(2): 1-945.

Thusu, B., and O. S. Vigran. 1985. Middle-Late Jurassic (Late Bajocian-Tithonian) Palynomorphs. Palynostratigraphy of north-east Libya. *Journal of Micropaleontology* 4: 113-120.

Vautrin, H. 1934. Contribution à l'étude de la série jurassique dans la chaîne de l'Anti-Liban et plus particulièrement dans l'Hermon (Syrie). *Comptes Rendus de l'Académie des Sciences* 198: 1438-1440.

Vidier, J. P., D. Marchand, A. Bonnot, and D. Fortwengler. 1993. The Callovian and Oxfordian of the Boulonnais area in northern France: New biostratigraphic data. *Acta Geologica Polonica* 43(3-4): 169-182.

Weir, J. 1925. Brachiopoda, Lamellibrachiata, Gastropoda and Belemnites. In B. K. N. Wyllie and W. R. Smellie (eds.), *The Collection of Fossils and Rocks from Somaliland.* Monographs of the Geological Department of the Hunterian Museum, Glasgow University, 1(6): 79-110.

Weir, J. 1929. *Jurassic Fossils from Jubaland*, East Africa. Collected by V. G. Glenday. Monographs of the Geological Department of the Hunterian Museum, Glasgow University, 3: 1-63.

Chapter Seven

Brachiopods from the Jurassic (Callovian) of Hamakhtesh Hagadol (Kurnub Anticline), Southern Israel

ABSTRACT

The Callovian Zohar and Matmor formations in the Negev, southern Israel, consisting of marls, shales and limestones, have yielded 13 brachiopod species (2 rhynchonellids, 11 terebratulids), referred to 12 genera, of which one genus and five species are new: *Apatecosia inornata, Bihenithyris mediocostata, Digonella boylani* sp. nov., *Dissoria bretti* sp. nov., *Burmirhynchia jirbaensis, Kutchithyris landeri* sp. nov., *Pleuraloma triangulatum, Polyplectella debriani* gen. et. sp. nov., *Ptyctothyris daghaniensis, Somalirhynchia africana, Striithyris saudiarabica, S. telemi* sp. nov., and *Zeilleria* sp. The brachiopods described herein from Hamakhtesh Hagadol (Kurnub Anticline), comprise a fauna located at the northernmost part of the Indo-African Faunal Realm within the Jurassic Ethiopian Province.

This study is a continuation of a long-term project undertaken by us in northern Sinai (Feldman 1986, 1987; Feldman and Owen 1988, 1993; Feldman *et al.* 1982, 1991) and Israel in order to taxonomically study and revise the brachiopod faunas; this will aid in establishing the history of brachiopod species and their evolution within the Ethiopian Province. We expect that the data compiled as a result of these studies will enable us to interpret the biogeographic history of the Ethiopian Province as well as gain insight into the structure and palaeoecology of its marine communities (see, for example, Feldman and Brett 1998).

Previous studies of Ethiopian Province brachiopods from northern Sinai (Feldman 1987; Feldman and Owen 1988; Feldman et al. 1982, 1991; Hegab 1988, 1989, 1991a, 1991b, 1992, 1993), along with Cooper's (1989) monograph, complement the present paper on the brachiopods of Hamakhtesh Hagadol. The Ethiopian Province has been recognized from the Early Jurassic until the mid and possibly end of the Cretaceous by the presence of endemic taxa at the species, genus and family level. These endemics (e.g. *Somalirhynchia africana, Daghanirhynchia daghaniensis, Somalithyris bihendulensis, Striithyris somaliensis, Bihenithyris barringtoni, B. weiri*) have been noted, especially in the brachiopods, by Weir (1925) and Muir-Wood (1935). Arkell (1952, 1956) recognized endemic faunas in the ammonoid Cephalopoda and Kitchin (1912) in the trigoniacean and crassatellacean bivalves. A clearer picture of the endemism so characteristic of many of these faunas is now available. The present fauna lies at the northernmost part of the Indo-African Faunal Realm and may therefore be related to species of the Tethyan Realm. Upon completion of a systematic revision of Israeli Jurassic brachiopods, including those from the Majdal Shams area near Mount Hermon currently under study, we will be able to define faunal- and province-realm boundaries with greater accuracy.

Cooper's (1989) work on the Jurassic brachiopods of Saudi Arabia was based largely on collections made during the years 1933-1953 by field geologists of the Arabian-American Oil Company (Aramco) and the Kier-Kauffmann collections (1962). Cooper (1989) noted that whereas several rock types are recorded, some of the specimens cannot be assigned to specific parts of the column, but most have been referred to various ammonite zones. Consequently, it is not possible to determine the palaeoecological conditions under which they existed during the Jurassic. In Hamakhtesh Hagadol, however, collecting was accomplished under strict stratigraphical control and an eventual palaeoecological study of the brachiopods and their communities will be undertaken.

STRATIGRAPHICAL SETTING

Hamakhtesh Hagadol (Text-fig. 1) is a 15-km-long and 6-km-wide erosional cirque, carved within the axial culmination of the Kurnub Anticline, located in the northern Negev (southern Israel). A 206-m-thick Jurassic sequence, 4.5 km long and 1.5 km wide, is exposed within the elliptical core of the Kurnub structure. The Kurnub Anticline was first discovered by F. E. Wellings and L. Damesin in the 1930's on behalf of the Iraq Petroleum Company (IPC), and its existence was first noted in the literature by Blake (1935). These fossiliferous marls and limestones became better known following the work of Damesin and S.N. Nasr for the Petroleum (Palestine) Development Company (PDP), the results of which were published by Shaw (1947). The lithostratigraphical subdivision used in the present study is based on the very detailed survey of the section by Goldberg (1963). The subdivision of formations follows that of Hirsch and Roded (1997).

Hudson (1958) published the first detailed palaeontological study of the Jurassic of Hamakhtesh Hagadol. He recognized the Callovian, Divesian, Argovian and Sequanian stages, proposed a stratigraphical subdivision, and composed a faunal list of bivalves, gastropods, ammonites and brachiopods from the section. Reiner (1968) studied Callovian gastropods from the lower part of the section and Hirsch (1979) wrote a monograph on the gastropods and bivalves found within the entire section. Rosenfeld and Honigstein (1991) studied the ostracodes and Conway (1990) described the palynomorphs from nearby boreholes. While studying the brachiopods of this region Feldman and Brett (1998) recognized epi- and endobiontic organisms on Jurassic crinoids from Hamakhtesh Hagadol and have extended the range of the ichnogenus *Tremichnus* up by almost 100 million years. The Callovian ammonite biozonation of the sequence at

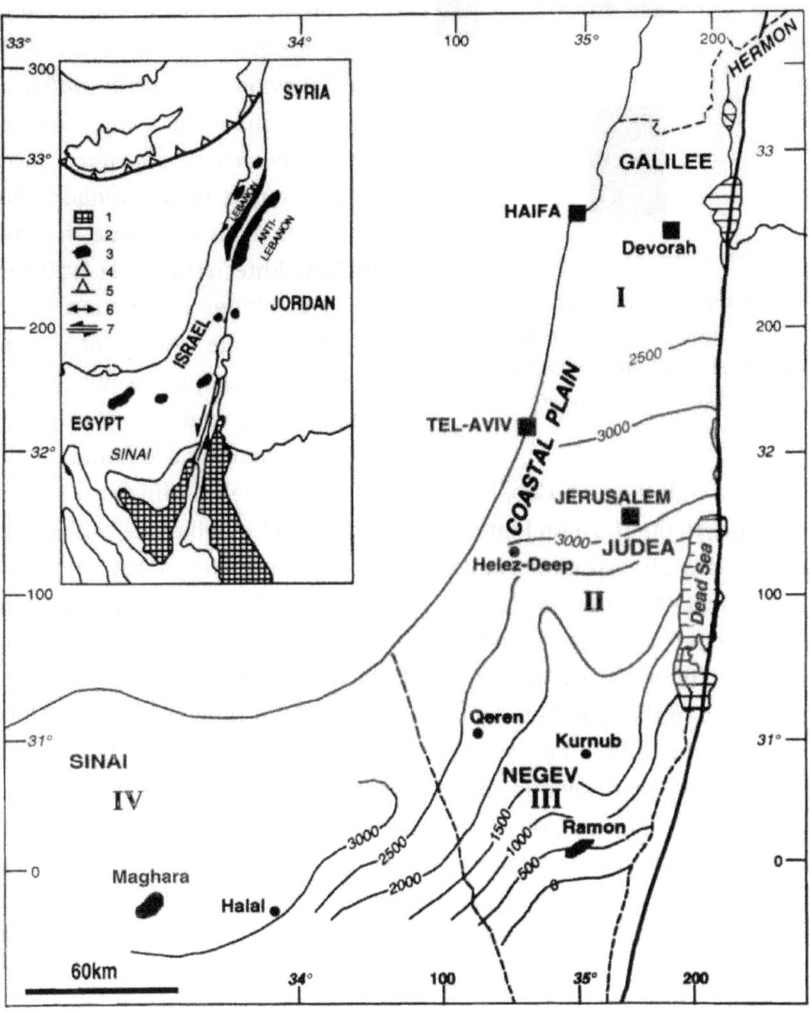

FIGURE 1: Location map of Jurassic exposures in the Levant. 1, Precambrian; 2, Palaeozoic-Triassic; 3, Jurassic; 4, Mount Hermon; 5, Alpine thrust-front; 6, Neogene sinistral transform, I, Galilee High; II, Judean Embayment; III, Negev High; IV, Sinai Deep, with extension of Lower Oxfordian Majdal Shams shales (shaded area) and Jurassic isopachs (from Hirsch et al. 1998).

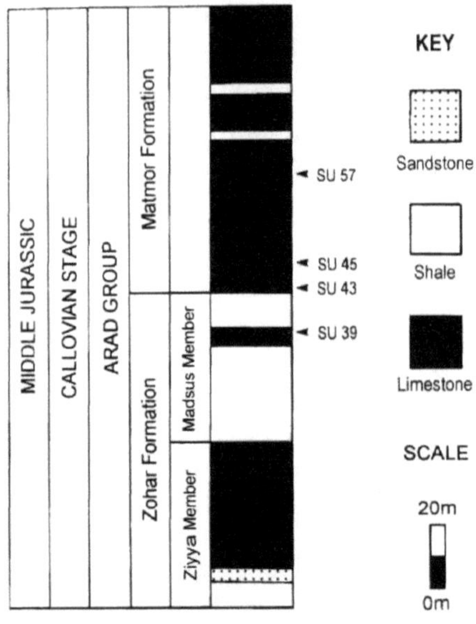

FIGURE 2: Simplified stratigraphical columnar section of the Jurassic sequence at Hamakhtesh Hagadol, Negev, Israel (after Hirsch and Roded 1997). Key fossiliferous subunits are numbered.

Hamakhtesh Hagadol was refined by Gill and Tintant (1975), Lewy (1983), Gill et al. (1985) and more recently by Cariou et al. (1997). Cariou et al. (1997) and Hirsch et al. (1998) considered the section at Hamakhtesh Hagadol to be a nearly continuous Middle to Upper Callovian sequence, unique in the Levant, comprising the *Erymnoceras coronatum* Zone (Middle Callovian), *Peltoceras athleta* Zone (Upper Callovian) and *Orionoides* (*Poculisphinctes*) *poculum* Subzone (= lower part of the *lamberti* Zone). The Callovian ammonite succession in Hamakhtesh Hagadol (Arabic Province) facilitates correlation with the standard scale stages of Submediterranean Europe. Elsewhere in the Middle East, coeval exposures of the Zohar Formation are found at Abu El Darag (Suez Gulf, Egypt), Gebel El-Maghara and Gebel El-Minshera (northern Sinai, Egypt), in the Hermon Limestone Formation (Antilebanon), and in the Tuwaiq Mountain Limestone Formation (Saudi Arabia). Coeval sequences of the Upper Callovian Matmor Formation (nearly 100 m thick) are only found in the Negev boreholes at Qeren, Sherif, Rehkme, Boqer and the Hallal borehole in Sinai, Egypt, all of which are part of the platform of the Higher Negev (Picard and Hirsch 1987). In the Gebel El-Maghara section (Sinai Deep) and in the Hermon exposures of the Judean Embayment (Hirsch *et al.* 1998), the Late Callovian is reduced to a few dm of section covered by Oxfordian shales (Cariou *et al.* 1997).

The 206-m-thick section at Hamakhtesh Hagadol (Text-fig. 2) was subdivided into 69 subunits (numbered 5-74) by Goldberg (1963). The sequence consists of the Ziyya and Madsus members of the Zohar Formation (Coates et al. 1963) and of the Matmor Formation (Goldberg, 1963, emend. Hirsch and Roded 1997).

Zohar Formation

Ziyya Member. The Ziyya Member of the Zohar Formation (Goldberg 1963, emend. Hirsch and Roded 1997; subunits 5-31) is a 56-m-thick sequence of shale, sandstone, marl and limestone overlying the hard limestone (subunit 4) that represents the top of the Halamish Member (Goldberg and Friedman 1974). This limestone unit is intermittently exposed in the river bed near the junction of Nahal Matmor and Nahal Hatira. Three cycles are recognized in the Ziyya Member. The lower cycle (14 m) consists of shales (subunit 5), a sandstone intercalation (subunits 6-7) rich in echinoids and thalassinoids, and a limestone unit (subunit 8) with abundant bivalves and occasional ammonites. The next cycle (17 m) consists of marl (subunit 9) and well-bedded marls and limestones (subunits 10-14) yielding abundant bivalves, gastropods and ammonites. The top cycle (25 m) consists of upward thinning marls (subunits 15, 16, 18, 20, 22, 24) with thin limestone interbeds (subunits 17, 19, 21, 23). Subunits 17 and 19 contain *Zoophycus*, indicating deepening, followed by alternating marl and limestone layers (subunits 25-31), all rich in fossils. The brachiopod *Somalirhynchia africana* was recovered from subunits 20-24. The Ziyya Member corresponds to the *Eligmus-Erymnoceras* beds of Hudson (1958) of Mid Callovian age (*coronatum* Zone) (Gill and Tintant 1975; Gill et al. 1985).

Madsus Member. The Madsus Member of the Zohar Formation (Goldberg, 1963 emend. Hirsch and Roded 1997; subunits 32-42) is 50 m thick, the lower 20 m of which consists of alternating marls and limestones (subunits 32-33) mostly covered by scree, and an upper 30 m thick shale sequence (subunits 34-42) separated by a 3.6-m-thick limestone unit (subunit 39). Scree (9 m thick) covers the lowermost section (subunit 32) followed by 11 m of marl (subunit 33) with thin limestone interbeds, rich in fossils, including ammonites and the brachiopods *Somalirhynchia africana* and *Pleuraloma triangulatum*. It is followed by rather sterile shales (subunits 34-38) topped by a calcarenite (subunit 39) in which the brachiopod *Ptyctothyris daghaniensis* occurs. Overlying these units are shales (subunits 40-42) that contain *Pleuraloma triangulatum* and *Somalithyris arabica* (subunit 40) and *Parathyridina plicatoides* (subunit 42). The Madsus Member correlates with the *Eligmus-Grossouvria* beds (Hudson, 1958), to which a Late Callovian age (*athleta* Zone) is assigned (Gill and Tintant 1975; Lewy 1983).

Matmor Formation

The 100 m thick Matmor Formation consists of a lower and upper sequence. The lower sequence (30 m; subunits 43-52) is composed of alternating fossiliferous limestones and marls, correlative with Hudson's (1958) *Somalirhynchia-Putealiceras* beds of Late Callovian age (*athleta* Zone, Gill and Tintant 1975; Lewy 1983). This lower sequence yields abundant bivalves and brachiopods, gastropods, and some ammonites. Subunit 43 is an extremely fossiliferous limestone from which have been recovered *Bihenithyris mediocostata*, *Burmirhynchia jirbaensis*, *Kutchithyris landeri* sp. nov., *Somalirhynchia africana*,

Striithyris saudiarabica, and *S. telemi*, sp. nov. The limestones of subunit 46 yield *Dissoria bretti* sp. nov. whereas in subunit 47 are found *Burmirhynchia jirbaensis, Ptyctothyris daghaniensis*, and *Striithyris telemi*, sp. nov. The white limestone of subunit 48 yields *Digonella boylani* sp. nov. The 70-m-thick upper sequence (subunits 53-74) consists mostly of alternating limestones and marls, correlative with Hudson's (1958) *Shuqraia* and *Cladocoropsis* beds. The 15-m lower beds of the upper sequence (subunits 53-58) still belong to the *athleta* Zone. Based on ammonite occurrences, most of the upper sequence (subunits 60-75) can be assigned to the Late Callovian *poculum* Subzone (*lamberti* Zone).

SYSTEMATIC PALAEONTOLOGY

Abbreviations. AMNH, American Museum of Natural History, New York; NHM, Natural History Museum, London; GSI, Geological Survey of Israel, Jerusalem. All measurements are in millimetres (mm); L, maximum length of shell; W, maximum width of shell; T, maximum thickness of shell.

Superfamily RHYNCHONELLOIDEA Gray, 1848
Family RHYNCHONELLIDAE Gray, 1848
Subfamily TETRARHYNCHIINAE Ager, 1965

Genus BURMIRHYNCHIA Buckman, 1918

Burmirhynchia jirbaensis Muir-Wood, 1935
Plate 1, figures 1-6

1935 *Burmirhynchia jirbaensis* Muir-Wood, p. 91, pl. 8, fig. 3a-c.

Type species. Burmirhynchia jirbaensis Muir-Wood, 1935.

Material. Four articulated specimens.

Remarks. Both the examples figured here (Plate 1, figures 1-6; Text-fig. 3) represent the species *Burmirhynchia jirbaensis* Muir-Wood, 1935, and give some indication of the variability of the species (Table 1) largely from their dorsal outline, which ranges from transversely oval to broadly triangular, perhaps a little less acutely triangular than the holotype. Both examples, however, show similar costation and a slightly inflated umbo with an incurved beak and

TABLE 1: Measurements of *Burmirhynchia jirbaensis* Muir-Wood, 1935.

Specimen	L	W	T	Subunit
NHM 1009	23.0	22.8	17.8	47
NHM 1010	21.9	22.0	16.5	53-54

well-exposed delthyrium. The median fold on the dorsal valve is more clearly defined than in the holotype, but the anterior margin shows a similar, broadly trapezoidal sulcation of the ventral valve with a broad arched or arcuate anterior commissure. Variation extends to a slightly more inflated dorsal umbo in the case of the two examples figured here (NHM 1009, 1010) but we, nevertheless, consider these two examples to be conspecific with Muir-Wood's species *Burmirhynchia jirbaensis* from the so-called "Argovian" of Somalia (formerly British Somaliland).

Stratigraphical occurrence. Matmor Formation (Upper Callovian), Subunits 47; 53-54, Hamakhtesh Hagadol, Negev, Israel.

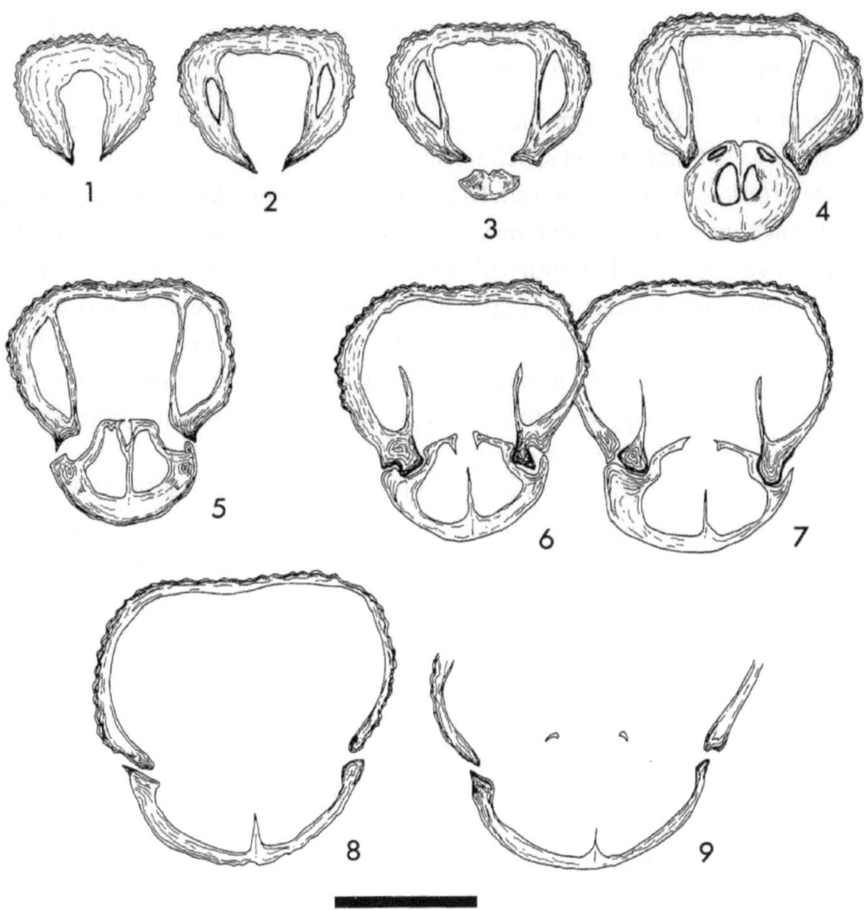

FIGURE 3: A series of nine transverse serial sections through the umbo of a specimen of *Burmirhynchia jirbaensis* Muir-Wood, 1935 from the Callovian of Hamakhtesh Hagadol, southern Israel. Numbers in parentheses represent distance between sections: 1(0.30); 2(0.80); 3(0.30); 4(0.20); 5(0.30); 6(0.30); 7(0.40); 8(0.70); 9(0.30); Scale bar = 1 cm; AMNH 46556.

Genus SOMALIRHYNCHIA Weir, 1925

Somalirhynchia africana Weir, 1925
Plate 1, figures 7-15

1925 *Somalirhynchia africana* Weir, p. 80, pl. 12, fig. 20-23
1935 *Somalirhynchia africana* Weir; Muir-Wood, p. 94, pl. 10, fig. 7a-c, text-fig. 7-8.
1965 *Somalirhynchia africana* Weir; Ager, p. H614, fig. 497, 9a-b.
1967 *Somalirhynchia africana* Weir; Dubar, p. 30, pl. 2, fig. 5a-b.
1974 *Somalirhynchia africana* Weir; Abbate et al., p. 439, pl. 39, fig. 4.
1989 *Somalirhynchia africana* Weir; Cooper, p. 58, pl. 12, fig. 37-41.
1991 *Somalirhynchia africana* Weir; Feldman et al., p. 7, fig. 4a-c.
1993 *Somalirhynchia africana* Weir; Shi and Grant, p. 104, pl. 8, fig. 15; pl. 10, fig. 9-11; pl. 12, fig. 8.

Type species. Somalirhynchia africana Weir, 1925, p. 80, pl. 12, fig. 20-23.

Material. 11 articulated specimens.

Description. Medium sized shells (Table 2) subtriangular in outline, strongly dorsibiconvex with maximum width attained past midlength; beak erect and massive with small hypothyrid pedicle foramen; deltidial plates disjunct. Anterior commissure strongly uniplicate; lateral commissure deflected ventrally; interareas large, somewhat concave. Ventral sulcus originates just past midlength, widening anteriorly into a long tongue with 6 costae. Dorsal valve high, domelike, with strong median fold with 7 costae that begins near midlength and becomes strongly elevated anteriorly.

Remarks. This is one of the most morphologically variable rhynchonellid species so far encountered within the Jurassic of Israel, Saudi Arabia and Somalia. There have been many specimens described as new species collected from an identical stratigraphical level, sometimes from the same bed, which would fit quite comfortably into a series of variations of the form originally described by Weir (1925). They vary in size from large to medium and a shell ornament that varies from coarse to deeply incised to fine and rounded. Some are described as having a pronounced median dorsal fold and deep sulcus, whereas others can be flat to incipiently biconvex. Our specimen, figured here, is smaller than the one figured originally by Weir (1925, pl. 12, fig. 20-23) but matches the characters described by that author. It is considerably smaller than the specimen figured by us from subunits at Gebel El-Maghara, Sinai, as *Somalirhynchia africana* Weir 1925 (Feldman et al. 1991, figs a-c) and agrees in overall morphology with the specimen figured in the same publication as figs d-f. However, the costation of this specimen is more rounded and not so deeply incised as the specimen figured here.

TABLE 2: Measurements of *Somalirhynchia africana* Weir, 1925.

Specimen	L	W	T	Subunit
NHM 1011	23.9	25.9	21.6	43
NHM 1012	25.0	30.4	19.0	43
NHM 1013	24.8	26.8	22.1	43
NHM 1014	25.9	25.1	18.4	43
AMNH 46559	26.0	27.0	18.0	43
AMNH 46560	22.1	25.0	15.9	43
AMNH 46561	27.6	27.0	18.1	43
AMNH 46562	24.8	28.1	16.9	43

Stratigraphical occurrence. Matmor Formation (Upper Callovian), Subunits 33, 43, Hamakhtesh Hagadol, Negev, Israel.

Superfamily TEREBRATULOIDEA Gray, 1840
Family TEREBRATULIDAE Gray, 1840
Subfamily LISSAJOUSITHYRIDINAE Cooper, 1983

Genus APATECOSIA Cooper, 1983

Apatecosia inornata Cooper, 1989
Plate 1, figures 16-18

Type species. Cererithyris nutiensis Bague, 1955; p. 219, fig. 2a-c.

Material. 14 articulated specimens. Dimensions of figured specimen (NHM 1015): L, 30.8; W, 23.7; T, 16.5.

Description. Medium to large sized, evenly biconvex, subpentagonal in outline. Umbo massive, beak suberect with rounded or poorly defined beak ridges. The foramen is large and circular. Deltidial plates obscured. The anterior commissure is plain to just uniplicate in the early stages of growth, becoming slightly biplicate in later stages and maturity, a shallow sulcus developing anteriorly on the dorsal valve. Shell surface with evenly spaced faint concentric growth lines.

Internal characters. Unknown.

Remarks. Cooper (1983, p. 53) described the genus *Apatecosia* originally from the *Macrocephalites macrocephalus* Zone of La Cude, near Velars, Cote d'Or, France. In his paper on the Jurassic brachiopods of Saudi Arabia he described (Cooper 1989, p. 69) two new species, *A. varians* and *A. inornata*, which he assigned to the genus. Examples of both species were collected from the Tuwaiq Mountain Formation on the west side of the Tuwaiq Mountains. The specimens described here as *Apatecosia inornata* are from subunits 39 and 47 in Hamakhtesh Hagadol, Negev, Israel. Fourteen further specimens collected from the same horizon and locality now in the Feldman, Hirsch and Owen collection at present in The Natural History Museum, London.

PLATE 1

Although the internal characters of the species described here are unknown, Cooper (1983, pl. 32, fig. 18-23, 28-29) illustrated brachial loops exposed by dissection of examples of the type species *Apatecosia nutiens* (Bague). They show a broad loop extending to approximately half the length of the dorsal valve, with long, inwardly curving crural processes and a high arcuate transverse band.

The internal characters of the species described here as *Apatecosia inornata* are known only from dissections made by Cooper and figured by him (Cooper, 1989, pl. 20, fig. 5) and compare favourably with loop dissections of the type species *Apatecosia nutiens* (Cooper, 1983, pl. 32, fig. 18-23, 28-29).

Stratigraphical occurrence. Zohar and Matmor formations (Upper Callovian), Subunits 39, 47, Hamakhtesh Hagadol, Negev, Israel.

Genus BIHENITHYRIS Muir-Wood, 1935

Bihenithyris mediocostata Cooper, 1989
Plate 1, figures 19-23

Type species. Bihenithyris mediocostata Cooper, 1989.

Material. 11 articulated specimens.

Description. Medium sized (Table 3), pentagonal, biconvex *Bihenithyris.* Dorsal valve sometimes less acutely convex than ventral. Umbo massive, suberect, truncated by large circular labiate foramen with distinct mesothyrid beak ridges. Maximum width of shell approximately just anterior to midvalve. Dorsal valve with two well-developed folds flanking a deep median sulcus that forms a marked episulcation with the corresponding ventral fold at the anterior margin.

EXPLANATION OF PLATE 1

FIGS 1-6. *Burmirhynchia jirbaensis* Muir-Wood, 1935, Matmor Formation (Upper Callovian, subunits 47, 53-54), Negev, Israel. 1-3, dorsal, lateral, anterior views, NHM 1009; 4-6, dorsal, lateral, anterior views, NHM 1010.

FIGS 7-15. *Somalirhynchia africana* Weir, 1925, Matmor Formation (Upper Callovian, subunits 33, 43), Negev, Israel. 7-9, dorsal, lateral, anterior views, NHM 1011; 10-12, dorsal, lateral, anterior views, NHM 1012; 13-15, dorsal, lateral, anterior views, NHM 1013.

FIGS 16-18. *Apatecosia inornata* Cooper, 1989, Zohar and Matmor formations (Upper Callovian, subunits 39, 47), Negev, Israel; dorsal, lateral, anterior views, NHM 1015.

FIGS 19-23. *Bihenithyris mediocostata* Cooper, 1989, Matmor Formation (Upper Callovian, subunit 43), Negev, Israel; 19-21, dorsal, lateral, anterior views, NHM 1016; 22-23, anterior, dorsal views, NHM 1017.

FIGS 24-26. *Kutchithyris landeri* sp. nov.; Matmor Formation (Upper Callovian, subunit 43), Negev, Israel; lateral, anterior, dorsal views, holotype, NHM 1022.

FIG 27. *Ptyctothyris daghaniensis* Muir-Wood, 1935, Zohar Formation (Upper Callovian, subunit 40), Negev, Israel; dorsal, lateral views, NHM 1023, x1.5.

All figures x1 except where indicated.

TABLE 3: Measurements of *Bihenithyirs mediocostata* Cooper, 1989.

Specimen	L	W	T	Subunit
NHM 1016	24.3	20.2	14.5	43
NHM 1017	23.6	20.4	14.6	43
NHM 1018	23.0	18.9	14.0	43
NHM 1019	25.1	20.0	14.0	43
NHM 1020	23.3	19.1	13.9	43
NHM 1021	22.0	18.1	13.0	43
AMNH 46563	26.5	20.0	16.0	43
AMNH 46564	25.4	19.0	15.5	43
AMNH46565	22.6	21.0	13.7	43
AMNH 46566	23.0	18.9	14.0	43
AMNH 46567	22.2	18.0	13.1	43

Remarks. This represents one of the most common terebratulid species occurring at this horizon in Hamakhtesh Hagadol. Its oval to subpentagonal outline and episulcate anterior commissure are similar characters to those described here as the species *Ptyctothyris daghaniensis* and it is tempting to assume that this form represents a juvenile of that species but, with very little variation, it appears to maintain its size and outline throughout a series of eleven specimens from the same locality. So far, little is known about the internal structures of this species and it may prove in time, and with more material, that both forms are congeneric. From similarities of external morphology we consider *Bihenithyris mediocostata* to be a senior synonym of the species described by Muir-Wood (1935, p. 138, pl. 12, fig. 3a-c) and closely resembles specimens figured by Dubar (1967, pl. 3, fig. 14a-c, 16a-c).

Stratigraphical occurrence. Matmor Formation (Upper Callovian), Subunit 43, Hamakhtesh Hagadol, Negev, Israel.

Genus KUTCHITHYRIS Buckman, 1918

Kutchithyris landeri sp. nov.
Plate 1, figures 24-26

Derivation of name. After Bernard Lander, Touro College, for his outstanding contributions to post-secondary education, worldwide.

Holotype. NHM 1022; L, 31.1; W, 25.8; T, 19.2.

Material. 4 articulated specimens.

Diagnosis. Medium sized, acutely biconvex, incipiently biplicate *Kutchithyris*.

Description. Typically subpentagonal in general outline. Evenly biconvex, lateral profile elongate oval. Umbo short, massive, foramen large, circular. Beak ridges rounded, interarea poorly defined. Dorsal valve with shallow lateral sulci and slightly deeper median sulcus developing anteriorly to form clearly defined but incipient biplication of the anterior commissure. Both valves show well-developed unevenly spaced concentric growth lines becoming more numerous and closer at the shell margins.

Internal structures. Unknown due to lack of duplicate material for serial sectioning.

Remarks. With the exception of the umbo, which is well developed and truncated by a large circular foramen, this species resembles one described by Cooper (1989, p. 98, pl. 26, fig. 28-30) as *Kutchithyris*? from the Tuwaiq Mountain Formation (Upper Callovian), Saudi Arabia, but differs from that species in its more elongate-pentagonal outline, more massive umbo and less marked biplication of the anterior commissure. It bears some faint resemblance to a specimen described by Cooper (1989, pl. 27, fig. 7-9) from the Dhruma Formation (Bathonian), Saudi Arabia, but differs from *Kutchithyris* sp. 1 in having a less marked pentagonal outline, less acute biconvexity, shorter, less massive umbo and more pronounced biplicate anterior margin. Our species can be more closely compared to *Kutchithyris parnesi* Feldman et al. (1991, p. 19, fig. 14A-C) but differs from that species in its more massive umbo, narrower width and more marked biplicate anterior commissure.

Stratigraphical occurrence. Matmor Formation (Upper Callovian), Subunit 43, Hamakhtesh Hagadol, Negev, Israel.

Genus PTYCTOTHYRIS Buckman, 1918

Ptyctothyris daghaniensis Muir-Wood, 1935
Plate 1, figure 27; Plate 2, figures 1-2

1935 *Ptyctothyris*? *daghaniensis* Muir-Wood, p. 122, pl. 13, fig. 2a-b.
1991 *Ptyctothyris*? *daghaniensis* Muir-Wood; Feldman et al., p. 25, fig. 17-18.

Type species. Terebratula stephani Davidson, 1877.

Material. 13 articulated specimens.

Description. Large (Table 4), oval, biplicate terebratulid. Biconvex with dorsal valve less acutely convex or flatter than ventral valve. Umbo large, incurved; foramen large, circular, labiate. Beak ridges permesothyrid, symphytium obscured. Anterior commissure broadly paraplicate; dorsal sulcus deepens anteriorly to meet a marked ventral fold. Shell surface with numerous concentric growth lines.

TABLE 4: Measurements of *Ptyctothyris daghaniensis* Muir-Wood, 1935.

Specimen	L	W	T	Subunit
NHM 1023	46.8	35.5	26.8	40
NHM 1024	38.8	30.7	9.4	40
AMNH 46568	38.1	32.0	8.1	40

PLATE 2

Remarks. The assignation of this species to *Ptyctothyris* is based largely upon comparison with species hitherto broadly assigned to the genus by Muir-Wood (1935) and Feldman *et al.* (1991, p. 25). Although Muir-Wood was able to compare her serial section of the species with those of the type species, it would appear that she was not convinced of a close match of the two and thus remained apprehensive. The external morphology of the specimens examined by us, both from localities in the Negev and from Gebel El-Maghara, northern Sinai, has convinced us that the characters comply almost exactly with those illustrated in the Treatise on Invertebrate Paleontology (Moore, 1965, p. H786, fig. 3a-c) and is broader than any of the specimens described subsequently. Our comparison is based on size and general outline, beak and umbonal features, lateral profile and anterior paraplication.

Stratigraphical occurrence. Zohar Formation (Upper Callovian), Subunit 40, Hamakhtesh Hagadol, Negev, Israel.

Genus STRIITHYRIS Muir-Wood, 1935

Striithyris saudiarabica Cooper, 1989

Plate 2, figures 3-5

Type species. Striithyris somaliensis Muir-Wood, 1935.

Material. 14 articulated specimens. Dimensions of figured specimen (NHM 1025): L, 21.1; W, 16.0; T, 12.3.

Diagnosis. Small, narrowly ovate *Striithyris* with incipient sulciplication.

EXPLANATION OF PLATE 2

FIGS 1-2. *Ptyctothyris daghaniensis* Muir-Wood, 1935, Zohar Formation (Upper Callovian, subunit 40), Negev, Israel; anterior view, NHM 1023, x 1.5.

FIGS 3-5. *Striithyris saudiarabica* Cooper, 1989, Matmor Formation (Upper Callovian, subunit 43), Negev, Israel; dorsal, lateral, anterior views, NHM 1025.

FIGS 6-8. *Striithyris telemi* sp. nov. Matmor Formation (Upper Callovian, subunits 43-44, 47), Negev, Israel; dorsal, lateral, anterior views, holotype, NHM 1029.

FIGS 9-13. *Pleuraloma triangulatum* Cooper, 1989, Zohar Formation (Upper Callovian, subunits 33, 40), Negev, Israel; 9-11, dorsal, lateral, anterior views, NHM 1032; 12-13, dorsal, anterior views, NHM 1033

FIGS 14-16. *Dissoria bretti* sp. nov. Matmor Formation (Lower Oxfordian, subunit 54), Negev, Israel; anterior, lateral, dorsal views, holotype, NHM 1035.

FIGS 17-19. *Digonella boylani* sp. nov. Matmor Formation (Upper Callovian, subunit 48), Negev, Israel; dorsal, lateral, anterior views, holotype, NHM 1036.

FIGS 20-25. *Polyplectella debriani* gen. et sp. nov. Matmor Formation (Upper Callovian, subunit 42), Negev, Israel; 20-22, dorsal, lateral, anterior views, NHM 1037, x 1.5; 23-25, dorsal, lateral, anterior views, holotype, NHM 1038, 1.5.

FIGS 26-28. *Zeilleria* sp., Matmor Formation (Upper Callovian, subunit 43), Negev, Israel; dorsal, lateral, anterior views, NHM 1039.

All figures x1 except where indicated.

Description. As the species was adequately described by Cooper (1989, p. 94, pl. 30, fig. 8-12) it is considered sufficient to give an abbreviated description herein. Small to medium (Table 5), elongate-oval, narrow; maximum width anterior to midvalve. Anterior and lateral margins rounded; posterolateral margins straight, forming acute angle. Anterior commissure uniplicate in young becoming sulciplicate in adult stage. Beak short, labiate; foramen large, permesothyrid. Shell surface evenly capillate. Shell moderately biconvex with ventral valve less acutely convex than dorsal valve. A shallow sulcus develops anteriorly on the well-formed dorsal fold to meet the wider and deep sulcus of the ventral valve at the commissural margin.

TABLE 5: Measurements of *Striithyris saudiarabica* Cooper, 1989.

Specimen	L	W	T	Subunit
NHM 1025	19.5	14.7	11.7	43
NHM 1026	22.2	17.8	13.9	43
NHM 1027	23.1	19.8	15.1	43
NHM 1028	23.5	20.0	15.7	43
AMNH 46569	22.1	18.0	12.3	43
AMNH 46570	22.7	18.4	13.0	43
AMNH 46571	19.2	15.8	11.1	43

Remarks. The type species *Striithyris somaliensis* is considerably larger than the species described here as *S. saudiarabica*, and is more robust. It has a distinct oval outline unlike our species which, like the specimens figured by Cooper (1989, pl. 30, fig. 8-12), are considered to be subpentagonal in general outline and have steeper flanks. Cooper's specimens are shown to have a deeper dorsal sulcus than the example figured here but this is probably a variable character. A single ventral valve, similar in form to the specimen illustrated here, was figured by Dubar (1967, pl. 4, fig. 10a-b) from the Kimmeridgian of Tazerdunet, Algeria as *Striithyris somaliensis* Muir-Wood.

Stratigraphical occurrence. Matmor Formation (Upper Callovian), Subunit 43, Hamakhtesh Hagadol, Negev, Israel.

Striithyris telemi sp. nov.
Plate 2, figures 6-8

Derivation of name. After Peter B. Telem for numerous discussions, valuable advice and much appreciated assistance.

Holotype. NHM 1029.

Material. 3 articulated specimens.

Diagnosis. Broadly triangular, finely striated, biconvex *Striithyris.*

Description. Unlike *Striithyris saudiarabica* Cooper, 1989, this species is flatter and distinctly triangular in general outline, the greatest width being about midway between the beak and the anterior margin. The convexity of the dorsal valve is greater than that of the ventral valve. The umbo is short with a sharp incurved beak and a proportionately large, circular foramen. The beak ridges are permesothyrid and fairly distinct. The dorsal valve is almost flat with a faint median fold and an incipient sulcus which is bounded on either side by two poorly defined carinae that become more highly developed anteriorly. The anterior commissure is sulciplicate, the ventral fold being more marked at the anterior extremity. Both valves are ornamented by numerous fine closely spaced striae, interrupted by very faint concentric growth lines. The shells are medium sized (Table 6).

TABLE 6: Measurements of *Striithyris telemi* sp. nov.

Specimen	L	W	T	Subunit
NHM 1029 holotype	14.0	14.1	10.0	43
NHM 1030 paratype	18.6	19.7	11.0	44
NHM 1031 paratype	19.5	17.0	10.0	47

Remarks. This may possibly be a wide variant of *S. saudiarabica* as it occurs at the same stratigraphical horizon and has been collected from subunit 43 with that species. However, two well-preserved specimens have been recovered from two other subunits and, although larger than the holotype, conform in detail to the description given here.

Stratigraphical occurrence. Matmor Formation (Upper Callovian), Subunits 43, 44, 47, Hamakhtesh Hagadol, Negev, Israel.

Genus PLEURALOMA Cooper, 1989
Pleuraloma triangulatum Cooper, 1989
Plate 2, figures 9-13

1989 *Pleuraloma triangulatum* Cooper, p. 87, pl. 27, fig. 13-15.

Type species. Pleuraloma triangulatum Cooper, 1989.

Material. 6 articulated specimens.

Description. Medium sized (Table 7; Text-fig. 4), equibiconvex *Pleuraloma*, broadly oval in dorsal outline and averaging 27.0 mm in length, 21.25 mm in width and 14.0 mm in thickness (Table 2). The ventral valve is massive, with a poorly developed umbo dominated by a large, labiate permesothyrid foramen. The beak ridges are rounded and the symphytium not exposed. The dorsal valve is evenly convex, smooth with ornamentation consisting of very faint concentric growth lamellae. The anterior commissure is rectimarginate. Both valves are marginally costate or polyplicate with a flattened appearance.

TABLE 7: Measurements of *Pleuraloma triangulatum* Cooper, 1989.

Specimen	L	W	T	Subunit
NHM 1032	29.0	23.3	20.7	40
NHM 1033	27.4	22.0	16.2	33
NHM 1034	25.6	20.7	17.1	33
AMNH 46572	28.0	21.9	17.7	33
AMNH 46573	31.6	23.9	20.9	33
AMNH 46574	20.9	16.2	12.3	33

Remarks. There can be no doubt that this species belongs to Cooper's (1989) genus *Pleuraloma* but we contend that it fits into the range of variation not recognized by Cooper but extends between that of *P. robustum* and *P. abruptum*. Specimens collected from Hamakhtesh Hagadol, in the Negev, show variation of morphology within these limits. All the forms described as separate species by

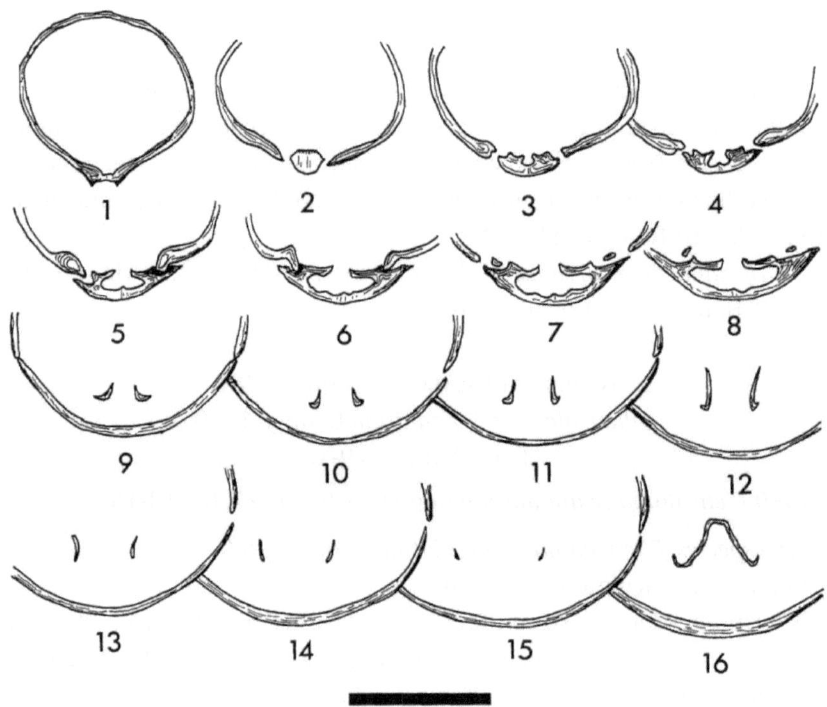

FIGURE 4: A series of 16 transverse serial sections through the umbo of a specimen of *Pleuraloma triangulatum* Cooper, 1989 from the Callovian of Hamakhtesh Hagadol, southern Israel. Numbers in parentheses represent distance between sections: 1(2.7); 2(0.3); 3(0.4); 4(0.2); 5(1.3); 6(0.2); 7(0.3); 8(0.2); 9(0.3); 10(0.3); 11(0.2); 12(0.2); 13(0.2); 14(0.1); 15(0.2); 16(0.3). Scale bar = 1 cm; AMNH 46557.

Cooper (1989, pp. 84-87) were collected from the Tuwaiq Mountain Formation at approximately the same level within the Callovian as our specimens.

Stratigraphical occurrence. Zohar Formation (Upper Callovian), Subunits 33, 40, Hamakhtesh Hagadol, Negev, Israel.

Genus DISSORIA Cooper, 1989

Dissoria bretti sp. nov.
Plate 2, figures 14-16

Derivation of Name. After Carlton E. Brett, University of Cincinnati, for his substantial contributions to the palaeontological literature.

Holotype. NHM 1035.

Material. 8 articulated specimens.

Diagnosis. Uniplicate, elongate-oval *Dissoria.*

Description. Medium sized (Table 8), unevenly oval to pyriform with greatest width attained at about two thirds length of shell. Ventral valve with a massive umbo, labiate beak dominated by a large, circular foramen; permesothyrid, symphytium not exposed. The dorsal valve is flatter with a slightly constricted anterior margin. Lateral profile shows an almost evenly biconvex shell with a slightly more inflated ventral umbo. Anterior commissure with broad, moderately extended uniplication. Ornamentation consists of faint, concentric growth lamellae.

Internal characters. Unknown.

Remarks. In general outline, this species closely resembles a specimen described and figured by Muir-Wood (1935, p. 117, pl. 12, fig. 9a-c) as *Heimia*? *incurvirostrum* but lacks the sulciplicate commissure of that species, having a well-marked uniplicate margin, a less evenly oval dorsal outline and flatter or less steep flanks. It differs from *Dissoria costata* and *D. tribulis* of Cooper (1989, pl. 25, fig. 1-4) in its lack of anterior costation and in its less robust and unevenly oval-pyriform general outline. It also differs from *Dissoria obscura* (Cooper, 1989, pl. 25, fig. 28-30) in general outline but resembles this species in having a marked uniplicate anterior margin. We nevertheless consider our species *Dissoria uniplicata* sp. nov. to be closely related to both *Heimia*? *incurvirostrum* Muir-Wood and *Dissoria obscura* Cooper.

Stratigraphical occurrence. Matmor Formation (Upper Callovian), Subunit 54, Hamakhtesh Hagadol, Negev, Israel.

TABLE 8: Measurements of *Dissoria bretti* sp. nov.

Specimen	L	W	T	Subunit
NHM 1035 holotype	26.5	18.3	15.0	45
GSI M6149 paratype	26.4	19.3	13.8	48

Superfamily ZEILLERIACEA Allan, 1940
Family ZEILLERIDAE Allan, 1940

Genus DIGONELLA Muir-Wood, 1934

Digonella boylani sp. nov.
Plate 2, figures 17-19; text-figure 5

Derivation of Name. After Stanley L. Boylan, Touro College, for his constant support and encouragement.

Holotype. NHM 1036.

Material. 18 articulated specimens. Dimensions of holotype: L, 21.8; W, 17.2; T, 17.0.

Diagnosis. Medium sized, subquadrate, acutely biconvex *Digonella.*

Description. Almost parallel sided subquadrate to spade-shaped, evenly biconvex. Maximum width is attained at approximately midlength. The ventral

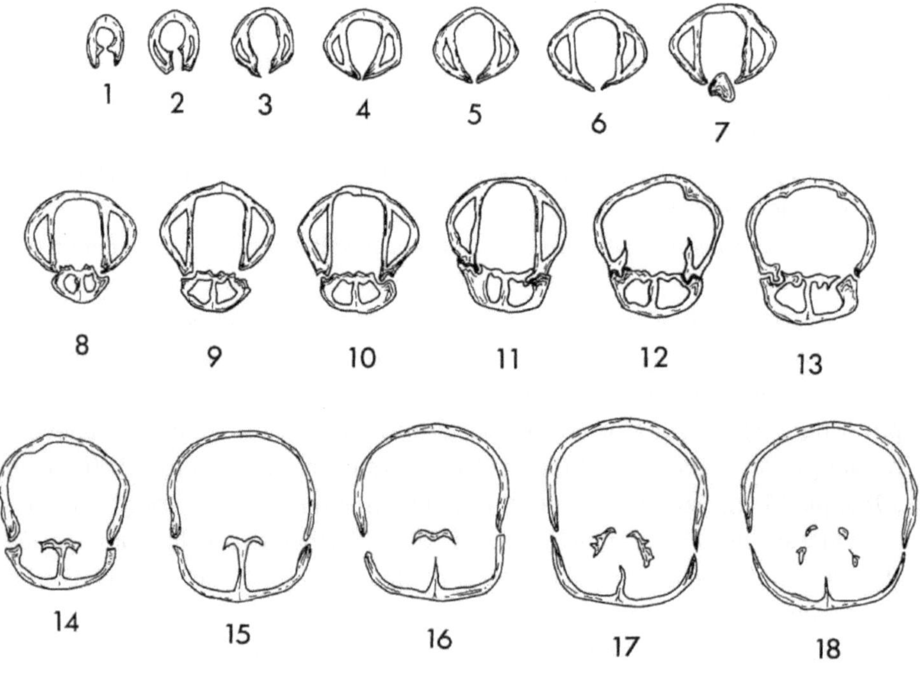

FIGURE 5: A series of 18 transverse serial sections through the umbo of a specimen of *Digonella boylani* sp. nov. from the Callovian of Hamakhtesh Hagadol, southern Israel. Numbers in parentheses represent distance between sections: 1(0.2); 2,(0.3); 3(0.5); 4,(0.4); 5 (0.6); 6(0.4); 7(0.4); 8(0.3); 9(0.2); 10(0.2); 11(0.4); 12(0.3); 13(0.2); 14(0.6); 15(0.4); 16(0.3); 17(0.2); 18(0.3). Scale bar = 5 mm; AMNH 46558.

umbo is slightly produced with a suberect beak. The beak ridges are distinct, foramen large, mesothyrid. The deltidial plates are conjunct. The interarea is wide and well defined. A median septum on the dorsal valve is clearly visible and extends to well over two-thirds the valve length. The anterior commissure is ligate. Ornamentation consists of faint, concentric growth lamellae becoming more marked toward the anterior margin.

Internal characters. In the ventral valve the dental lamellae are long, supporting deeply inserted, inwardly directed, hinge teeth that articulate strongly with the sockets. In the dorsal valve a strong, high median septum supports flat or horizontal hinge plates with well-developed inner and outer socket ridges. A shallow, central indentation marks the junction of the two elongate-triangular hinge plates with the septum as seen in transverse section (Text-fig. 5). The descending branches of the brachial loop are given off ventrally from hooked-shaped bases at the distal end of the hinge trough. Long, narrow spines develop along the descending branches of the loop, projecting laterally.

Remarks. In her original description of the genus *Digonella*, Muir-Wood (1934, p. 550) gave the stratigraphical range as Middle Jurassic (Bathonian). The specimen described and figured here is one of 18 specimens collected from Hamakhtesh Hagadol and is younger than the morphologically similar specimens described by Cooper (1989, p. 118, pl. 32, fig. 26-35) as *Rugitela primeria* that were collected from the Marrat Formation (*Bouleiceras* Zone), Saudi Arabia (lower Lias equivalent) and is probably the earliest recorded species of *Rugitela*. *Digonella boylani* sp. nov. bears a strong resemblance to *Rugitela primeria* Cooper but differs in its subquadrate general outline, steeper subparallel flanks and more marked ligation of the anterior commissure. The brachial loop is supported for a greater distance anteriorly than in *Rugitela primeria*.

Stratigraphical occurrence. Matmor Formation (Upper Callovian), Subunit 48, Hamakhtesh Hagadol, Negev, Israel.

Genus POLYPLECTELLA gen. nov.

Type species. Polyplectella debriani sp. nov., by monotypy.

Derivation of name. Greek, in reference to pronounced polyplications (*poly*, many; *plek*, twisted) on the valve margins.

Diagnosis. Medium sized plicate terebratellid, oval to pentagonal in general outline.

Description. Almost evenly convex valves, but ventral valve becoming more inflated in late growth stage. Umbo large, erect, with circular foramen and distinct mesothyrid beak ridges. Valves developing marked polyplication marginally. Anterior commissure uniplicate. A distinct median septum is visible on the dorsal valve, extending about two-thirds the length of the shell. The presence of dental plates visible within the umbo of a damaged specimen confirms the systematic position of this new taxon in the Zeilleridae.

Polyplectella debriani sp. nov.

Plate 2, figures 20-25

Derivation of Name. After Debra J. Belowich and Brian A. Feldman for many hours of dedicated assistance in the field and laboratory.

Holotype. NHM 1038. Dimensions of holotype (NHM 1038): L 42.8; W, 34.7; T, 31.1; dimensions of paratype (NHM 1037): L, 40.5; W, 36.6; T, 20.9.

Material. 2 articulated specimens.

Description. As for the genus.

Remarks. Only two specimens have been recorded from the Jurassic beds of Hamakhtesh Hagadol and both are figured here. One of these (NHM 1037) is noticeably flatter than the second specimen (NHM 1038), which is damaged but is considered to be an older individual belonging to the same species. Both specimens have much in common with a specimen described by Cooper (1989, p. 99, pl. 30, fig. 37-40) as an undetermined terebratulacean genus and species. It has a similar but more elongate general outline, uniplicate anterior commissure, and distinct beak ridges as seen in our specimens figured here. Cooper referred to shell ornament as "faint costae" but the specimen figured by him seems to possess faint plication of the valves in keeping with our specimens. Cooper's specimen is said to have been collected from the Lower Dhruma Formation (*Ermoceras* Zone) Upper Bajocian equivalent, Saudi Arabia.

Stratigraphical occurrence. Matmor Formation (Upper Callovian), Subunit 42, Hamakhtesh Hagadol, Negev, Israel.

Zeilleria sp.

Plate 2, figures 26-28

Material. 8 articulated specimens.

Description. Shells small (Table 9), oval to almost subpentagonal in general outline. Umbo short, foramen large. Beak suberect, beak ridges distinct, permesothyrid. Interarea proportionately extensive, slightly concave. Deltidial plates exposed. Lateral profile oval, evenly biconvex. Anterior commissure rectimarginate, ligate.

Internal structures. Unknown due to lack of suitable material for sectioning.

TABLE 9: Measurements of *Zeilleria* sp.

Specimen	L	W	T	Subunit
NHM 1039	13.0	9.9	8.8	43
NHM 1040	10.6	11.0	8.8	43
NHM 1041	10.3	9.2	7.6	43
NHM 1042	10.7	10.0	7.9	43
NHM 1043	10.3	8.9	8.2	43
AMNH 46575	10.7	9.6	6.8	43
AMNH 46576	10.6	9.3	7.0	43
AMNH 46577	9.5	8.8	5.8	43

Remarks. The general morphology of this species with its broadly oval outline and tapering anterior suggests a close affinity with Zeilleridae from the Upper Bathonian to Upper Callovian beds. Distinguishing features, such as a slightly extended hinge line and comparatively broad interarea bounded by distinct beak ridges distinguish it from either *Rugitela* or *Mycerosia*, both of which have been described by Cooper from beds of a similar age in Saudi Arabia (see Cooper, 1989; pp. 117-118, pl. 32, fig. 1-17). Although a total of eight specimens have been recovered from Hamakhtesh Hagadol there is no reliable comparative material from the same horizon recorded from any other locality known to the authors. The species described here is broadly referred to the genus *Zeilleria* until more material and stratigraphical information is obtained.

Stratigraphical occurrence. Matmor Formation (Upper Callovian), Subunit 47, Hamakhtesh Hagadol, Negev, Israel.

ACKNOWLEDGEMENTS

This study was supported by a research grant from the National Geographic Society (to HRF). Feldman wishes to thank Dr. Y. Mimran, former Director of the Geological Survey of Israel, for providing facilities for two field seasons during his tenure as visiting scientist in connection with this work. We gratefully acknowledge the comments and suggestions of Dr. M. R. Sandy, University of Dayton, and an anonymous reviewer both of whom were responsible for improving the manuscript. Thanks also to Sarah Long, The Natural History Museum, London, and Susan Klofak, American Museum of Natural History, New York, for assistance in specimen preparation.

REFERENCES

Abbate, E., G. Ficcarelli, C. Pirini, A. Radrizzani, D. Salvietti, and A. Turi. 1974. Jurassic sequences from the Somali coast of the Gulf of Aden. *Revista Italiana di Paleontologia e Stratigraphia* 80: 409-478.

Ager, D. V. 1965. Mesozoic and Cenozoic Rhynchonellacea. H597-H625. In R. C. Moore (ed.), *Treatise on Invertebrate Paleontology. Part H. Brachiopoda.* Lawrence, KS: Geological Society of America and the University of Kansas Press.

Allan, R. S. 1940. A revision of the classification of the terebratuloid Brachiopoda. *Records of the Canterbury Museum,* 4: 267-275.

Arkell, W. J. 1952. Jurassic ammonites from Jebel Tuwayq, Central Arabia. *Philosophical Transactions, Royal Society of London,* Series B, 236: 241-313.

------. 1956. *Jurassic geology of the world.* Edinburgh: Oliver and Boyd.

Bague, M. 1955. Contribution a l'étude des Brachiopodes du Bathonien Supérieure et du Callovien de la Côte d'Or. *Bulletin Scientifique de Bourgogne,* 15: 213-240.

Blake, G. S. 1935. *The stratigraphy of Palestine and its building stones.* Jerusalem: Printing and Stationary Office.

Buckman, S. S. 1918. The Brachiopoda of the Namyau Beds, northern Shan States, Burma. *Memoirs of the Geological Survey of India, Palaeontologica Indica, New Series,* 3, 1-299. [Date on title page is 1917, footnote to Corrigenda sheet from the Geological Survey states: "Print off orders regarding this memoir were given to the Press in September, 1917, but owing to shortage of paper, copies were not actually issued till July, 1918"; Shi and Grant, 1993, p. 144]

Cariou, E. J. P., L. Bassoullet, L. Grossowicz, and F. Hirsch. 1997. Le Callovo-Oxfordien du Sud Levant: données biostratigraphiques nouvelle (ammonites, foraminifères) endémesme et corréations stratigraphiques. In Société Géologique de France, (ed.), *Réunion Spécialisée APF-SGF "De la Biostratigraphie à la Paléobiogéographie,"* Lyon 27-28 novembre 1997, p. 21.

Coates, J., J. P. Gottesman, M. Jacobs, and E. Rosenberg. 1963. Gas discoveries in the western Dead Sea region. 21-36. *Proceedings of the Sixth World Petroleum Congress, Frankfurt am Main,* Hanseatische Druckanstalt, Hamburg-Wandsbek, 1: 943.

Conway, B. 1990. Paleozoic-Mesozoic palynology of Israel, II. Palynostratigraphy of the Jurassic succession in the subsurface of Israel. *Bulletin of the Geological Survey of Israel* 82: 1-39.

Cooper, G. A. 1983. The Terebratulacea (Brachiopoda), Triassic to Recent: a study of the brachidia (loops). *Smithsonian Contributions to Paleobiology* 50: 1-445.

------. 1989. Jurassic brachiopods of Saudi Arabia. *Smithsonian Contributions to Paleobiology* 65: 1-213.

Davidson, T. 1876-1878. Supplement to the Jurassic-Triassic species. 243-479. In T. A. Davidson, Monograph of the British Fossil Brachiopoda. *Monograph of the Palaeontographical Society* 4: 145-241.

Dubar, G. 1967. Brachiopodes Jurassiques du Sahara Tunisien. *Annales de Paléontologie* 53: 1-71.

Feldman, H. R. 1986. Cladistic analysis of a Jurassic rhynchonellid brachiopod genus from the Middle Eastern Ethiopian Province. Abstracts of the Geological Society of America Northeastern Section Meeting, Kiamesha Lake, New York, 18, 16.

------. 1987. A new species of the Jurassic (Callovian) brachiopod *Septirhynchia* from northern Sinai. *Journal of Paleontology* 61: 1156-1172.

Feldman, H. R., and C. E. Brett. 1998. Epi- and endobiontic organisms on Late Jurassic crinoids columns from the Negev Desert, Israel: implications for co-evolution. *Lethaia* 31: 57-71.

Feldman, H. R., and E. F. Owen. 1988. *Goliathyris lewyi*, new species (Brachiopoda, Terebratellacea) from the Jurassic of Gebel El-Minshera, northern Sinai. *American Museum Novitates* 2908: 1-12.

------. 1993. Vicariance biogeography of the Middle Eastern Ethiopian Province: implications for the Paleozoic. Abstracts of the Geological Society of America Annual Meeting, Boston, Massachusetts, 25, 55.

Feldman, H. R., F. Hirsch, and E. F. Owen, E. F. 1982. A comparison of Jurassic and Devonian brachiopod communities: Trophic structure, diversity, substrate relations and niche replacement. *Journal of Paleontology, Supplement 2*, 56: 9-10.

Feldman, H. R., E. F. Owen, and F. Hirsch. 1991. Brachiopods from the Jurassic of Gebel El-Maghara, northern Sinai. *American Museum Novitates* 3006: 1-28.

Gill, G. A., and H. Tintant. 1975 Les ammonites Calloviennes du sud d'Israel. Stratigraphie et relations paleogeographiques. *Société Géologique de France, Comptes Rendus des Scéances* 4: 103-106.

Gill, G. A., J. Thierry, and H. Tintant. 1985. Ammonites Calloviennes du sud d'Israel: systematique, biostratigraphie et paleobiogeographie. *Geobios* 6: 705-751.

Goldberg, M. 1963. Reference section of Jurassic sequence exposed in Hamahktesh Hagadol (Kurnub Anticline). Detailed binocular sample description, including field observations. Unpublished Internal Report, Oil Division, *Geological Survey of Israel.*

Goldberg, M., and Friedman, G. M. 1974. Paleoenvironments and paleogeographic evolution of the Jurassic System in southern Israel. *Geological Survey of Israel* 61: 1-44.

Gray, J. E. 1840. *Synopsis of the contents of the British Museum*, 42nd edition. London: Oxford University.

Gray, J. E. 1848. On the arrangement of the Brachiopoda. *Annals and Magazine of Natural History, Series 2*, 2: 435-440.

Hegab, A. A. 1988. Analysis of growth patterns in *Eudesia* (Brachiopoda) and its potential use for the identification of different species of the genus. *Bulletin of the Faculty of Science, Assiut University* 17: 25-36.

------. 1989. New occurrence of Rhynchonellida (Brachiopoda) from the Middle Jurassic of Gebel El-Maghara, northern Sinai. *Journal of African Earth Sciences* 9: 445-453.

------. 1991a. New genus *Praeudesia* (Brachiopoda) from the Jurassic outcrops of Gebel Maghara, northern Sinai, Egypt. *Bulletin of the Faculty of Science, Assiut University* 20: 1-17.

------. 1991b. The occurrence of genus *Flabellothyris* (Brachiopoda) from the Jurassic of northern Sinai. *Bulletin of the Faculty of Science, Assiut University* 20: 39-49.

------. 1992. Terebratulida (Brachiopoda) from the Jurassic of Gebel El-Maghara, northern Sinai. *Proceedings of the 8th Symposium of Phanerozoic Developments in Egypt* 8: 33-42.

------. 1993. *Eudesia* (Brachiopoda) community from the Bathonian of Gebel El-Maghara (northern Sinai): their morphologic adaptation, ontogenetic variation and paleoecology. *Palaeontographica* 229: 1-14.

Hirsch, F. 1979. Jurassic bivalves and gastropods from northern Sinai and southern Israel. *Israel Journal of Earth-Sciences* 28: 128-163.

Hirsch, F., and R. Roded. 1997. The Jurassic stratigraphic nomenclature in Hamahktesh Hagadol, northern Negev. *Geological Survey of Israel, Current Research* 10: 1014.

Hirsch, F., Bassoullet, J.-P., É. Cariou, B. Conway, B., H. R. Feldman, L. Grossowicz, A. Honigstein, E. F. Owen, and A. Rosenfeld. 1998. The Jurassic of the southern Levant. Biostratigraphy, palaeogeography and cyclic events. 213-235. In S. Carasquin-Soleau and É. Barrier, (eds.), Peri-Tethys memoir 4: epicratonic basins of Peri-Tethyan platforms, *Mémoires du Muséum national d'Histoire Naturelle* 179: 1-294.

Hudson, R. G. S. 1958. The Upper Jurassic faunas of southern Israel. *Geological Magazine* 95: 415-425.

Kitchin, F. L. 1912. *Palaeontological work: England and Wales: summation of programme for* 1911. *Memior of the Geological Survey of Great Britain and Museum of Practical Geology*, 59-60.

Lewy, Z. 1983. Upper Callovian ammonites and Middle Jurassic geological history of the Middle East. *Bulletin of the Geological Survey of Israel* 76: 1-56.

Moore, R. C. 1965. *Treatise on Invertebrate Paleontology. Part H. Brachiopoda.* Lawrence, KS: Geological Society of America and the University of Kansas Press.

Muir-Wood, H. M. 1934. On the internal structure of some Mesozoic Brachiopoda. *Philosophical Transactions, Royal Society, London,* Series B, 223: 511-567.

------. 1935. *The Mesozoic palaeontology of British Somaliland Part II of the Geology and Palaeontology of British Somaliland. Jurassic Brachiopoda,* 75-147 and 8-13. London: Government of the Somaliland Protectorate.

Picard, L., and F. Hirsch. 1987. *The Jurassic stratigraphy in Israel and the adjacent countries.* Jerusalem: The Israel Academy of Sciences and Humanities.

Reiner, W. 1968. Callovian gastropods from Hamahktesh Hagadol (southern Israel). *Israel Journal of Earth Sciences* 17: 171-198.

Rosenfeld, A., and A. Honigstein. 1991. Callovian-Oxfordian ostracodes from the Hamahktesh Hagadol section, Southern Israel. *Revista Española de Micropaleontología* 3: 133-148.

Shaw, S. H. 1947. *Southern Palestine geological map on a scale of 1:250,000 with explanatory notes.* Jerusalem: Palestine Government Printer.

Shi, X., and R. E. Grant. 1993. Jurassic rhynchonellids: internal structures and taxonomic revisions. *Smithsonian Contributions to Paleobiology* 73: 1-190.

Weir, J. 1925. Brachiopoda, Lamellibranchiata, Gastropoda and Belemnites. In The collection of fossils and rocks from Somaliland made by mssrs. B. K. N. Wyllie and W. R. Smellie. *Monographs of the Geological Department of the Hunterian Museum, Glasgow University* 3: 1-63, especially 1-5.

CHAPTER EIGHT

A NEW SPECIES OF *COENOTHYRIS* (BRACHIOPODA) FROM THE TRIASSIC (UPPER ANISIAN—LADINIAN) OF ISRAEL

ABSTRACT

Coenothyris oweni new species is described from the Lower Member (Upper Anisian-Ladinian) of the Triassic Saharonim Formation (Upper Anisian-Lower Carnian) at Har Gevanim, Makhtesh Ramon, southern Israel. The Saharonim Formation was deposited under normal, calm, shallow marine conditions as part of the ingression of the Saharonim Sea. The presence of *Coenothyris* along with characteristic conodonts, ostracodes, foraminiferans, bivalves, cephalopods, gastropods, echinoderms and vertebrate remains is 1) indicative of the Sephardic Province; 2) diagnostic of the Middle Triassic series of Israel; and 3) important in differentiating the Sephardic Province from the Germanic Muschelkalk and Tethyan Realm faunas to the north and correlating the Triassic rocks in the Negev.

INTRODUCTION

The discovery of Carnian fossils near Latakia, Syria, in 1915 (Picard and Flexer, 1974) was the first indication of the existence of Triassic rocks in the Levant. Wyllie et al. (1923) reported on Triassic outcrops at Wadi Hisban and Zarqa Ma'in. Cox (1924) described a Triassic fauna from Transjordan collected by the Turkish Petroleum Company, Awad (1946) noted the occurrence of Marine Triassic (Muschelkalk) deposits in the Sinai and Negev (Gebel Areif en-Naqa) deserts, and Shaw (1947) summarized the research done by the British Petroleum Company in southern Israel during World War II. Information on the fauna and bedrock of the Middle Eastern Triassic was made available after World War II when oil companies and governments published the results of their field surveys and wildcat drillings (Picard and Flexer, 1974). Marine Triassic outcrops can be found in southern Israel (Makhtesh Ramon; described by Lerman, 1960), Egypt (Har Arif, Gebel Areif en-Naqa; Druckman, 1974b), and in Jordan (Wadi Hisban, Wadi Zarqa and Zarqa Ma'in; Druckman, 1974b). Boreholes in Israel, Sinai and Jordan (Parnes, 1986, fig. 1) have yielded cores of Triassic rocks (Bartov et al., 1980; Bender, 1968; Bentor et al., 1965; Druckman,

1974a, 1974b, 1976; Garfunkel and Derin, 1985; Zak, 1957, 1963, 1964). Triassic rocks in Israel, mostly marine but with occasional progradation of terrestrial environments, range in thickness from 500-1100 m (Parnes, 1986). Continental Triassic rocks are exposed or penetrated by boreholes in southwest and central Sinai (Druckman et al., 1970), whereas to the south and southeast the Triassic section is truncated by the regional Early Cretaceous unconformity (Druckman, 1974b). Goldberg (1964) first suggested that part of the Triassic sequence in Israel be correlated using subsurface techniques, and correlation of surface outcrops with subsurface sections was proposed by Druckman (1966). Halperin (l968) and Rosenberg (1969) prepared electrolog correlations of the Triassic rocks in Israel and Druckman (1974a) constructed a reference section for the subsurface. There have been numerous biostratigraphic studies on the subsurface Triassic rocks of southern Israel [Avnimelech (1958), Glickson (1964), Gerry (1967), Horowitz (1974), and Hirsch and Gerry (1974)].

The Triassic sequence at Makhtesh Ramon and Gebel Areif en-Naqa was described by Bentor and Vroman (1951, 1952), mapped and correlated with the section at Gebel Areif en-Naqa by Zak (1964), and studied in detail both stratigraphically and tectonically by Nevo and Zak (1955), Zemel et al. (1956), and Zak (1957, 1963). Biostratigraphic and chronostratigraphic analyses of the sections at Makhtesh Ramon, Har Arif, and Gebel Areif en-Naqa were made by Eicher (1947), Brotzen (1956), Parnes (1957, 1962), Kummel (1960), Sohn (1963, 1968), Sohn and Reiss (1964), Hirsch (1972), and Hirsch and Gerry (1974).

FIGURE 1: Location map of collecting localities in Makhtesh Ramon, southern Israel (Israel grid coordinates 1370/9998-1379/9994; 30°35'N, 34°55'W). *Coenothyris* beds in the Fossiliferous Limestone Member, Saharonim Formation, are denoted by an "X" (Modified from Feldman, 1987).

GEOGRAPHIC OCCURRENCE AND STRATIGRAPHY

The brachiopods described herein were collected at Har Gevanim in Makhtesh Ramon (Israel grid coordinates between 1370/9998 and 1379/9994; 30°35'N, 34°55'W), a large (40 km long, 8 km wide), northeast trending, erosional cirque described by Picard (1951) as a complex erosion funnel, in southern Israel (Fig. 1). Makhtesh Ramon, the central breached part of the Ramon anticlinorium in which sedimentary rocks of Cretaceous to Triassic age are exposed, is the largest and southernmost of three breached domes in the Negev of southern Israel (Sohn, 1968), the other two being Hamakhtesh Hagadol (Makhtesh Hathira; Kurnub Anticline) and Hamakhtesh Haqatan (Makhtesh Hazera).

The stratigraphic sequence is divided into the Negev Group (Yamin and Zafir formations; Weissbrod, 1969, 1976) and Ramon Group (Ra'af, Gevanim, Saharonim and Mohilla formations; Zak, 1963) and consists of carbonates, gypsum, sulphates, sandstones, siltstones, clays, that is largely clastic in the lower part of the section, predominantly calcareous in the middle, and more evaporitic in the upper part (Parnes, 1975, 1986). The ages range from Scythian (Lower Triassic) to Carnian-Norian (Upper Triassic). The brachiopod shells were recovered from the Fossiliferous Limestone Member of the Triassic Saharonim Formation (Anisian-Ladinian) (Fig. 2). The Fossiliferous Limestone Member consists of limestone beds alternating with shale; many of the limestones are fossiliferous and some of the shales are calcareous. The limestone beds consist of biomicrites and biosparites, dark gray, black and thinly bedded (in most cases ranging from 0.4-1.5 m in thickness) whereas the soft shale and calcareous shale layers are greenish yellow and dark gray (Druckman, 1974b). The contact of the Fossiliferous Limestone Member, conformably overlying the Gevanim Formation, is marked at the type locality (Makhtesh Ramon, Har Gevanim) by the first fossiliferous limestone layer above the last variegated shale and siltstone bed. The middle and upper part of the section consists of micritic concretionary and micritic "helminthoid" limestone with clay and marl interbeds, some dolomite and gypsum intercalations, some hemispheroidal, laminated stromatolitic limestones and, in places, intraformational conglomerates and endolithic breccias (Parnes, 1986). Some of the limestones contain flat pebbles, mostly dolomitic. At the base are some sandstone layers whereas in the upper part of the sequence there are some sandy limestones and marls. The upper contact with the Mohilla Formation, also conformable, is marked by the upper "helminthoid" bed.

This study is a result of a long-term project in the biogeography, paleoecology and taxonomy of Mesozoic faunas, especially brachiopods, in the Middle East (Feldman et al., 1982, 1991, 2000, 2001; Feldman, 1986, 1987; Feldman and Owen, 1988, 1993; Feldman and Brett, 1998; Hirsch et al., 1998). Endemic faunas of the Jurassic Ethiopian Provinces have been under investigation for many years and now the Triassic brachiopods of the Negev, belonging to the Sephardic Province, are being revised in order to establish the early history of various brachiopod species and their evolution within the province.

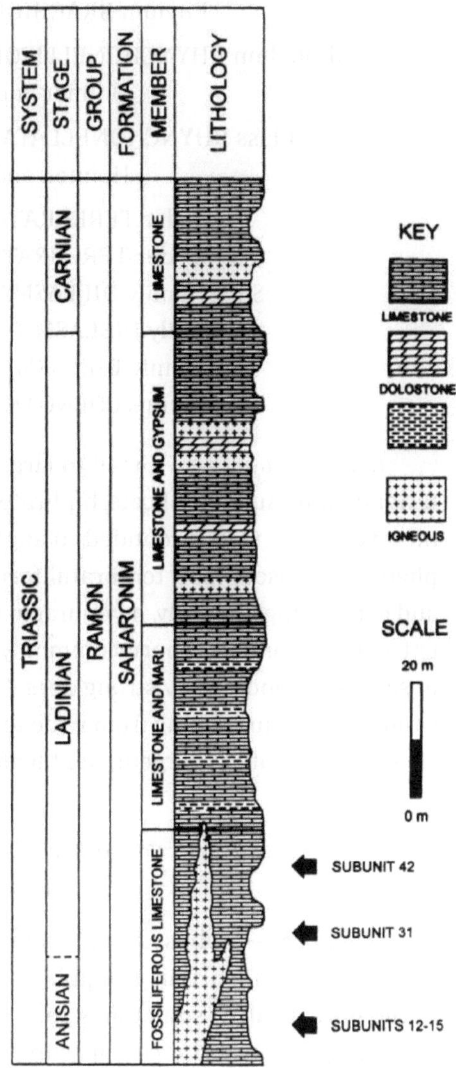

FIGURE 2: Generalized columnar section of the Triassic Saharonim Formation at Har Gevanim, Makhtesh Ramon, southern Israel. For detailed stratigraphy see Druckman (1974b; Parnes, 1986). Arrows represent occurrences of *Coenothyris* beds at marked subunit intervals. Subunit designations are after Zak (1964) and Parnes (1975).

SYSTEMATIC PALEONTOLOGY

The following abbreviations are used: AMNH = American Museum of Natural History, New York; NHM = Natural History Museum, London; GSI = Geological Survey of Israel, Jerusalem; HU = The Hebrew University, Jerusalem; IG = Museum of the State Geological Survey, Warsaw; MUZ PIG = Museum of the Polish Geological Institute, Warsaw; SMF = Senckenberg-Museum, Frankfurt am Main; USNM = United States National Museum, Smithsonian Institution, Washington, D.C.; ZPALWr. = Paleozoology Department, Institute of Zoology, Wroclaw University; L = length; W = width; T = thickness.

Phylum BRACHIOPODA Duméril. 1806

Subphylum RHYNCHONELLIFORMEA Williams, Carlson, Brunton, Holmer and Popov, 1996

Class RHYNCHONELLATA Williams, Carlson, Brunton, Holmer, and Popov. 1996

Order TEREBRATULIDA Waagen, 1883
Suborder TEREBRATULIDINA Waagen, 1883
Superfamily DIELASMATOIDEA Schuchert, 1913
Family DIELASMATIDAE Schuchert, 1913
Subfamily DIELASMATINAE Schuchert, 1913
Genus COENOTHYRIS Douvillé, 1879

Revised diagnosis.—Small to large sized, astrophic, sulco- to biconvex; anterior commissure uniplicate, biplicate or rectimarginate; beak erect to strongly incurved, beak ridges rounded to angular, mesothyrid to permesothyrid, symphytium exposed; loop terebratuliform; long, high crural processes thickened and converging ventrally, myophragm thin, long; pedicle collar commonly present, cardinal process short, commonly bilobate; hinge plates ventrally concave in section, supported by strong dorsal septum; crural bases prominent, demarcating septalium ranging from wide to deep, narrow to shallow; adductor muscle scars elongate, divergent; dental plates absent.

Coenothyris Oweni new species
Figures 3-5

Diagnosis.—Globose form is somewhat similar to *Coenothyris vulgaris,* but differs from *C. vulgaris* in its equibiconvex valves, lack of any umbonal sulcation of the dorsal valve (sometimes seen in young *C. vulgaris)* and its consistently rectimarginate anterior commissure in all ontogenetic stages.

Description.—Shells small to medium sized (Table 1; Fig. 6), astrophic, with rounded beak ridges but in some shells beak ridges angular. Outline mainly elongate ranging from transversely suboval to subpyriform; neanic shells ventribiconvex in lateral profile, become biconvex in ephebic stage. Ventral beak mostly incurved but some specimens display strongly incurved umbo; dorsal beak fits under anterior end of symphytium. Foramen medium sized, circular and mesothyrid. Greatest width attained at or just anterior to midlength and greatest thickness at about the middle. Anterior commissure rectimarginate. Shell smooth with external sculpture consisting of numerous, closely spaced very fine concentric growth lines (approximately 10 per 5 mm near anterior commissure) on juveniles as well as adults; no evidence of color banding or capillae.

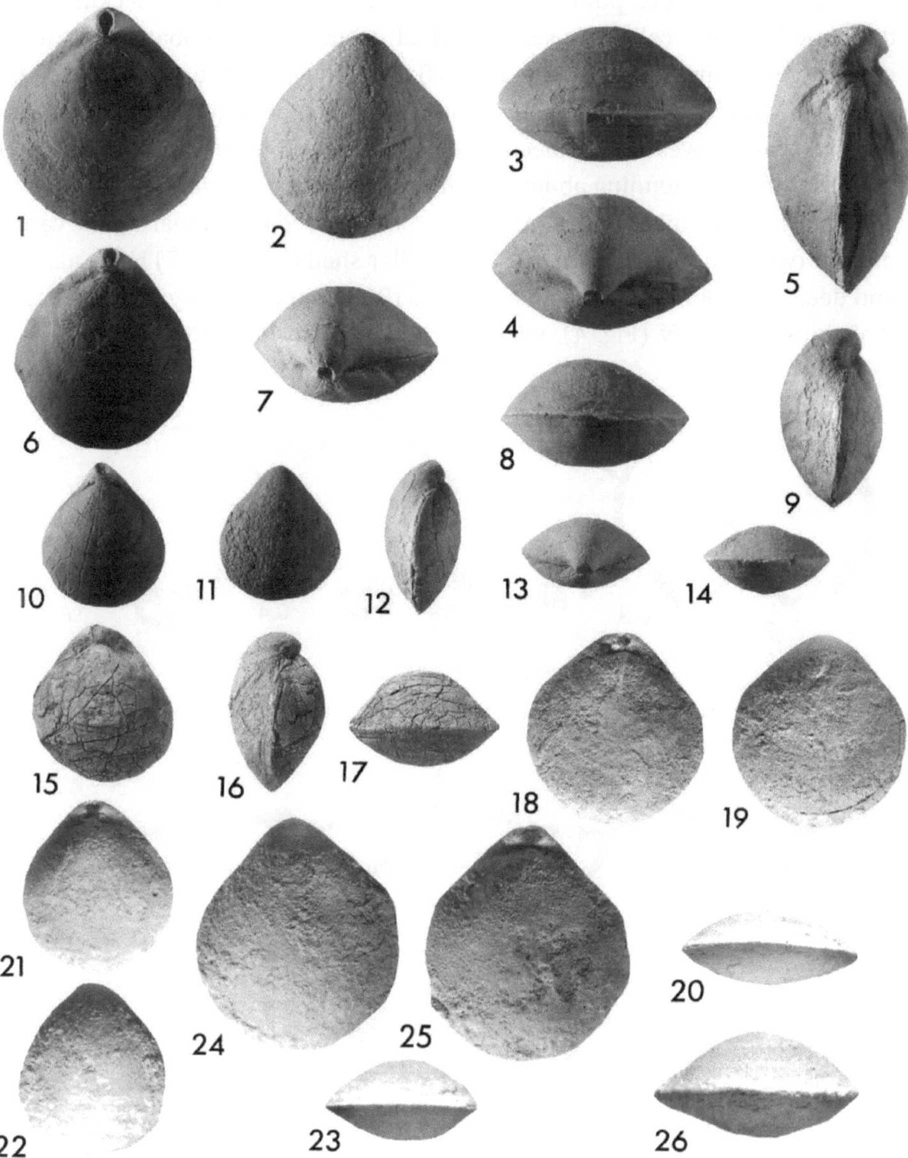

FIGURE 3: *Coenothyris oweni* n. sp. *1-5*, Dorsal (x1.8), ventral, anterior, posterior (x1.6), lateral (x2.2) views, holotype, Subunit 42, AMNH 46490; *6-9*, dorsal (x1.8), posterior, anterior (x2), lateral (x1.7) views, paratype, Subunit 42, NHM 996; *10-14*, dorsal, ventral, lateral, posterior, anterior views, (x2), Paratype, Subunit 31, NHM 997; *15-17*, dorsal, lateral, anterior views (x1.5), sectioned specimen (same as fig. 4), paratype, Subunit 42, NHM 1007; *18-20,* dorsal, ventral, anterior views (x5) of juvenile paratype, Subunit 42, GSI M8122; *21-23*, dorsal, ventral, anterior views (x6) of juvenile paratype, Subunit 42, AMNH 46514: *24-26*, ventral, dorsal, anterior views (x5.4) of juvenile paratype, Subunit 31, AMNH 46515.

Hinge teeth short, pointed, directed medially; dental plates lacking. Well developed pedicle collar present; cardinal process low, bilobate, with deep concavity separating lobes; hinge plates flat to convex in section, supported by strong, low (in juveniles) median septum forming septalium. The septum, visible externally in weathered specimens, extends one-third of valve length. Loop terebratuliform, extending about one-third length of dorsal valve.

Internal variability in the species can be observed by comparing the cardinalia of two sectioned specimens. The smaller shell (NHM 1007) has a narrow and deep septalium (sensu Shi and Grant, 1993), supported by a high septum that thins anteriorly (Fig. 4), whereas the larger shell (AMNH 46516) has a

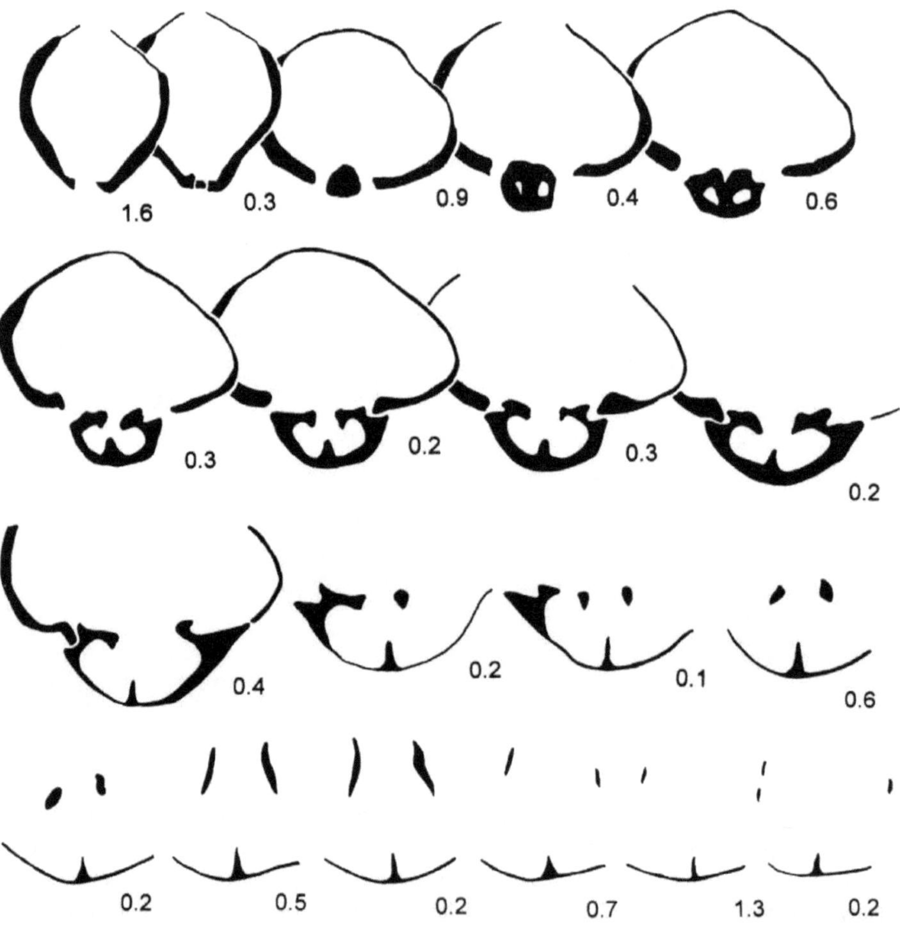

Figure 4: Transverse serial sections of *Coenothyris oweni* n. sp., (x3), paratype, Subunit 42, NHM 1007, (L=18.1; W=16.3; T=9.5). Numbers show distance in mm between sections; first section made at a distance of 3.6 mm from beak.

shallow and wide septalium (Fig. 5); the low septum does not support the hinge plates in the same way as it does in NHM 1007. Also, in the larger specimen the terminal ends of the descending branches of the loop are not triangular in outline and there is no marked high arched transverse band.

Etymology.—Named in honor of Dr. Ellis F. Owen, paleontologist emeritus, The Natural History Museum, London, for his many significant and valuable contributions to the study of Mesozoic Brachiopoda.

Types.—Holotype, AMNH 46490, from Subunit 42 of the Saharonim Formation. Lower Ladinian, southern Israel; paratypes, AMNH 46514, 46515 (from Subunit 31), AMNH 46516 (from Subunit 42), all in the collection of the American Museum of Natural History, New York, Saharonim Formation, Lower Ladinian, southern Israel; Paratypes NHM 996, 997 (from subunit 31), NHM 1007 (from Subunit 42), Saharonim Formation, Lower Ladinian, southern

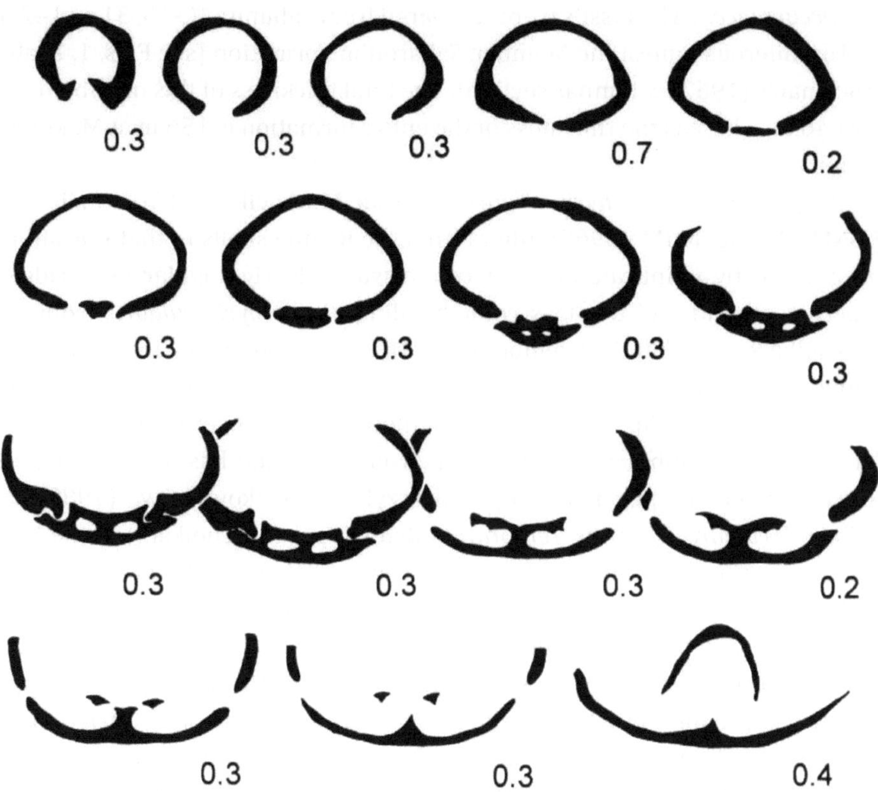

FIGURE 5: Transverse serial sections of *Coenothyris oweni* n. sp., (x3), AMNH 46516, Subunit 42, (L=20.3; W=18.5; T=10.8). Numbers show distance in mm between sections; first section made at a distance of 4.0 mm from beak.

Israel, in the collection of the Natrual History Museum, London: Paratype GSI M 8122 from Subunit 42, in the collection of the Geological Survey of Israel, Jerusalem, Saharonim Formation. Lower Ladinian, southern Israel.

Other material examined.—Specimens collected from Subunit 12-15, in the collection of the Geological Survey of Israel, Jerusalem, Saharonim Formation. Upper Anisian, southern Israel: GSI M8124 (33 shells); Specimens collected from Subunit 31, in the collection of the Geological Survey of Israel, Jerusalem, Saharonim Formation, Lower Ladinian, southern Israel: GSI M8125 (158 shells); in the collection of the American Museum of Natural History, New York: AMNH 46517-46542 (26 slabs with 2,807 articulated shells, two free ventral valves, one free dorsal valve); Specimens collected from Subunit 42, in the collection of the Geological Survey of Israel, Jerusalem, Saharonim Formation, Lower Ladinian: GSI M8126 (195 shells); Specimens collected from Subunit 31, in the collection of the Hebrew University. Jerusalem, Saharonim Formation, Lower Ladinian, southern Israel: HU 35325 (5 shells); H 35327.

Occurrence.—The fossils were recovered from subunits 12-15, 31 and 42 of the Fossiliferous Limestone Member, Saharonim Formation [see Figs. 1, 2; also Druckman's (1967) columnar section]. The total thickness of this member is almost 40 m, whereas the thickness of the entire formation is 156 m at Makhtesh Ramon.

Discussion.—*Coenothyris vulgaris* from the Muschelkalk of Toulon, France (USNM 319165; NHM 95968) differs from the Ramon shells in that it is larger, has a distinctly uniplicate anterior commissure, clearly angular beak ridges, is permesothyrid and displays color banding. *Coenothyris vulgaris* from the Muschelkalk of Tulau, Wurtemberg (NHM 22696) differs in its larger size, uniplicate anterior and greater convexity. *Coenothyris vulgaris* from Gotha, Germany (AMNH 45369) is uniplicate, although not as severely as shells from Toulon, is more gibbous than the shells from Israel and has minute pustules on well preserved specimens. Popiel-Barczyk and Senkowiczowa (1989) described *C. vulgaris* from the *Terebratula* Bed (Upper Muschelkalk) of the Holy Cross Mountains (MUZ IG 1362 II 109b), central Poland, that has a triangular, ovate shell, is more globose, displays a distinct sulcation of the anterior commissure and has a depression bisecting the ventral valve. Shells of *C. vulgaris* from Bukowie, at the northeastern border of the Holy Cross Mountains, described by Senkowiczowa and Popiel-Barczyk (1996), have a delicate wavy striation radially arranged on the lateral margins of the valves, and are distinctly sulcate (MUZ PIG 1362 II 131). *Coenothyris vulgaris* from Silesia (ZPALWr. Br/28/I/67, ZPALWr. Br/28/I/89, ZPALWr. Br/28/I/96) is markedly different internally from *C. oweni* n. sp. (see Usnarska-Talerzak, 1988 for a comparison of transverse serial sections).

Coenothyris antra (SMF 50070) from Stara Planina Mountain in the Yugoslavian Carpartho-Balkanides (Radulovic et al., 1992) differs from the Ramon shells in its strongly bisulcate anterior commissure, folded shell surface and carinate beak. *Coenothyris radulovici* from the same locality (SMF 50075, 50076) differs in its elongate oval outline and varying uniplicate to recti marginate anterior commissure. The anterior commissure of *C. oweni* n. sp. is invariably rectimarginate.

Popiel-Barczyk and Senkowiczowa (1989) have collected *Coenothyris cycloides* from the *Terebratula* Bed (Upper Muschelkalk) of the Holy Cross Mountains (IG1362 II 110 [fig. 10]; IG1362 II 111) that differs externally from *C. oweni* n. sp. in its more rounded shell outline and suberect to straight ventral beak. Also, *C. cycloides* is a smaller shell whose length is not greater than 14 mm (see Kirchner, 1934), whereas the shells of *C. oweni* n. sp. range in length up to 27.6 mm (e.g., AMNH 46491). Internally *C. cycloides* has a more narrow and shallow septalium (see Popiel-Barczyk and Senkowiczowa, 1989, fig. 10).

PALEOECOLOGIC AND PALEOBIOGEOGRAPHIC SIGNIFICANCE

The discovery of *Coenothyris oweni* n. sp. and associated fauna in the Ladinian Saharonim Formation of Makhtesh Ramon, southern Israel, sheds new light on the paleoecology of the genus. In the Balaton Highland, Hungary, Palfy and Torok (1992) described a depositional setting of foreslope patch-reef deposits (Köveskál), high energy environment with redeposition by slumping (Aszófo), low energy local basinal development (Felsoörs), and subtidal soft stratum environment (Iszkaszentgyörgy). *Coenothyris* faunas from the Mecsek Mountains described by Palfy and Torok (1992) suggest that the sediments were deposited in quiet shallow water just below wave base on a semi-lithified substrate. Radulovic et al. (1992) mentioned *C. vulgaris* from marly limestones in the Stara Planina Mountains of the Yugoslavian Carpatho-Balkanides along with *C. aeuta* from Upper Carnian sandstones with an associated fauna of *Cardita, Nucula, Sphaerocodium* and *Glomospira?,* as well as *C. radulovici* from Upper Norian biomicrosparites with an associated fauna of *Aulotorfus, Vidlicia,* and *Megalodon* sp. In Israel the *Coenothyris* fauna existed under normal, calm shallow marine conditions as part of the transgression of the Saharonim Sea that prevailed in the area during Upper Anisian to Lower-Middle Ladinian times (Druckman, 1974a). Druckman (1974b) noted that the occurrence of a diverse benthic and planktic fauna, along with the absence of evaporites and dolomites, suggests that deposition of the Saharonim Formation occurred in an open shelf environment of normal salinity at least beneath wave base.

TABLE 1: Measurements (in millimeters) of articulated specimens of *Coenothyris oweni*, n. sp., Fossiliferous Limestone Member (Upper Anisian-Ladinian), Saharonim Formation (Upper Anisian-Lower Carnian), Makhtesh Ramon, southern Israel. Shells collected from each subunit ranged from well preserved to highly weathered.

Specimen Number	(L)	(W)	(T)
Subunit 42			
AMNH 46490	24.5	23.2	14.3*
AMNH 46491	27.6	23.7	15.1**
AMNH 46492	23.3	19.6	11.9
AMNH 46493	20.9	20.1	10.7
AMNH 46494	18.1	16.1	9.3
NHM 998	15.3	14.0	12.3
NHM 999	14.1	10.6	13.3
NHM 1000	12.9	6.3	5.6
GSI M8115	8.4	8.1	4.1
GSI M8116	5.5	4.6	2.5
Subunit 31, top			
GSI M8117	25.4	24.0	15.9
GSI M8118	24.2	21.6	14.0
GSI M8119	20.2	19.9	12.9
NHM 1001	18.3	17.9	8.7
NHM 1002	15.9	13.3	8.4
NHM 1003	13.5	12.2	6.9
AMNH 46495	12.9	11.2	6.4
AMNH 46496	11.1	9.9	5.4
AMNH 46497	10.4	10.5	5.0
AMNH 46498	4.7	4.5	2.2
Subunit 31, bottom			
AMNH 46499	11.9	10.4	4.3
AMNH 46500	10.2	9.7	5.8
AMNH 46501	8.4	7.4	3.7
AMNH 46502	7.5	6.9	3.2
AMNH 46503	7.4	7.0	2.9
AMNH 46504	6.9	6.1	3.0
AMNH 46505	5.6	5.0	2.8
AMNH 46506	5.3	4.8	2.0
AMNH 46507	4.9	4.6	2.1
AMNH 46508	3.7	3.7	1.5
Subunits 12-15			
AMNH 46509	25.5	22.0	15.1
AMNH 46510	25.2	19.4	12.3
AMNH 46511	22.9	19.0	13.6
AMNH 46512	22.5	18.6	12.0
AMNH 46513	21.6	17.4	13.5
NHM 1004	19.6	15.9	10.5
NHM 1005	17.9	15.7	9.3
NHM 1006	14.8	12.5	7.9**
GSI M8120	14.9	12.2	7.1
GSI M8121	5.5	4.5	2.8

* = holotype; ** = estimated

The marine Middle Triassic series of Israel belongs to the Sephardic Province which sharply differs in its cephalopod composition from coeval "normal" or Panthalassan faunas. Extreme abundance and diversity of nautiloids, as in the Triassic of Israel, is diagnostic of shallow marine environments according to Bucher (personal commun., 1996). The Sephardic Province is also known from Spain, and shows some affinities in its abnormal faunal composition and shallow depositional environments with the Germanic Muschelkalk. The Saharonim Formation shows affinities to the Germanic Muschelkalk in that it consists primarily of fossiliferous, bioturbated, and stromatolitic limestones, marls, shales, occasionally sandstones with plant remains and reptile bones, lithographic limestones with fish remains and occasionally gypsum intercalations (Feldman et al., 2000). Vertebrate remains found in the Saharonim Formation include *Hybodus, Nothosaurus, Placodus,* and *Psephosaurus.* The Sephardic Province Muschelkalk facies contains the following characteristic taxa: *Pseudofurnishius murcianus, Sephardiella mungoensis* (conodonts), *Gevanites, Israelites, Iberites, Protrachyceras hispanicum* (ammonites), *Myophoria, Gervillia, Anodontophora* (bivalves). According to Hirsch (1992) their equivalents in the Germanic Muschelkalk are endemic neogondolellid conodont and ceratidid ammonite taxa and endemic species of cosmopolitan bivalve genera. The presence of *Coenothyris oweni* n. sp. in the Triassic of Israel is useful in correlating the Triassic rocks in the Negev and helps differentiate the Sephardic Province from the Germanic Muschelkalk and the Tethyan Realm faunas to the north.

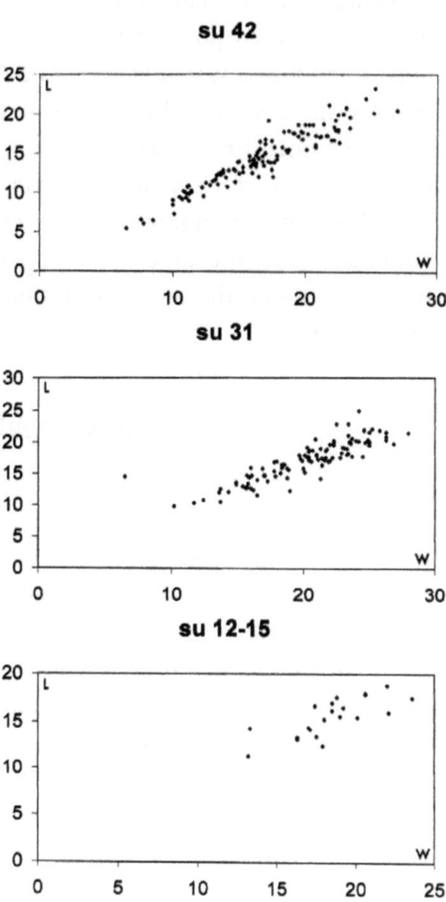

FIGURE 6: Scattergrams of length (L) versus width (W) in *Coenothyris oweni* n. sp. from the Triassic Saharonim Formation, Makhtesh Ramon, southern Israel. Measurements (in mm) are based on 22 specimens from Subunits (su) 12-15, GSI M8124, 110 specimens from Subunit (su) 31, GSI M8l25, and 119 specimens from Subunit (su) 42, GSI M8126.

EVOLUTION

The presence of *Coenothyris oweni* n. sp. in the Triassic (Ladinian) of southern Israel provides additional data for interpreting the evolution of lineages within the genus. Recently Radulovic et al. (1992) extended the range of the genus *Coenothyris,* previously known only from Middle Triassic sediments, into the Upper Triassic. They described the Carnian *C. acuta* and the Norian *C. radulovici* from Stara Planina Mountain in the Yugoslavian Carpatho-Balkanides and concluded that, based on the stratigraphic position and morphology of these Upper Triassic genera, they represent an evolutionary lineage descended from the widespread Middle Triassic *C. vulgaris.* A cladistic analysis of these species, and others in the genus, would clarify their evolutionary relationships since stratigraphic position should not be considered as *a priori* evidence for character evaluation (Schaeffer et al., 1972). Schaeffer et al. contend that all organisms should be treated without initial regard to stratigraphic position and that only after a cladistic hypothesis has been proposed should the worker examine the correlation between stratigraphic position and relative primitiveness.

ACKNOWLEDGMENTS

The research was supported by the National Geographic Society and carried out during my tenure as a visiting scientist at the Geological Survey of Israel, Jerusalem. I thank the following for comments, discussion and suggestions: C. E. Brett (University of Cincinnati), H. Bucher (Université Claude Bernard), N. Eldredge (AMNH), F. Hirsch (GSI), N. Landman (AMNH), R. H. Lindemann (Skidmore College), N. D. Newell (AMNH), E. F Owen (NHM), M. R. Sandy (University of Dayton). P Rollings and A. Modell (both AMNH) deserve thanks for assisting in photographic work. S. Klofak (AMNH) and S. Long (NHM) helped in the preparation of specimens. I would like to express my appreciation to F. Alvarez (Universidad de Oviedo) and P. E. Isaacson (University of Idaho) for critical review and improvement of the manuscript.

REFERENCES

Avnimelech, M. 1958. Triassic in the deep boring at Kfar Yeruham (Rekhme), northern Negev. *Research Council of Israel Bulletin*, Section G, Geo-sciences, 7G: 173-175.

Awad, G. H. 1946. On the occurrence of Marine Triassic (Muschelkalk) deposits in Sinai. *Bulletin de L'Institut d'Égypte* 27: 397 27.

Bartov, Y., Z. Lewy, G. Steinitz, and I. Zak. 1980. Mesozoic and Tertiary stratigraphy, paleogeography and structural history of the Gebel Areif en-Naqa area, eastern Sinai. *Y. K. Bentor Honourary Volume, Israel Journal of Earth-Sciences* 29: 114-139.

Bender, F. 1968. *Geologie van Jordanien*. Berlin: Berlin Gebrueder Borntraeger; also 1974, Geology of Jordan, translated by M. K. Khdeir et al. *Beitrage zur Regionalen Geologie der erde, Supplementary [English] Edition* 7: 1-196.

Bantor, Y. K., and A. Vroman. 1951. *The geological map of the Negev, Avdat Sheet. Scale 1:100,000, with explanatory notes.* Tel Aviv: Israel Defense Forces. [In Hebrew]

------. 1952. A new occurrence of the Marine Triassic in Israel. *Research Council of Israel Bulletin*, Section G, Geosciences, G 1: 1-98.

Bentor, Y. K., U. Golani, L. Picard, A. Vroman, and I. Zak. 1965. *The Geological Map of Israel, 1:250,000 (2 sheets)*. Tel Aviv: Survey of Israel.

Brotzen, F. 1956. Stratigraphical studies on the Triassic vertebrate fossils from Wadi Raman Israel. *Arkiv for Mineralogi och Geologi* 2: 191-217.

Cox, L. R. 1924. A Triassic fauna from the Jordan Valley. *Annals and Magazine of Natural History* 14: 52-96.

Douvillé, H. 1879. Sur quelques genres de brachiopodes Terebratulidae et Waldheimiidae. *Société Géologique de France Bulletin* 7: 251-277.

Druckman, Y. 1966. Triassic Project. *Institute for Petroleum Research and Geophysics (IPRG)* 1018: 38-44.

------. 1967. Reference section of the Ramon Group (Triassic) for the subsurface of the northern Negev. *Institute for Petroleum Research and Geophysics (IPRG)* 1023: 21-22.

------. 1974a. Triassic paleogeography of southern Israel and Sinai Peninsula. Symposium, Wien, May 1973. *Schriftenreihe der Erdwissenschaftlichen Kommissionen. Oesterreichische Akademie der Wissenschaften* 2: 79-86.

------. 1974b. Stratigraphy of the Triassic sequence in southern Israel. *Geological Survey of Israel Bulletin* 64: 1-94.

------. 1976. The Triassic in southern Israel and Sinai: a sedimentological model of marginal, epicontinental, marine environments. Unpublished Ph.D. dissertation, Hebrew University, Jerusalem. [Hebrew, with English summary]

Druckman, Y., T. Weissbrod, and A. Horowitz. 1970. The Budra Formation: a Triassic continental deposit in southwestern Sinai. *Geological Survey of Israel* 1048: 1-20.

Duméril, A. M. C. 1806. *Zoologique analytique ou méthode naturelle de classification des animaux*. Paris: Allais.

Eicher, D. B. 1947. Micropaleontology of the Triassic of north Sinai. *Bulletin de L'Institut d'Égypte*, 28: 87-92.

Feldman, H. R. 1986. Cladistic analysis of a Jurassic rhynchonellid brachiopod genus from the Middle Eastern Ethiopian Province. *Abstracts, Geological Society of America* Northeastern Section Meeting, Kiamesha Lake, New York, 18: 16.

------. 1987. A new species of the Jurassic (Callovian) brachiopod *Septirhynchia* from northern Sinai. *Journal of Paleontology* 61: 1156-1172.

Feldman, H. R., and C. E. Brett. 1998. Epi- and endobiontic organisms on Late Jurassic crinoid columns from the Negev Desert, Israel: Implications for coevolution. *Lethaia* 31: 57-71.

Feldman, H. R., and E. F. Owen. 1988. *Goliathyris lewyi,* new species (Brachiopoda, Terebratellacea), from the Jurassic of Gebel El-Minshera, northern Sinai. *American Museum Novitates* 2908: 1-12.

------. 1993. Vicariance biogeography of the Middle Eastern Ethiopian Province: Implications for the Paleozoic. *Abstracts, Geological Society of America* Annual Meeting, Boston, Massachusetts, 25: 55.

Feldman, H. R., F. Hirsch, and E. F Owen. 1982. A comparison of Jurassic and Devonian brachiopod communities: trophic structure, diversity, substrate relations and niche replacement. *Third North American Paleontological Convention Proceedings,* Montreal, Canada, 1: 169-174.

Feldman, H. R., E. F Owen, and F. Hirsch. 1991. Brachiopods from the Jurassic of Gebel El-Maghara, northern Sinai. *American Museum Novitates* 3006: 1-28.

------. 2001. Brachiopods from the Jurassic (Callovian) of Hamakhtesh Hagadol (Kurnub Anticline), southern Israel. *Palaeontology* 44: 637-658.

Feldman, H. R., R. Gellis, E. F. Owen, F. Hirsch, A. Rosenfeld, and M. R. Sandy. 2000. Paleobiogeography of Triassic brachiopods from the Sephardic Province in the Middle East. *Abstracts Geological Society of America* Northeastern Section Meeting, New Brunswick, New Jersey, 32: 16.

Garfunkel, Z., and B. Derin. 1985. Permian-Early Mesozoic tectonism and continental margin formation in Israel and its implications for the history of the eastern Mediterranean. In J. E. Dixon and A. H. F. Robertson (eds.), *The Geological Evolution of the Eastern Mediterranean,* 187-201. Oxford: Blackwell Scientific Publications.

Gerry, E. 1967. Paleozoic and Triassic Ostracoda from outcrops and wells in southern Israel. The Israel Institute of Petroleum, Report 1/67: 1-9.

Glickson, M. R. 1964. Palynological investigations in "Makhtesh Qatan 2," Boring in the Negev, Israel. *Israel Journal of Earth-Sciences* 13: 16-26.

Goldberg, M. 1964. Problems in the Jurassic stratigraphy of Israel. *Israel Journal of Earth-Sciences* 13: 169-171.

Halperin, D. 1968. Gas and oil possibilities in southern Israel. Lapidoth, Israel Oil Prospectors Corporation Ltd., Internal Report, 29 p.

Hirsch, F. 1972. Midde Triassic conodonts from Israel, southern France and Spain. *Mitteilungen der Geselschaft für Geologie und Bergbaustudenten* 21: 811-828.

------. 1992. Circummediterranean Triassic eustatic cycles. *Israel Journal of Earth-Sciences* 40: 29-38.

Hirsch, F., and E. Gerry. 1974. Conodont and ostracode biostratigraphy of the Triassic in Israel. Symposium, Wien, May 1973. *Schriftenreihe der Erdwissenscahftlichen Kommissionen. Oesterreichische Akademie der Wissenschaften* 2: 107-114.

Hirsch, F., J.-P Bassoullet, E. Cariou, B. Conway, H. R. Feldman, L. Grossowicz, A. Honigstein, E. F. Owen, and A. Rosenfeld. 1998. The Jurassic of the southern Levant. Biostratigraphy, palaeogeography anel cyclic events, p. 213-235. In S. Carasquin-Soleau and E. Barrier (eds.), Peri-Tethys Memoir 4: Epicratonic Basins of Peri-Tethyan Platforms. Mémoires du Muséum National d'Histoire Naturelle, Paris 179: 1-294.

Horowitz, A. 1974. Triassic miospores from southern Israel. *Review of Paleobotany and Palynology* 17: 175-207.

Kirchner, H. 1934. Die Fossilien del Würzburg Trias; Brachiopoda. *Neus Jahrbuch für Mineralogie, Geologie und Paläontologie,* Abteilung B, 71: 11-136.

Kummel, B. 1960. Middle Triassic nautiloids from Sinai, Egypt and Israel. *Bulletin of the Museum of Comparative Zoology* 123: 285-302.

Lerman, A. 1960. Triassic pelecypods from southern Israel and Sinai. *Bulletin of the Research Council of Israel,* Section G, Geo-sciences 9G: 1-60.

Nevo, A., and I. Zak 1955. Preliminary report on geological survey of Har 'Arif, Geological Survey of Israel, unpublished report. [Hebrew]

Palfy, J., and A. Torok. 1992. Comparison of Alpine and Germanotype Middle triassic brachiopod faunas from Hungary, with remarks on *Coenothyris vulgaris* (Schlotheim 1820). *Annales Universitatis Scientiarum Budapestinensis de Rolando Eotvos nominatae, secto geologica* 29: 303-323.

Parnes, A. 1957. The Triassic in Makhtesh 'Arif. *Israel Geological Society Bulletin* 4: 9-12.

------. 1962. Triassic ammonites from Israel. *Geological Survey of Israel Bulletin* 33: 1-77.

------. 1975. Middle Triassic ammonite biostratigraphy in Israel. *Geological Survey of Israel Bulletin* 66: 1-35.

------. 1986. Middle Triassic cephalopods from the Negev (Israel) and Sinai (Egypt). *Geological Survey of Israel Bulletin* 79: 1-59.

Picard, L. 1951. Geomorphogeny of Israel, Pt. I, The Negev. *Research Council of Israel Bulletin,* Section G, Geo-sciences, G1: 5-32.

Picard, L., and A. Flexer. 1974. *Studies on the Stratigraphy of Israel: The Triassic.* Tel Aviv: The Israel Institute of Petroleum.

Popiel-Barczyk, E., and H. Senkowiczowa. 1989. Representatives of the genus *Coenothyris* Douvillé, 1879, from the Terebratula Bed (Upper Muschelkalk) of the Holy Cross Mountains, central Poland. *Acta Geologica Polonica* 39: 93-112.

Radulovic, V., D. Urosevic, and N. Banjac. 1992. Upper Triassic brachiopods from the Yugoslavian Carpatho-Balkanides (Stara Planina Mountain). *Senckenbergiana Lethaea* 72: 61-67.

Rosenberg, E. 1969. Oil and gas prospects in the Triassic and Paleozoic section in the northern and central Negev. *Naphta Geological Office Note* 277: 1-14.

Schaeffer, B., M. K. Hecht, and N. Eldredge. 1972. Paleontology and Phylogeny. *Evolutionary Biology* 6: 31-46.

Schuchert, C. 1913. Class 2. Brachiopoda, p. 355-420. In K. A. von Zittel (translated and edited by C. R. Eastman), *Text-book of Palaeontology, Volume I.* London: Macmillan & Co., Ltd.

Senkowiczowa, H., and E. Popiel-Barczyk. 1996. Some Terebratulida (Brachiopod) from the Muschelkalk sediments in the Holy Cross Mountains. *Polish Institute of Geology, Geological Quarterly* 40: 443-466.

Shaw, S. H. 1947. Southern Palestine, geological map on a scale of 1:250,000 with explanatory notes. Government of Palestine, Jerusalem.

Shi, X., and R. E. Grant. 1993. Jurassic rhynchonellids: internal structures and taxonomic revisions. *Smithsonian Contributions to Paleobiology* 73: 1-190.

Sohn, I. G. 1963. *Middle Triassic Marine ostracoeles in Israel. U.S. Geological Survey Professional Paper, 475-C:* C58-59.

------. 1968. Triassic Ostracodes from Makhtesh Ramon. *Israel Geological Survey of Israel Bulletin* 44: 1-71.

Sohn, I. G., and Z. Reiss. 1964. Conodonts and foraminifera from the Triassic of Israel. *Nature* 201: 1209.

Usnarska-Talerzak, K. 1988. Morphology and postembryonic development of *Coenothyris vulgaris* (Schlotheim) (Brachiopoda, Middle Triassic). *Acta Palaeontologica Polonica* 33: 169-202.

Waagen, W. H. 1883. Salt Range Fossils, Pt. 4(2), Brachiopoda. *Memoirs of the Geological Survey of India*, Palaeontologica Indica, series 13, I fascicule, 2: 391-546.

Weissbrod, T. 1969. The Paleozoic of Israel and adjacent countries, pt. I: The subsurface Paleozoic strmigraphy of southern Israel. *Geological Survey of Israel Bulletin* 47: 1-35.

------. 1976. The Permian in the Near East, p. 200-214. In H. Falke (ed.), *The Continental Permian in Central, West and South Europe. NATO Advanced Study Institute Series 22, D.* Dordrecht, Holland: Reidel Publishing Group.

Williams, A., S. J. Carlson, C. H. C. Brunton, L. E. Holmer, and L. Popov. 1996. A supra-ordinal classification of the Brachiopoda. *Philosophical Transactions of the Royal Society, London,* B 351: 1171-1193.

Wyllie, B. K. N., K. A. Campbell, and G. M. Lees. 1923. Petroleum prospects in Syria, Palestine and Transjordan. Anglo-Palestine Oil Corporation report.

Zak, I. 1957. The Triassic in Makhtesh Ramon. Unpublished MA thesis, Hebrew University, Jerusalem. [Hebrew, with English summary]

------. 1963. Remarks on the stratigraphy and tectonics of the Triassic of Makhtesh Ramon. *Israel Journal of Earth-Sciences* 12: 87-89.

------. 1964. The Triassic in Areif en-Naqa, Sinai. *Geological Survey of Israel,* Report 64/ 1: 1-28. [Hebrew, with English summary]

Zemel, E. M., U. Wurtzburger, and Y. Bartura. 1956. The geology of Makhtesh 'Arif. Geological Survey of Israel, Internal Report. [Hebrew]

CHAPTER NINE

PALEOECOLOGY, TAPHONOMY, AND BIOGEOGRAPHY OF A *COENOTHYRIS* COMMUNITY (BRACHIOPODA, TEREBRATULIDA) FROM THE TRIASSIC (UPPER ANISIAN–LOWER LADINIAN) OF ISRAEL

ABSTRACT

A brachiopod community from the Fossiliferous Limestone Member (Upper Anisian-Lower Ladinian) of the Triassic Saharonim Formation at Har Gevanim, Makhtesh Ramon, southern Israel, is dominated by the terebratulid *Coenothyris oweni* Feldman. The community shows evidence of time-averaging and is largely composed of a single cohort of juvenile mortality of one spatfall. The Saharonim Formation was deposited under normal, calm, relatively shallow marine conditions as part of the global Anisian-Ladinian transgression. One horizon, varying in thickness between 1 and 1.5 cm, represents an autochthonous obrution deposit of juvenile *Coenothyris* brachiopods and 10 bivalve genera that were rapidly buried by pulses of clay in the form of flocculated mud. Other faunal constituents of the Saharonim Formation include conodonts, ostracodes, foraminiferans, bivalves, cephalopods, gastropods, echinoderms, and vertebrate remains that belong to the Sephardic Province and are diagnostic of the Middle Triassic series of Israel. The faunal composition and shallow depositional environment of the strata studied are useful in correlating the Triassic rocks in the Negev with those in Europe and help to differentiate the Sephardic Province from the Germanic Muschelkalk and the Alpine Tethyan faunas to the north.

INTRODUCTION

Triassic rocks were first recognized in the Middle East as a result of the discovery of Carnian fossils in 1915 near Latakia in Syria (Picard and Flexer, 1974). Cox (1924) described a Triassic fauna from Transjordan collected by the Turkish Petroleum Company, Awad (1946) wrote on the occurrence of Marine Triassic (Muschelkalk) deposits in the Sinai and Negev (Gebel Areif en-Naqa) deserts, and Shaw (1947) summarized the research done by the British Petroleum Company in southern Israel during World War II. In general,

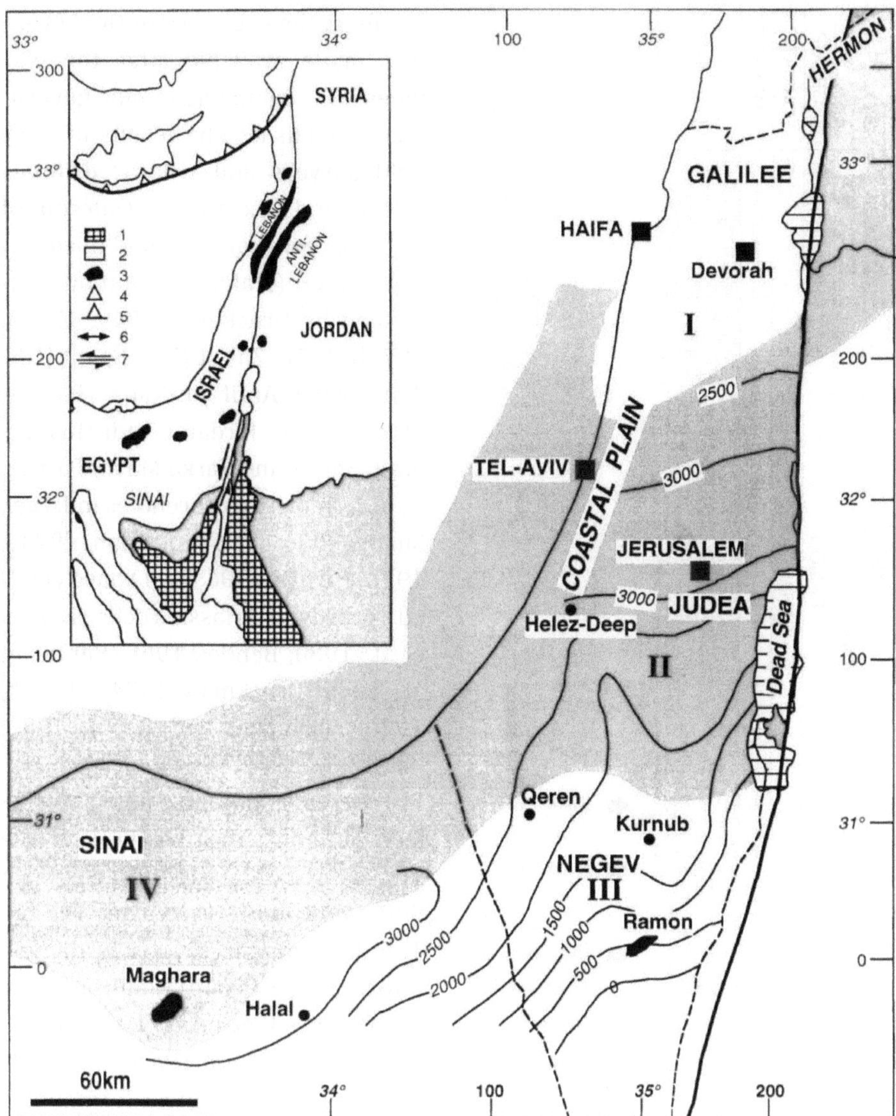

FIGURE 1: Location map of Precambrian-Oxfordian exposures in the Levant. 1, Precambrian; 2, Paleozoic-Triassic; 3, Jurassic; 4, Mount Hermon; 5, Alpine thrust-front; 6, 7, Neogene sinistral transform; I, Galilee High; II, Judean Embayment; III, Negev High; IV, Sinai Deep; with extension of Lower Oxfordian Majdal Shams shales (shaded area) and Jurassic isopachs (from Hirsch et al., 1998). "Ramon" signifies the Triassic Ramon Group, an anticlinorium that was breached through several phases of erosion, beginning in the Late Tertiary (Miocene-Pliocene), resulting in erosional cirques (Parnes, 1986) Tectonic movements since the Late Mesozoic, including reactivation of faults, resulted in the development of WSW-ENE- to S-N-trending reverse and thrust faults that today consist of several secondary monoclinal and domal buckles with an associated fault system (Zak, 1963).

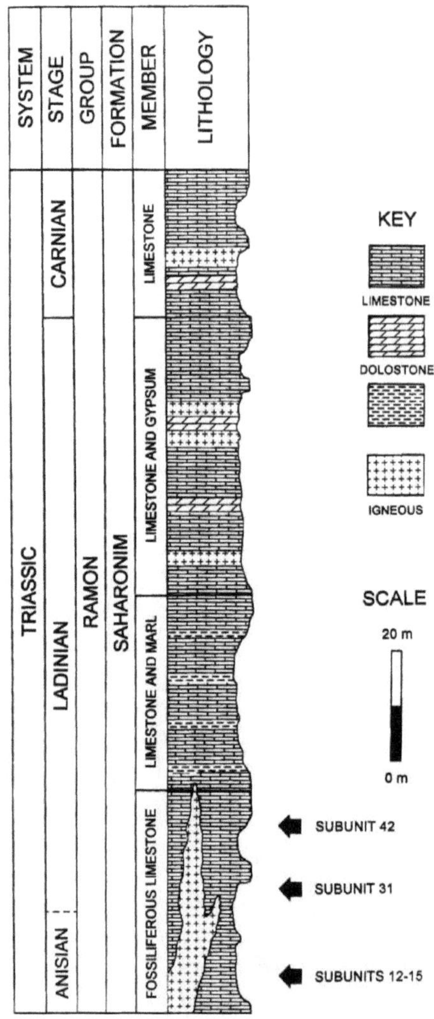

FIGURE 2: Generalized columnar section of the Triassic Saharonim Formation at Har Gevanim, Makhtesh Ramon, southern Israel. For detailed stratigraphy, see Druckman (1974b). Arrows represent occurrences of *Coenothyris oweni* Feldman at marked subunit intervals. Subunit designations are after Zak (1964) in Parnes (1975).

research on the Triassic rocks and fossils in the countries of the Levant was made available after the war when oil companies and governments published the results of their field surveys and wildcat drillings (Picard and Flexer, 1974). Outcrops of Triassic rocks can be found in southern Israel (Makhtesh Ramon; described by Druckman, 1969, 1974a, 1974b, 1976; Zak, 1957), Egypt (Har Arif, Gebel Areif en-Naqa; Parnes, 1986), and in Jordan (Wadi Husban, Wadi Zarka, and Zarka Ma'in; Parnes, 1986). Numerous boreholes in Israel, Sinai, and Jordan (Druckman, 1974a, 1976; Parnes, 1986: fig. 1) have yielded records of Triassic rocks (Bartov et al., 1980; Bender, 1968; Bentor et al., 1965; Druckman, 1974a, 1974b, 1976; Garfunkel and Derin, 1985; Zak, 1957, 1963, 1964).

Based on borehole data and surface outcrops, the thickness of the Triassic rocks in Israel ranges from 500 to 1100m (Druckman, 1974a). The stratigraphic section is divided into the Negev Group (Yamin and Zafir formations; Weisbrod, 1969, 1976) and Ramon Group (Ra'af, Gevanim, Saharonim, and Mohilla formations; Zak, 1963). The section consists of carbonates, sulfates, sandstones, siltstones, and clays; it is largely clastic in the lower part, more carbonate-rich in the middle, and more evaporitic in the upper part (Druckman, 1969, 1974a, 1976), and ranges in age from Scythian (Early Triassic) to Carnian-Norian (Late Triassic).

The material described in this paper was collected at Har Gevanim in Makhtesh Ramon (N30°35', W34°55'), a large (40 km long, 8 km wide), northeast-trending structure described by Picard (1951) as a complex erosion funnel, part of the Ramon anticline, in southern Israel (fig. 1). Above the Triassic sediments exposed in the study area, there are Jurassic-Cretaceous beds that cover large parts of the interior of Makhtesh Ramon, as well as alkaline trachybasaltic to trachysyenitic dikes and sills and larger intrusive bodies such as an essexitic laccolith and nordmarkite boss (a plutonic body having an areal extent of less than 40 square miles and a roughly circular outline in map view) (Mazor, 1955; Picard and Flexer, 1974). The shells were recovered from the Fossiliferous (lower) Member of the Saharonim Formation (Upper Anisian-Lower Ladinian) at Har Gevanim, Makhtesh Ramon, southern Israel (fig. 2). The Fossiliferous Member consists of fossiliferous limestone beds alternating with calcareous shales.

This study is an outgrowth of the author's long-term project in the taxonomy, biogeography, and paleoecology of Triassic and Jurassic faunas, particularly brachiopods, in the Middle East (Feldman et al.. 1982, 1991; Feldman, 1986, 1987, 2002; Feldman and Owen, 1988, 1993; Feldman and Breu, 1998; Hirsch et al., 1998). Endemic faunas of the Jurassic Ethiopian provinces have been under investigation for many years and now the Triassic brachiopods of the Negev, belonging to the Sephardic Province, are being revised in order to establish the early history of various brachiopod species and their evolution within the province.

ABBREVIATIONS

AMNH = Division of Paleontology, American Museum of Natural History
GSI = Geological Survey of Israel, Jerusalem
NHM = Department of Palaeontology, Natural History Museum, London

PALEOECOLOGY AND TAPHONOMY

A *Coenothyris* brachiopod community from the Middle Triassic (Upper Anisian-Lower Ladinian) southern Israel was collected from the lower part of subunit 31 (Druckman, 1969; 1974), Saharonim Formation, Makhtesh Ramon. This community (hereafter referred to as the *Coenothyis* Bed) is unique in that:

1. It comprises solely juveniles; no ephebic specimens occur in the population, whereas in stratigraphic horizons above and below (subunit 12-15, the top of subunit 31, and subunit 42) there are normal ontogenetic gradations of the brachiopods ranging from neanic to gerontic forms.

2. It represents an obrution deposit which, according to Taylor and Brett (1996), are of the most common *Lagerstätten* that remain poorly documented compared to the famous soft-bodied occurrence such as the Burgess Shale and Mazon Creek faunas that are widely cited in modern studies (Nitecki, 1979; Baird et al., 1985; Whittington and Conway Morris, 1985; Gould, 1989: Butterfield, 1990; Allison and Briggs, 1991; Collins, 1996).

TABLE 1: Faunal Constituents of the *Coenothyris oweni* Community, Saharonim Formation (Upper Anisian-Lower Ladinian), Makhtesh Ramon, Southern Israel

Taxon	Number	% Composition
Brachiopods		
Coenothyris oweni Feldman	2399	81.54
Bivalves		
Myophoria germanica (Hohenstein)··	72	2.44
Modiolus sp.*	239	8.12
Pseudoplacunopsis fissistriata (Winkler)**	11	0.37
Pleuromya cf. *mactroides* (Schlotheim)**	205	6.96
byssate pteriid**	5	0.17
Parallelodon sp.*	6	0.20
*Daonella?**	1	<1
Arcid indet.*	1	<1
Pecten indet.*	2	<1
Anodontophora munsteri (Wissman)	1	<1
Total bivalves	543	18.46
Epifaunal bivalves	293	53.96
Infaunal bivalves	250	46.06

* = epifaunal; ** = infaunal.

Coenothyris accounts for 100% of the brachiopods in the community, whereas 10 bivalve genera constitute the remaining fauna (table 1). The exposures of Anisian-Ladinian rocks in the Negev and Sinai are discontinuous, making it difficult to collect and study in situ shells laterally for any great distance. Brett et al. (1986) noted this difficulty in locating and tracing simple smothered layers from one outcrop to the next in the Middle Devonian Hamilton Group of New York, although it was possible in many cases. Brett (personal commun.) noted that it may be possible to trace certain smothered layers for considerable distances. Because these layers may preserve original patchy distributions, it may be difficult to find them in given localities, but with careful scrutiny of large areas it is often possible. This suggests that mud obrution sometimes forms a regional sediment blanket rather than being localized fallout from a point input source. Normal populations of *Coenothyris,* that is, those in which there is a complete gradation in size from juvenile to adult, occur in Gebel Areif en-Naqa, Egypt, but no obrution beds containing solely juveniles have been observed there yet.

OBRUTION BED

The *Coenothyris* Bed in the lower part of subunit 31 of the Saharonim Formation (figs. 3, 4) can best be characterized as an obrution bed; that is, very rapid burial of intact brachiopods that resulted in exceptional preservation. A rapid burial event of this type can record brief moments in time or "snapshots" of seafloor communities (sensu Brett, 1995; Taylor and Brett, 1996; Brett et al., 1997), and may form important marker horizons in event stratigraphy (Brett et al., 1997). Ager (1974) noted that the characteristics of tempestites, or storm deposits, are that they show evidence of violent disturbance of preexisting sediments followed by their rapid redeposition, all in a shallow water environment. Here, however, there is no indication that preexisting sediments were disturbed. In any case, Ager's (1974) definition for tempestites may be too restrictive in demanding evidence for violent disturbance of preexisting sediments. Mud tempestites may be a more appropriate term for indicating

FIGURE 3: Slab showing bivalve pavement colonized by *Coenothyris oweni* Feldman, subunit 31, AMNH 46517. Abbreviations: co, *Coenothyris*; pl, *Pleuromya*; mo, *Modiolus;* my, *Myophoria*; pa, *Parallelodon*. Note that this slab has relatively few brachiopods (compare with fi g. 4). Scale bar = 4 cm.

FIGURE 4: Slab of juvenile specimens of *Coenothyris* that represents an obrution deposit. Note brachiopods in a horizontal to subhorizontal attitude relative to the sediment-water interface, many in dorsal side up position. Large bivalves under the brachiopods are *Pleuromya*. Subunit 31, AMNH 46518. See text for further discussion.

the rapid settlement of sediment in relatively low energy settings after a storm (Brett, personal commun.). The *Coenothyris* obrution bed would seem to fall into this category. In most cases there is no direct evidence the preexisting sediments were disturbed, only evidence of rapid redeposition. It is probable that in most cases redeposition was involved. The only other case might be flood washout of muds following the flooding of rivers, but these would be a form of storm deposit in most instances. The brachiopods here need not have been disturbed if they lie below stormwave base; they will still get the backwash of storm-suspended sediments. The evidence here does suggest deposition below stormwave base. Stanley (1994) illustrated terebratulid brachiopods in the Early Mesozoic Shalypayco Formation of northern and central Peru that resemble the *Coenothyris* Bed in richness but differ in that the orientations and packing suggest storm deposition.

The burial of immature *Coenothyris* in the Saharonim Formation best conforms to Johnson's (1960) model 1, in which a life assemblage is buried suddenly with no evidence of postmortem transport or exposure (see also Johnson, 1997: fig. 7.7). This community can best be categorized as having a taphonomic grade of "A" based on Brandt's (1989) scheme in which there is <10% fossil breakage, >90% articulation, and <10% corrasion/orientation. Johnson's (1960) model 2 describes a more gradual accumulation and burial of shells winnowed by brief postmortem exposure and transport (see, for example, Stanley's [1994: fig. 17] illustration of a bed of Upper Triassic terebratulids where orientations and packing suggest storm deposition; note the high proportion of disarticulated valves), whereas model 3 describes an assemblage altered by long postmortem exposure and appreciable transportation from the original habitat.

The *Coenothyris* Bed does not seem to be hiatal (condensed) or a lag accumulation, as shells are all well preserved and it is not traceable laterally for any great distance. Brett and Baird (1993) considered that a first line of evidence for hiatal concentrations is stratigraphic and that in passive continental margins and epeiric seas these beds display a remarkable lateral persistence, even across major changes in facies of the adjacent beds, although so too are some obrution beds. However, not only condensed beds may show large lateral persistence. Brett and Baird (personal commun.) have traced out many single event obrution beds in the Ordovician-Silurian and Devonian with considerable precision. Indeed, the fact that this occurrence appears localized may require explanation. They suggested (unpublished manuscript) that two mechanisms of mud redeposition may be indicated: (1) widespread layers may indicate hemipelagic sedimentation from detached sediment plumes, and (2) localized occurrences may reflect point source bottom flows (true mud turbidites or distal tempestites).

There is also no evidence of mixtures of fossils representing two or more usually distinct biofacies as described by Kidwell and Bosence (1991) on the obrution horizon (although there is shell mixing in the subsurface horizon), nor is there any indication of corroded or abraded shell fragments typical of environmentally condensed shells beds and simple composite beds. All of the brachiopods are articulated, although not all are in life position. Further indication that the beds are not lag deposits is the lack of any reworked fossil material derived by significant erosional truncation of older strata (Kidwell, 1991).

Seilacher (1990) noted that some groups are more susceptible to catastrophic smothering than others by virtue of their general organization, and that sudden mud sedimentation can kill and preserve victims simultaneously. He also mentioned that the echinoderms are one such group because their ambulacral system communicates with the ambient sea water and becomes easily clogged by fine sediment. Although some brachiopods have good sediment shedding mechanisms, they may constitute another such group since they are extremely sensitive to suspended sediment clogging the lophophore. Also, they are sessile and therefore unable to escape burial as can trilobites and crinoids. Datillo (2004) reported on trace fossil evidence that the plectambonitoid brachiopod *Sowerbyella rugosa* from the Ordovician of Kentucky was able to move upward through the sediment to reach the sediment-water interface by snapping its valves. He further noted that if plectambonitoids were able to snap their valves to move through the sediments, strophomenoids might have been able to accomplish the same maneuver. Here, however, the biconvex *Coenothyris* shells differ morphologically from the concavo-convex-geniculated *Sowerbyella rugosa* and were unlikely to have been able to either shed or move through the sediment. If the *Coenothyris* brachiopods were nonpedunculate and able to have assumed a near-vertical, semiinfaunal position to escape catastrophic burial as were the Ordovician *Sowerbyella* shells, they might have escaped burial by "valve snapping."

Coenothyris shells preserved are in perfect condition, with >99.9% of the shells articulated and only three free valves evident (out of 2399 shells studied in the obrution bed). The shells lay directly on the buried surface, in this case a bivalve pavement, and were buried suddenly. Johnson (1989) described an exterminated *Pentamerus* population from the Silurian of Norway that was terminated not solely by the physical disruption of the sea by storm waves, but also by passive suffocation by fine clastic sediments falling out of suspension. Johnson's populations tended to have a high percentage of whole individuals, as is the case with the *Coenothyris* Bed described herein. He concluded that the termination of an immature population was due to an exceptionally severe storm.

OBRUTION EVENT IN RELATION TO DEPOSITIONAL SEQUENCE

The obrution deposit represented by the accumulation and preservation of *Coenothyris* occurs in the lower part of the Fossiliferous Limestone Member, Saharonim Formation. Brett and Seilacher (1991) noted that obrution deposits are most frequently recognized in the late transgressive to early highstand deposits of many successions. According to Druckman (1974b), the Fossiliferous Limestone Member was deposited as part of the transgression that prevailed in the area during Upper Anisian to Lower-Middle Ladinian times. Brett and Seilacher (1991) noted that this empirical relationship is somewhat paradoxical because transgressive sections are relatively the most sediment-starved and consequently should contain the least number of burial events. They noted that many condensed sections, however, may accumulate as a stack of very infrequent large depositional events and that major storms may carry siliciclastic or allodapic carbonate sediments into basins that are otherwise almost completely sediment starved. Reworking is not a factor in the preservation of organisms trapped in these deposits because of the deep position of the sea floor.

The *Coenothyris* Bed under discussion here is an example of a sediment-starved shell bed that was smothered.

ALLOCHTHONOUS VERSUS AUTOCHTHONOUS BURIAL

Based on examination of the brachiopod and bivalve fauna preserved in the *Coenothyris* Bed, there is no evidence of preferential alignment of valves or hydrodynamic sorting (see figs. 3-5), two criteria noted by Brett and Baird

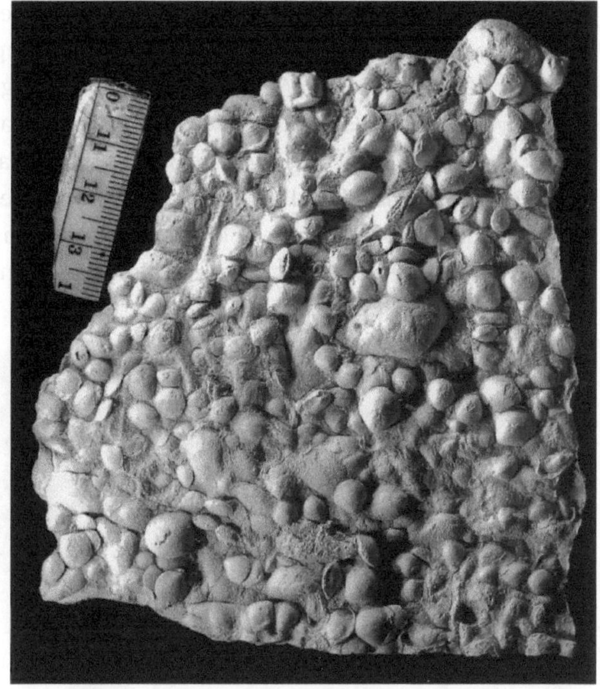

FIGURE 5: Slab of juvenile specimens of *Coenothyris*. Note the large *Pleuromya*, just off-center, under which lie disarticulated *Myphoria*. Some shells are slightly exfoliated, but most are almost perfectly preserved. Subunit 31, AMNH 46524. See text for further discussion.

(1993) for recognizing, at least locally, some degree of current reorientation. The elongated shape of the epifaunal *Modiolus* would make that bivalve shell a sensitive indicator of any current that would most likely result in at least a weak alignment of shells. The infaunal bivalves *(Pleuromya, Myophoria)* are preserved articulated, but with few specimens in burrows. However, in two cases, both observed in the field, these infaunal bivalves were buried within their burrows, thus ruling out postmortem transport (Brett and Baird, 1993; Kranz, 1974). Most *Coenothyris* shells (89.9%) are preserved in life orientations. Moreover, Alexander (1986) provided evidence that the decay of brachiopod pedicles after death would result in postmortem instability unless they were buried very rapidly. Thus, the obrution bed represents an autochthonous deposit, that is, a deposit of taxa buried in their habitat.

ENVIRONMENT OF DEPOSITION

Druckman (l974b) noted that the occurrence of a diverse benthic and nektic macrofauna, along with the absence of evaporites and dolomites, suggests that deposition of the Saharonim Formation occurred in an open-shelf environment of normal salinity. He further suggested that the complete absence of scouring within the carbonates or signs of channeling and ripple marks implies that most of the member was deposited at least beneath wave base (at a paleolatitude of within 10°N) and may have been deposited even at a depth of between 100 and 200 m. Westermann (1996) suggested that the cephalopod *Monophyllites* (common in the lower part of the Saharonim Formation although not recovered from the *Coenothyris* Bed) indicates a deep ocean environment. However, the marine Middle Triassic series of Israel belongs to the Sephardic Province that sharply differs in its cephalopod composition from coeval "normal" or Panthalassan faunas. Extreme abundance and diversity of nautiloids as

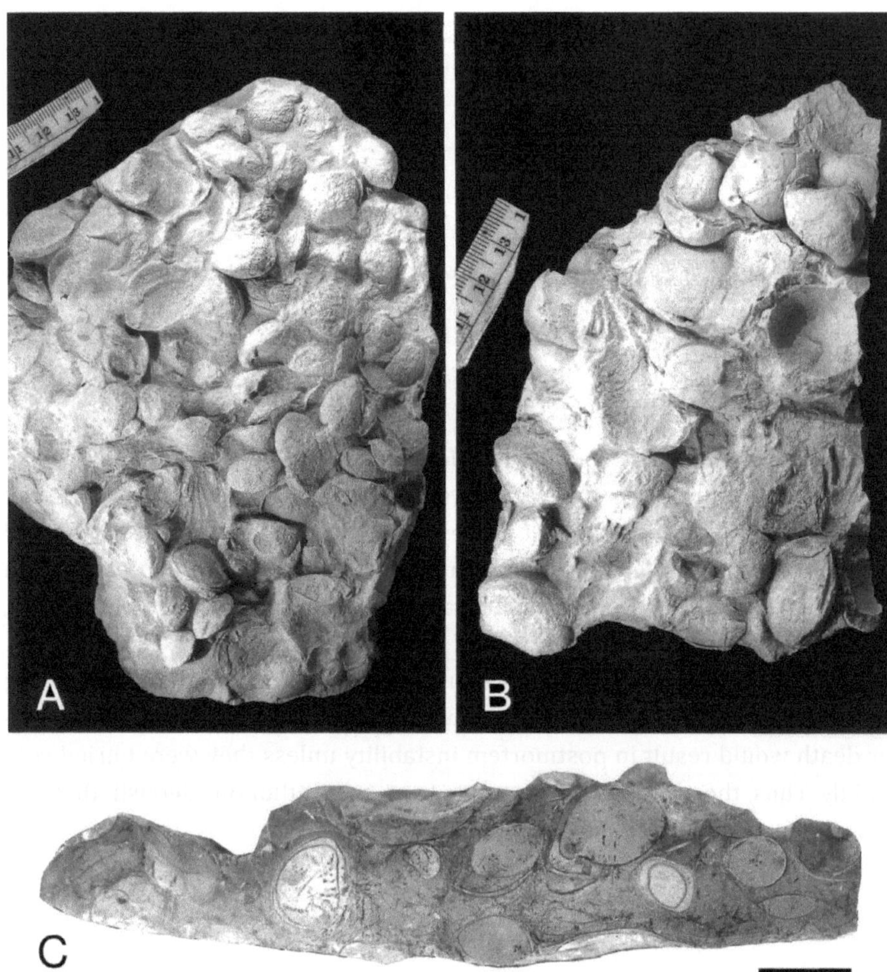

FIGURE 6: **A, B.** Slabs with normal populations of *Coenothyris* that range from ephebic to gerontic shells. **A,** Subunit 42, AMNH 46522; **B,** Subunit 42, NHM 1008. C. Cross section of *Coenothyris* horizon consisting of jumbled adult specimens similar in appearance to the Triassic German Muschelkalk. Subunit 42, AMNH 46522. Scale bar = 2 cm.

in the Triassic of Israel are diagnostic of shallow marine environments according to Bucher (personal commun.). Bucher further noted that the Sephardic Province is also known from Spain and shows some affinities in its abnormal faunal composition and shallow depositional environments with the Germanic Muschelkalk (fig. 6A-C). Triassic phylloceratids such as *Monophyllites* are not, according to Bucher, indicative of deep water, but are generally ubiquitous in ammonoid-bearing beds. In Israel, cephalopods are common only in the deeper, more transgressive, basins where there is an opening to the Tethys.

Simple layered bottom horizons may be sequentially arranged along a proximal-distal axis, and simple layers, of which the *Coenothyris* Bed is typical, are more common in proximal upslope settings (Brett et al., 1986). However, concretionary and pyritic beds are typical of more distal downslope environments. The depth of the *Coenothyris* Bed may to be similar to that inferred for Johnson's (1989) *Pentamerus* beds from Norway. He noted that suffocated populations tend to have a high percentage of whole individuals and that whether by fallout below or active disruption above wave base, only the most severe storms appear to have affected these deeper, more offshore sea-floor zones.

The burial increment (sensu Kidwell, 1991) consists of unfossiliferous sediment ranging in thickness from 1.5 to 2 cm that was rapidly deposited in a period that ranged from a few hours to a few days. There is no evidence of fining-upward size gradation or planar and cross lamination that are indicative of deposition from a waning current (Brett, 1990). Brett noted that the burial layer most commonly consists of barren structureless mudstone or siltstone. Here it consists of marl in the lower part and pure clay in the upper part with no evidence of gradation between the two layers. The most likely reason for rapid burial here was that the muds were probably flocculated, causing the suspended particles to behave hydrodynamically as silts or fine sands rather than as muds, as discussed by Brett (1990) and Brett and Taylor (1997). This seems most plausible given the high clay content of the sediment (22% clay in the lower part of the bed, with the remainder consisting of carbonate, no pyrite in the clay fraction, no quartz, mica, or silt, and no indication of volcanics; 100% clay in the upper part, no carbonate, no fine sand or silt). In fact, clay matrix can be seen filling the spaces between the immature brachiopod shells (fig. 7A) in the upper part of the bed, but it is not clearly visible on the sectioned slabs (fig. 7B, C) due to the crumbly nature of the claystone; it spalled off during sectioning. As noted by Brett (1990), most of the best obrution *Lagerstätten* are preserved in and beneath layers of fine-grained sediments, for which extremely rapid sedimentation might seem an impossibility.

A small percentage (10.1 %) of the shells (of 2336 counted on bedding surfaces) are inverted, dorsal side up (see appendix), signifying some kind of physical perturbation of the environment during or prior to burial. Taylor and Brett (1996) indicated that environmental disturbances associated with the onset of a storm may have produced changes in salinity, oxygenation, and turbidity that adversely affected the benthos. However, most *Coenothyris* are in a horizontal to sub-horizontal attitude relative to the sediment-water interface, ventral side up, indicating clogging of the mantle and subsequent collapse and decay of the pedicle before final burial. Webby and Percival (1983) reported rapid decay after displacement of the soft parts of *Eodinobolus,* an inarticulate with no functional pedicle present during ontogeny (Norford and Steele, 1969), from the Ordovician of New South Wales. Here, however, decay of the pedicle occurred before displacement; otherwise, the pedicles would have kept the brachiopod shells attached in a vertical (= living) position. As noted by Boardman et al. (1987), the presence of a pedicle foramen in fossil brachiopods does not necessarily mean that the shell was attached to the substrate by the pedicle. They further stated that there are some individuals of living species that are attached by their pedicles when hard substrates are available, but are able to adapt a free-lying mode of life when only mud bottoms are present. Here, however, attachment was on a hard bivalve substrate.

In this case *Coenothyris* was smothered by a sudden pulse, or pulses, of clay sedimentation, as a sedimentary event deposit, and subsequently subjected to weak currents. This explains the small percentage of articulated shells dorsal side up. Druckman (1974b) proposed that cyclic alternation of fossiliferous limestone and shale layers on a meter scale in the Lower Saharonim Formation may be due to pulses of terrigenous clay that were supplied to the area over a steady "background" of carbonate precipitation or secretion. However, it seems possible that: (1) there was a change in salinity, and/or (2) there was a change in oxygen levels, and (3) there was an episode of sedimentation that clogged the mantles of the brachiopods, resulting in death by anoxia. This was followed by a period of weak current activity that flipped about 10% of the dead shells whose pedicle foramens were either weakened or decayed.

Some shells were buried rapidly, as evidenced by the layer of spar-filled brachiopods (see the lower half of fig. 7B), since rapid entombment does not allow for the decay of soft tissues with subsequent displacement by sediment. Also, the decay of organic matter may increase alkalinity and favor early mineralization. The brachiopods just above the spar-filled zone (fig. 7B, upper half) are mud-filled, indicative of individuals that had died earlier and decayed to become infilled with sediment; approximately 25% of the shells were spar filled. There are also some shells with geopetal structure (fig. 7B) that indicates the original orientation of the rock, with the spar-filled upper half representing the "up" position and the spar-mud boundary roughly parallel to the sediment-water interface.

Bioturbation by the burrowing bivalves, mainly by the relatively large *Pleuromya* and the medium-sized *Myophoria,* likely overturned the brachiopods after death. Well-preserved bivalves that formed the burial pavement were probably smothered by a previous influx of sediment too thick to

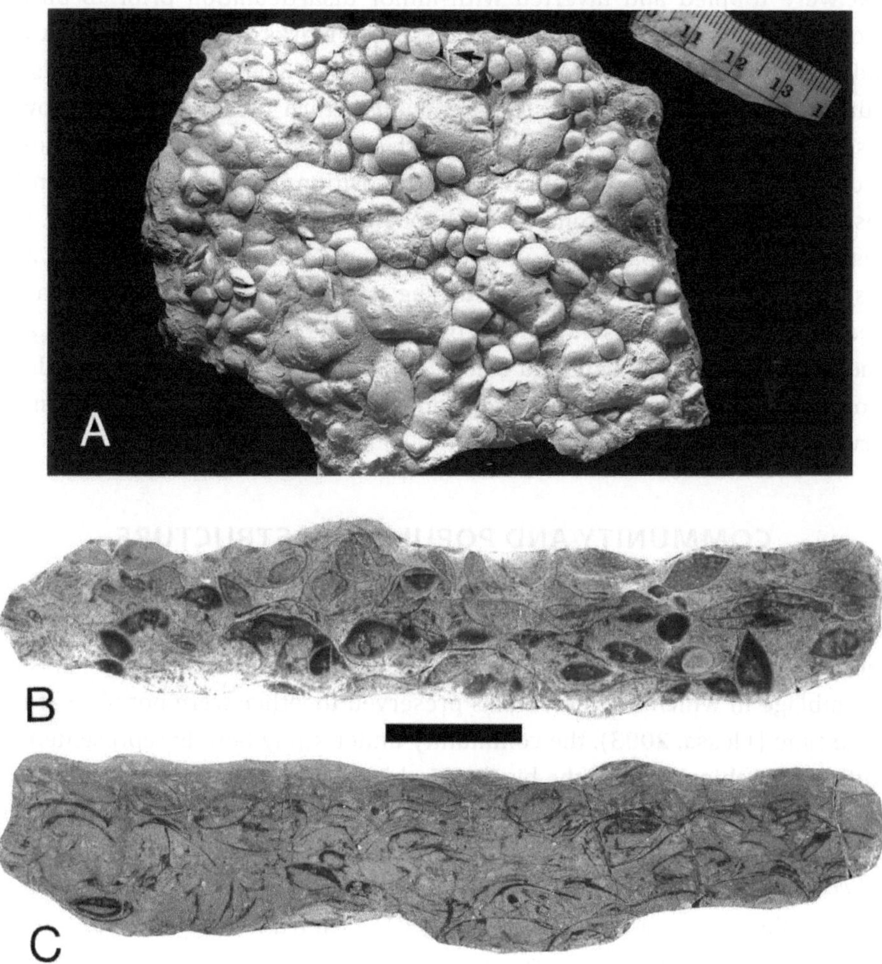

FIGURE 7: **A.** Slab of juvenile specimens of *Coenothyris.* Note rare disarticulated ventral valve with small hinge teeth (arrow). Subunit 31, AMNH 46519. **B.** Cross section of *Coenothyris* Bed showing spar-filled articulated shells (bottom half) indicative of rapid burial that did not allow for the decay of soft tissues with subsequent displacement by sediment. The brachiopods just above the spar-filled zone (upper half) are mud filled, indicative of individuals that had died earlier and decayed to become infilled with sediment; approximately 25% of the shells were spar filled. Subunit 31, AMNH 46520. **C.** Cross section of *Coenothyris* Bed showing bivalve hash (composed mainly of the bivalve *Myophoria)* that served as a pavement for brachiopod spat. Subunit 31, AMNH 4652 1. Scale bar = 2 cm.

allow them to escape. However, some of the infaunal forms (e.g., *Pleuromya, Myophoria)* may have been able to escape death and bioturbate the sediment after the settling of the *Coenothyris* spat. There is no other evidence of disturbance among *Coenothyris,* even minimal disturbance, similar to what Taylor and Brett (1996) reported for their low-level benthic forms (e.g., brachiopods) that were toppled and inverted with minor disarticulation prior to burial. There are no signs of current activity, no ripple marks, no scours of any kind, and no significant numbers of disarticulated shells (3 out of 3000); there is no indication of current activity strong enough to turn the shells upside down. Less than 1% of the bivalves that occur with *Coenothyris* are open, facing down into the sediment—an indication of death position. *Myophoria* valves can be observed (fig. 7B, C) in a concave-down position as a disarticulated hash, but this represents the pavement upon which the *Coenothyris* spat settled and lies directly under the obrution bed. The bed as a whole is time-averaged (see fig. 7C) and includes pavements of bivalve shells that were reworked and stacked. This process facilitated colonization by pedunculate brachiopods in providing a hard substrate for attachment. The shell bed may represent an interval of sediment starvation.

COMMUNITY AND POPULATION STRUCTURE

The *Coenothyris* population is composed of juvenile brachiopods that were rapidly buried after they settled on a shell pavement that consisted of 10 bivalve genera. Whereas there is clearly evidence of time-averaging, which is an assemblage in which the individuals preserved together were not alive at the same time (Flessa, 2003), the community under study here is represented by both the brachiopods and the bivalves (which were likely buried earlier than the brachiopods) found in the lowermost beds of subunit 31. The *Coenothyris* Bed is considered a local paleocommunity since the shelly parts of the original community were collected from a single bed at one outcrop that belong to within-habitat, time-averaged assemblages (Zuschin and Stanton, 2002). Some of the infaunal bivalves likely escaped immediate death by burrowing. The brachiopods represent a single cohort (table 2) of juvenile mortality of one spatfall. Some workers (e.g., Cloud, 1948) recognized that environmental factors could have produced a subnormal life-span among brachiopod populations. There is no evidence of stunting; the environment was well oxygenated, salinity was normal, and nutrients were abundant, as evidenced by a diverse fauna above and below the obrution bed (see table 3). The shells were preserved in a type 3 deposit (Brett et al., 1997), which is the most common type of obrution deposit, mud tempestites, which are typically preserved in lower-energy settings.

TABLE 2: Statistical Characterization of the *Coenothyris oweni* Cohort, Saharonim Formation (Upper Anisian-Lower Ladinian), Makhtesh Ramon, Southern Israel

	Length	Width	Thickness
Subunit 42			
OR	5.5-27.6	11.5-22.5	2.5-15.1
SD	3.94	3.67	2.58
Subunit 31, top			
OR	4.7-26.3	4.5-30.0	2.2-19.4
SD	3.55	2.88	3.07
Subunit 31, bottom			
OR	3.7-13.3	1.8-5.8	1.8-6.5
SD	2.01	1.65	1.12
Subunits 12-15			
OR	5.5-25.2	4.5-19.4	2.8-13.6
SD	4.01	3.54	2.74

OR, observed range; SO, standard deviation. All measurements in millimeters. Note that normal populations consisting of a morphocline ranging from neanic to mature forms are represented by the shells found in subunits 42, 31 (top) and 12-15, whereas the juvenile population with its noticeably smaller observed range (length 3.7-13.3; width 1.8-5.8; thickness 1.8-6.5) was recognized only in the lower part of subunit 31.

BIOGEOGRAPHY

The Saharonim Formation is similar to the Germanic Muschelkalk in that it consists primarily of fossiliferous, bioturbated, and stromatolitic limestones, marls, shales, occasional sandstones with plant remains and reptile bones, lithographic limestones with fish remains, and a few gypsum intercalations in the upper part of the formation (Hirsch, 1992). Vertebrate remains found elsewhere in the Gevanim and Saharonim formations, not in the same stratigraphic horizon as the *Coenothyris* Bed, include the shark *Hybodus,* the marine reptiles *Nothosaurus, Placodus,* and *Psephosaurus* (table 2). The Sephardic Province (Hirsch, 1972) Muschelkalk facies contains endemic or latitudinally restricted taxa such as: *Pseudofurnishius murcianus, Sephardiella mungoensis* (conodonts), *Gevanites epigonus, Israelites ramonensis, Protrachyceras hispanicum* (ammonites), *Myophoria coxi,* and *Gervillia joleaudi* (bivalves). According to Hirsch (1992), their equivalents in the Germanic Muschelkalk are equally restricted neogondolellid conodonts and various ceratidid ammonite taxa, as well as endemic species of cosmopolitan bivalve genera. There are distinct differences between the Sephardic Province faunas that can be recognized on the southern margin of the Tethys shelf and the Germanic (= Muschelkalk) Province and Tethyan Realm faunas to the north (see Hirsch, 1992: fig. 1; Marguez-Aliaga et al., 1986).

TABLE 3: Faunal Constituents of the Fossiliferous Limestone Member, Saharonim Formation (Upper Anisian-Lower Ladinian), Makhtesh Ramon, Southern Israel

Microfossils	Bivalves (continued)
Conodonts	*G.* aff. *alberti* (Goldfuss)
Pseudofurnishius murcianus	*G.* cf. *bouei* (Hauer)
Lonchodina mulleri	*Lima striata* (Schlotheim)
Enanthiognathus ziegleri	*Lima* sp.
Hibbardella mngnidentata	*Mysidioptera* cf. *vix-costata* (Stoppani)
Hindeodella sp.	*Pecten discites* (Schlothheim)
Ostracodes	*P. albertii* (Goldfuss)
Reubenella avnimelechi	*Schafhautlia* aff. *Mellingi* (Hauer)
Foraminifera	*Anodontophora munsteri* (Wissmann)
Indet. miliolids	*Pleuromya* cf. *mactroides* (Schlotheim)
	Gastropods
Megafossils	*Naticopsis* sp.
Brachiopods	*Zygopleura* spp.
Coenothyris oweni Feldman	*Omphaloptycha* sp.
Lingula sp.	Cephalopods
Dielasma julicum (Bittner)	*Mojsvaroceras* cf. *morloti* (Mojsisovics)
Echinoderms	*M.* cf. *augusti* (Mojsisovics)
Encrinus sp.	*Germanonautilus salinarius* (Mojsisovics)
Bivalves	*G. bidorsatus* (Schlotheim)
Leda cf. *fibula* Mansuy	*G.* cf. *advena* (Fritsch)
Palaeoneilo elliptica (Goldfuss)	*Indonautilus awadi* Kummel
Lyriomyophoria elegans (Dunker)	*"Ceratites"* spp.
Myophoria germanica (Hohenstein)	*Israelites ramonensis* Parnes
Costatoria coxi (Awad)	*Negebites zaki* Parnes
C. multicostata (Lerman)	*Protrachyceras wahrmani* Parnes
Neoschizodus laevigatus (Ziethen)	*P. curionii* Mojsisovics
Myophoriopsis cf. subundata (Schaurolh)	*P.* cf. *ladinum* Mojsisovics
Pseudoplacunopsis fissistriata (Winkler)	*P.* cf. *hispanicum* Mojsisovics
P. cf. *ostracina* (Schlotheim)	*P.* cf. *mascagnii* Tornquist
Placunopsis cf. *flabellum* Schmidt	*Proarcestes* sp.
Ostrea montis-caprilis Klipstein	*Monophyllites* cf. *sphaerophyllus* (Hauer)
Modiola cf. *raibliana* Bittner	Vertebrates
M. cf. *salzstettensis* Hohenstein	*Hybodus* sp.
Cassianella cf. *decussata* (Munster)	*Nothosaurus* spp.
Cassianella spp.	*Palcodus* sp.
Gervillia joleaudi Schmidt	*Psephosaurus picardi* Brotzen
	Psephosaurus sp.

Coenothyris vulgaris shells from the Muschelkalk of France and Germany differ from the Israeli shells in that it they are larger and have a distinctly uniplicate anterior commissure. *Coenothyris vulgaris* from the Upper Muschelkalk (*Terebratula* Bed) of the Holy Cross Mountains, central Poland, and specimens from Bukowie, at the northeastern border of the Holy Cross Mountains, also display a distinct sulcation of the anterior commissure. *Coenothyris acuta* from Stara Planina Mountain in the Yugoslavian Carpartho-Balkanides has a strongly biplicate anterior commissure, folded shell surface, and carinate beak, whereas *C. radulovici* from the same locality differs in its elongate oval outline and varying uniplicate to rectimarginate anterior commissure. The anterior commissure of *C. oweni* is invariably rectimarginate. There is a distinct morphologic similarity between the Israeli shells and those from the European Muschelkalk, indicating some proximity, the degree of which, however, is difficult to determine. *Coenothyris oweni* in the Triassic beds of southern Israel allows for an enhanced correlation with Muschelkalk strata in Europe and enables Tethyan Realm faunas to be more easily distinguished from faunal elements of the Sephardic Province.

CONCLUSIONS

A *Coenothyris* brachiopod community recovered from the Middle Triassic (Upper Anisian-Lower Ladinian) of southern Israel is unique in that it (1) is comprised solely of juveniles, and (2) represents an obrution deposit. *Coenothyris oweni* Feldman accounts for 100% of the brachiopods in the community, whereas 10 bivalve genera constitute the remaining fauna. The horizon is not a condensed bed or lag accumulation since it is not traceable laterally for any great distance, and there is no evidence of any reworked fossil material derived by significant erosional truncation of older strata. Brachiopods may constitute a group that is more susceptible to catastrophic smothering than others since they are extremely sensitive to suspended sediment that can clog the mantle. The depositional environment was that of a normal, calm, relatively shallow marine sea on an open shelf. An absence of scouring, channelling, or ripple marks implies deposition at least beneath wave base.

Juvenile *Coenothyris* were buried rapidly by pulses of clay sedimentation probably in the form of flocculated muds, causing suspended particles to act as silts or fine sands rather than as muds. Most of the shells are in a horizontal to subhorizontal attitude relative to the sediment-water interface, ventral side up, indicating mantle clogging and subsequent collapse and decay of the pedicle before final burial. Some brachiopods (about 10%) were overturned after burial by burrowing bivalves such as *Pleuromya* and *Myophoria* that also homogenized the sediment.

The community under study here shows evidence of time-averaging and is represented by both the brachiopods and the bivalves found in the lower part of subunit 31. Here we have a local paleocommunity since the shelly parts of the original community were collected from a single bed at one outcrop that belongs to within-habitat, time-averaged assemblages. The brachiopods represent a single cohort of juvenile mortality of one spatfall. There is no evidence of stunting, and the shells were preserved in a type 3 deposit, which is the most common type of obrution deposit, mud tempestites, which are typically preserved in lower energy settings.

Biogeographically, the *Coenothyris* community belongs to the Sephardic Province, also known from Spain, which is similar in its faunal composition and shallow depositional environments with the Germanic Muschelkalk. The presence of the brachiopod *Coenothyris* in southern Israel is useful in correlating the Triassic rocks in the Negev with those in Europe and helps differentiate the Sephardic Province from the Germanic Muschelkalk and the Tethyan realm faunas to the north.

ACKNOWLEDGMENTS

The work was supported by the National Geographic Society and carried out during my tenure as a visiting scientist at the Geological Survey of Israel, Jerusalem. I am especially grateful to C.E. Brett (University of Cincinnati, Cincinnati, Ohio) for stimulating comments, suggestions for improving the manuscript, and access to unpublished data. I thank the following for discussion and criticism of the text: C. Benjamini (Ben Gurion University, Beersheva, Israel), H. Bucher (Université Claude Bernard, Lyon, France), N. Eldredge (AMNH), F. Hirsch (GSI), N. Landman (AMNH), R.H. Lindemann (Skidmore College, Saratoga Springs, NY), N.D. Newell (AMNH), E.F Owen (NHM), and M.R. Sandy (University of Dayton). P. Rollings (AMNH) deserves thanks for assisting in photographic work, and S. Klofak (AMNH) helped in the preparation of specimens. Thanks also to two anonymous reviewers for their insightful comments and suggestions.

REFERENCES

Ager, D. V. 1974. Storm deposits in the Jurassic of the Moroccan High Atlas. *Palaeogeography, Palaeoclimatology, Palaeoecology* 15: 83-93.

Alexander, R. R. 1986. Life orientation and postmortem reorientation of Chesterian brachiopods shells by paleocurrents. *Palaios* 1: 303-311.

Allison, P. A., and D. E. G. Briggs. 1991. *Taphonomy: releasing the data locked in the fossil record*. New York: Plenum Press.

Awad, G. H. 1946. On the occurrence of Marine Triassic (Muschelkalk) deposits in Sinai. Bulletin de L'Institut d'Egypte 27: 397-427.

Baird, G. C., S. D. Sroka, S. W. Shabica, and T. L. Beard. 1985. Mazon Creek-type fossil assemblages in the U.S. midcontinent Pennsylvanian: Their recurrent character and paleoenvironmental significance. In A. Seilacher, W.E. Reif, and F. Westphal (editors), *Sedimentological, ecological and temporal patterns of fossil Lagerstätten*: *Philosophical Transactions of the Royal Society of London* B 311: 87-98.

Bartov, Y., Z. Lewy, G. Steinitz, and I. Zak. 1980. Mesozoic and Tertiary stratigraphy, paleogeography and structural history of the Gebel Areif en-Naqa area, eastern Sinai. *Y. K. Bentor Honourary Volume, Israel Journal of Earth Science* 29: 114-139.

Bender, F. 1968. *Geologie van Jordanien*. Berlin-Stuttgart: Berlin Gebrueder Borntraeger, 230 pp.; also 1974, *Geology of Jordan*, translated by M. K. Khdeir et al., *Beitrage zur Regionalen Geologie der Erde*, Supplementary [English] Edition 7: 1-196.

Bentor, Y. K., U. Golani, L. Picard, A. Vroman, and I. Zak. 1965. *The geological map of Israel*, 1: 250,000 (2 sheets). Tel Aviv: Survey of Israel.

Boardman, R. S., A. H. Cheetham, and A. J. Rowell (eds). 1987. *Fossil invertebrates*. Palo Alto, CA: Blackwell Scientific Publications.

Brandt, D. S. 1989. Taphonomic grades as a classification for fossiliferous assemblages and implications for paleoecology. *Palaios* 4: 303-309.

Brett, C. E. 1990. Obrution deposits. In D. E. G. Briggs and P. R. Crowther (eds.), *Palaeobiology: A synthesis*, 239-243. Oxford: Blackwell Scientific Publications.

------. 1995. Sequence stratigraphy, biostratigraphy, and taphonomy in shallow marine environments. *Palaios* 10: 597-616.

Brett, C. E., and G. C. Baird. 1993. Taphonomic approaches to temporal resolution in stratigraphy: Examples from Paleozoic marine mudrocks. In S. M. Kidwell and A. K. Behrensmeyer (eds.), *Taphonomic approaches to time resolution in fossil assemblages*: 250-274. Boston: The Paleontological Society.

Brett, C. E., G. C. Baird, and S. E. Speyer. 1997. Fossil Lagerstätten: Stratigraphic record of paleontological and taphonomic events. In C. E. Brett and G. C. Baird (eds.), *Paleontological events: stratigraphic, ecological, and evolutionary implications*, 3-40. New York: Columbia University Press.

Brett, C. E., and A. Seilacher. 1991. Fossil Lagerstäitten: a taphonomic consequence of event sedimentation. In G. Einsele, W. Ricken, and A. Seilacher (eds.), *Cycles and events in stratigraphy*: 283-297. Berlin: Springer-Verlag.

Brett, C. E., S. E. Speyer, and G. C. Baird. 1986. Storm-generated sedimentary units: Tempestite proximality and event stratification in the Middle

Devonian Hamilton Group of New York. In C. E. Brett (ed.), *Dynamic stratigraphy and depositional environments of the Hamilton Group (Middle Devonian) in New York State, part 1.* New York State Education Department, Bulletin 457: 129-156.

Brett, C. E., and W. L. Taylor. 1997. The *Homocrinus* beds: Silurian crinoid Lagerstätten of western New York and southern Ontario. In C. E. Brett and G. C. Baird (eds.), *Paleontological events: Stratigraphic, ecological, and evolutionary implications,* 181-223. New York: Columbia University Press.

Butterfield, N. J. 1990. Organic preservation of non-mineralizing organisms and the taphonomy of the Burgess Shale. *Paleobiology* 16: 272-286.

Cloud, P. E., Jr. 1948. Assemblages of diminutive brachiopods and their paleoecological significance. *Journal of Sedimentary Petrology* 18: 56-60.

Collins, D. 1996. The "evolution" of *Anomalocaris* and its classification in the arthropod class Dinocarida (nov.) and order Radiodonta (nov.). *Journal of Paleontology* 70: 280-293.

Cox, L. R. 1924. A Triassic fauna from the Jordan Valley. *Annals and Magazine of Natural History* 14: 52-96.

Datillo, B. F. 2004. A new angle on strophomenid paleoecology: trace-fossil evidence of an escape response for the plectambonitoid brachiopod *Sowerbyella rugosa* from a tempestite in the Upper Ordovician Kope Formation (Edenian) of northern Kentucky. *Palaios* 19: 332-348.

Douvillé, H. 1879. Sur quelques genres de brachiopodes Terebratulidae et Waldheimiidae. Société Géologique de France, Bulletin 7: 251-277.

Druckman, Y. 1969. The petrography and environment of deposition of the Triassic Saharonim Formation and the Dolomite Member of the Mohilla Formation in Makhtesh Ramon, central Negev (southern Israel). Geological Survey of Israel, Bulletin 49: 1-22.

------. 1974a. Triassic paleogeography of southern Israel and Sinai Peninsula. Symposium, Wien, May 1973. Oesteneichische Akademie der Wissenschaften Schriftenreihe der Erdwissenschaftlichen Kommissionen 2: 79-86.

------. 1974b. Stratigraphy of the Triassic sequence in southern Israel. Geological Survey of Israel, Bulletin 64: 1-94.

------. 1976. The Triassic in southern Israel and Sinai: a sedimentological model of marginal, epicontinental, marine environments. Unpublished Ph.D. dissertation, Hebrew University, Jerusalem, 188 pp. [In Hebrew, English summary.]

Duméril, A. M. C. 1806. *Zoologique analytique ou méthode naturelle de classification des animaux.* Paris: Allais.

Feldman, H. R. 1986. Cladistic analysis of a Jurassic rhynchonellid brachiopod genus from the Middle Eastern Ethiopian Province. Abstracts of the Geological Society of America Northeastern Section Meeting, Kiamesha Lake, New York 18: 16.

------. 1987. A new species of the Jurassic (Callovian) brachiopod *Septirhynchia* from northern Sinai. *Journal of Paleontology* 61: 1156-1172.

------. 2002. A new species of *Coenothyris* (Brachiopoda) from the Triassic (Ladinian) of Israel. *Journal of Paleontology* 76: 34- 42.

Feldman, H. R., and C. E. Brett. 1998. Epi- and endobiontic organisms on Late Jurassic crinoid columns from the Negev Desert, Israel: implications for coevolution. *Lethaia* 31: 57-71.

Feldman, H. R., F. Hirsch, and E. F. Owen. 1982. A comparison of Jurassic and Devonian brachiopod communities: trophic structure, diversity, substrate relations and niche replacement. Third North American Paleontological Convention Proceedings. Montreal, Canada 1: 169-174.

Feldman, H. R., and E. F. Owen. 1988. *Goliathyris lewyi,* new species (Brachiopoda, Terebratellacea), from the Jurassic of Gebel El-Minshera, northern Sinai. *American Museum Novitates* 2908: 1-12.

-------. 1993. Vicariance biogeography of the Middle Eastern Ethiopian Province: implications for the Paleozoic. *Abstracts of the Geological Society of America Annual Meeting, Boston, Massachusetts* 25: 55.

Feldman, H. R., E. F. Owen, and F. Hirsch. 1991. Brachiopods from the Jurassic of Gebel El-Maghara, northern Sinai. *American Museum Novitates* 3006: 1-28.

Flessa, K. W. 2003. Time-averaging. In D. E. G. Briggs and P. R. Crowther (edss), *Palaeobiology* 2: 292-296. Oxford: Blackwell Science.

Garfunkel, Z., and B. Derin. 1985. Permian-Early Mesozoic tectonism and continental margin formation in Israel and its implications for the history of the eastern Mediterranean. In J. E. Dixon and A. H. F. Robertson (eds.), *The geological evolution of the eastern Mediterranean*, 187-20l. Oxford: Blackwell Scientific Publications.

Gould, S. J. 1989. *Wonderful life: The Burgess Shale and the nature of history.* New York: Norton.

Hirsch, F. 1972. Middle Triassic conodonts from Israel, southern France and Spain. *Mitteilungen Gesellschaft Geologie Bergbaustudenten* 21: 811-828.

------. 1992. Circummediterranean Triassic eustatic cycles. *Israel Journal of Earth Sciences* 40: 29-38.

Hirsch, F., J-P. Bassoullet, É. Cariou, B. Conway, H. R. Feldman, L. Grossowicz, A. Honigstein, E. F. Owen, and A. Rosenfeld. 1998. The Jurassic of the southern Levant. Biostratigraphy, palaeogeography and cyclic events. In S. Carasquin-Soleau and É. Barrier (eds.), Peri-Tethys memoir 4: Epicratonic basins of Peri-Tethyan platforms. *Mémoires du Muséum National d'Histoire Naturelle* 179: 213-235.

Huxley, T. H. 1869. *An introduction to the classification of animals.* London: John Churchill and Sons.

Johnson, M. E. 1989. Tempestites recorded as variable *Pentamerus* layers in the Lower Silurian of southern Norway. *Journal of Paleontology* 63: 195-205.

------. 1997. Silurian event horizons related to the evolution and ecology of penatmerid brachiopods. In E. E. Brett and G. E. Baird (eds.), *Paleontological events: Stratigraphic, ecological, and evolutionary implications*, 162-180. New York: Columbia University Press.

Johnson, R. I. 1960. Models and methods for analysis of the mode of formation of fossil assemblages. Geological Society of America, Bulletin 71: 1075-1086.

Kidwell, S. M. 1991. The stratigraphy of shell concentrations. In P. A. Allison and D. E. G. Briggs (eds.), *Taphonomy: Releasing the data locked in the fossil record*, 211-290. New York: Plenum Press.

Kidwell, S. M., and P. Bosence. 1991. Taphonomy and time averaging of marine shelly faunas. In P. A. Allison and D. E. G. Briggs (eds.), *Taphonomy: releasing the data locked in the fossil record*, 116-209. New York: Plenum Press.

Kranz, P. M. 1974. The anastrophic burial of bivalves and its paleoecological significance. *Journal of Geology* 82: 237-265.

Marquez-Aliaga, A., F. Hirsch, and A. C. Lopez-Garrido. 1986. Middle Triassic bivalves from Hornos-Siles, Sephardic Province. *Neues Jahrbuch für Geologie und Palaeontologie*, Abhandlungen 173: 201-227.

Mazor, E. 1955. The magmatic occurrences in Makhtesh Ramon. *Geological Survey of Israel*, 29 pp. [unpublished report, in Hebrew]

Nitecki, M. H. (ed.). 1979. *Mazon Creek fossils.* New York: Academic Press.

Norford, B. S., and H. M. Steele. 1969. The Ordovician trimerellid brachiopod *Eodinobolus* from south-east Ontario. *Palaeontology* 12: 161-171.

Parnes, A. 1975. Middle Triassic ammonite biostratigraphy in Israel. *Geological Survey of Israel Bulletin* 66: 1-35.

------. 1986. Middle Triassic cephalopods from the Negev (Israel) and Sinai (Egypt). *Geological Survey of Israel Bulletin* 79: 1-59.

Picard, L. 1951. Geomorphogeny of Israel, Part I: The Negev. *Bulletin of the Research Council of Israel*, Geo-sciences, G 1: 5-32.

Picard, L., and A. Flexer. 1974. *Studies on the stratigraphy of Israel: the Triassic.* Tel Aviv: Israel Institute of Petroleum.

Popiel-Barczyk, E., and H. Senkowiczowa. 1989. Representatives of the genus *Coenothyris* Douvillé, 1879, from the Terebratula Bed (Upper Muschelkalk) of the Holy Cross Mountains, central Poland. *Acta Geologica Polonica* 39: 93-112.

Schuchert, C. 1913. Class 2. Brachiopoda. In K. A. von Zittel (translated and edited by C. R. Eastman), *Text-book of palaeontology* vol. 1, 2nd ed., 355-420. London: Macmillan.

Seilacher, A. 1990. Taphonomy of fossil-lagerstätten: Overview. In D. E. G. Briggs and P. R. Crowther (eds.), *Palaeobiology: A synthesis*, 266-270. Oxford: Blackwell Scientific Publications.

Shaw, S. H. 1947. *Southern Palestine, geological map on a scale of 1:250,000 with explanatory notes*. Jerusalem: Government of Palestine.

Stanley, G. D., Jr. 1994. Early Mesozoic carbonate rocks of the Pucara Group in northern and central Peru. *Palaeontographica* 233: 1-32.

Taylor, W. L., and C. E. Brett. 1996. Taphonomy and paleoecology of echinoderm *Lagerstätten* from the Silurian (Wenlockian) Rochester Shale. *Palaios* 11: 118-140.

Waagen, W. H. 1883. Salt Range fossils, part 4(2) Brachiopoda. Memoirs of the Geological Survey of India, *Palaeontologica Indica*, series 13, 1 fascicule 2: 391-546.

Weisbrod, T. 1969. The Paleozoic of Israel and adjacent countries, part I: The subsurface Paleozoic stratigraphy of southern Israel. *Geological Survey of Israel Bulletin* 47: 1-35.

------. 1976. The Permian in the Near East. In H. Falke (ed.), *The Continental Permian in central, west and south Europe*. NATO Advanced Study Institute Series 22: 200-214. Dordrecht, Holland: D. Reidel Publishing Group.

Webby, B. D., and I. G. Percival. 1983. Ordovician trimerellacean brachiopod shell beds. *Lethaia* 16: 215-232.

Westermann, G. E. G. 1996. Ammonoid life and habitat. In N. H. Landman, K. Tanabe, and R. A. Davis (eds.), *Ammonoid paleobiology*, 608-707. New York: Plenum Press.

Whittington, H. B., and S. Conway Morris. 1985. Extraordinary fossil biotas: their ecological and evolutionary significance: *Philosophical Transactions of the Royal Society of London, Series B: Biological Sciences* 311: 1-192.

Zak, I. 1957. The Triassic in Makhtesh Ramon. Unpublished M.A. thesis, Hebrew University, Jerusalem. [Hebrew, English summary]

------. 1963. Remarks on the stratigraphy and tectonics of the Triassic of Makhtesh Ramon. *Israel Journal of Earth Sciences* 12: 87- 89.

------. 1964. *The Triassic in Areif en-Naqa, Sinai*. Israel Geological Survey Report Geochemistry Department no. 6411, 28 pp., appendices. [In Hebrew, English summary.]

Zuschin M., and R. J. Stanton, Jr. 2002. Paleocommunity reconstruction from shell beds: a case study from the Main Glauconite Bed, Eocene, Texas. *Palaios* 17: 602-614.

Appendix

NUMBERED SLABS COLLECTED FROM OBRUTION BED (SUBUNIT 31, LOWER PART) OF FOSSILIFEROUS LIMESTONE MEMBER (UPPER ANISIAN-LOWER LADINIAN), SAHARONIM FORMATION (ANISIAN-CARNIAN) AT HAR GEVANIM, MAKHTESH RAMON, SOUTHERN ISRAEL

This list denotes the number of *Coenothyris oweni* found dorsal valve up (= overturned) and the number and percentage (in parentheses) found ventral valve up (in a horizontal to subhorizontal altitude relative to the sediment-water interface and consquently an inferred life position: see text for further discussion).

AMNH number	Dorsal valve up	Ventral valve up
46517, slab #1	21	144 (87.3%)
46518, slab #2	20	175 (89.7%)
46519, slab #3 [One free ventral valve]	22	81 (78.6%)
46520, slab #4	4	61 (93.8%)
46521, slab #5 (no preserved brachiopods)		
46523, slab #6	10	260 (96.3%)
46524, slab #7	10	192 (95.0%)
46525, slab #8 [One free dorsal valve]	11	93 (89.4%)
46526, slab #9	6	44 (88.0%)
46527, slab #10	24	139 (85.3%)
46528, slab #11	22	225 (91.0%)
46529, slab #12	1	13 (92.9%)
46530, slab #13	33	189 (85.1%)
46531, slab #14	17	208 (92.4%)
46532, slab #15	1	9 (90.0%)
46533, slab #16	2	14 (87.5%)
46534, slab #17	0	5 (100%)
46535, slab #18 (no preserved brachiopods)		
46536, slab #19	3	51 (94.4%)
46537, slab #20	5	22 (81.5%)
46538, slab #21 (no preserved brachiopods)		
46539, slab #22 (no preserved brachiopods)		
46540, slab #23	20	120 (87.5%)
46541, slab #24	4	52 (92.9%)
46542, slab #25	0	3 (100%)
Totals	236	2100 (89.9%)

CHAPTER TEN

A New Equatorial, Very Shallow Marine Sclerozoan Fauna from the Middle Jurassic (Late Callovian) of Southern Israel

ABSTRACT

Tropical Jurassic sclerozoan faunas are poorly known, yet they are critical to our understanding of Jurassic biogeography and the evolution of hard substrate communities. A diverse assemblage of hard substrate fossils is here described from subunits 53 and 54 of the upper Matmor Formation (Callovian) of Hamakhtesh Hagadol in the Negev Desert of southern Israel. This region was on, or very near, the equator during the Middle Jurassic. The fauna is dominated by flat, platter-shaped sponges (*Actostroma*?) and scleractinian corals (*Microsolena*), some of which have the depressed centers and raised rims of "microatolls" which form today in the shallowest subtidal zone associated with reef systems. The coral and calcareous sponge platters are encrusted on their top surfaces by one species of serpulid worm and many small coral and sponge recruits with narrow attachments and mushroom-shaped or conical skeletons growing upwards. The undersurfaces of the platters are encrusted by another serpulid species, at least two calcareous sponges, rare cyclostome bryozoans, oysters, plicatulid bivalves, and numerous thecideide brachiopods (the first known from the Jurassic of the region). The upper and undersurfaces of the platters are often bored by bivalves, forming the ichnospecies *Gastrochaenolites torpedo*. These borings were occasionally reoccupied by a nestling mytilid bivalve on the undersurfaces of the skeletal platters. This encrusting fauna of the Matmor Formation apparently lived in a shallow lagoon on the landward side of a coral reef. The fossils in the surrounding muddy sediments are primarily echinoids, oysters, and rhynchonellid and terebratulid brachiopods. This sclerozoan fauna is an evidence that Jurassic tropical hard substrate faunas were serpulid-rich and bryozoan-poor as predicted, but more diverse (at twelve species) than expected. This community was also ecologically divided into open and cryptic assemblages like its Jurassic equivalents in the temperate and subtropical waters of Europe and North America.

INTRODUCTION

Sclerozoans are animals which inhabit hard substrates, often by encrusting, boring, or nestling in cavities (Taylor and Wilson, 2002, 2003). Marine fossil sclerozoans are commonly found on hardgrounds (synsedimentarily-cemented seafloor sediments), rockgrounds (exposed surfaces of rocks lithified much earlier), and various biotic substrates including carbonate skeletons, wood, and other plant materials. Diverse sclerozoan communities are found in the marine fossil record throughout the Phanerozoic. The relatively constant physical parameters of this ecological niche, and the preservation of most sclerozoans *in situ* on these hard surfaces, has made them excellent subjects for studying community evolution (e.g., Wilson and Palmer, 1992) and the adaptations and competition between specific clades over time (e.g., McKinney, 1995).

The Jurassic saw a great increase in diversity and abundance of sclerozoan faunas worldwide (Taylor and Wilson, 2003). This was in large part due to the increase in carbonate hard substrates in shallow marine environments including hardgrounds (Palmer, 1982) and thick carbonate skeletons such as those of oysters, sponges and corals (Stanley and Hardie, 1998). There are dozens of systematic and paleoecological studies of Jurassic sclerozoan assemblages (see Taylor and Wilson, 2003), but almost all of them are in the relatively high northern paleolatitudes (N30°) of Europe and North America, or the equivalent southern paleolatitudes of India and Argentina. Prior to the present work there has been only one published study on a Jurassic sclerozoan fauna from tropical paleolatitudes, that of Feldman and Brett (1998) in Israel. If we are to understand the evolution of sclerozoan communities over time, we need more paleogeographic diversity in our studies to detect trends which may change with latitude and climate. Johnson and Baarli (1999) introduced larger questions of latitudinal diversity in the evolution of rocky-shore communities in an important summary paper, pointing out that we know little about paleolatitudinal gradients in critical intervals such as the Jurassic. The diverse and well-preserved Jurassic sclerozoan communities in Israel are thus important data points for the study of sclerozoan evolution.

This paper is a continuation of work undertaken in the last two decades that deals with the marine faunas of the Ethiopian Province. The Jurassic Period has been divided into two distinct faunal realms by biogeographers: the Boreal Realm, occupying the northern part of the northern hemisphere, and the much larger Tethyan Realm, which occupied the rest of the world (Arkell, 1956; Hallam, 1975). The Tethyan Realm (= the Tethys-Panthalassa Realm sensu Dommergues, 1987; see also Westermann, 2000) has been further subdivided into a number of faunal provinces, such as the Sub-Mediterranean, Mediterranean, Indo-Pacific and Ethiopian provinces, based on the amount of endemism displayed by the faunas in these areas. The Ethiopian Province is characterized by the presence of endemic taxa at the species, genus and family levels. It is recognizable from early in the Jurassic until the middle and possibly the end of the Cretaceous in India, Madagascar, and East Africa and, at the end of the Jurassic, in South America. Its first occurrence seems to be in the shallow seas within rifts formed during the break-up of Gondwanaland, but it apparently ends at some unknown southern margin because none of its species are known in the contemporaneous deposits of Antarctica and New Zealand. This long term study began with a taxonomic revision of the brachiopod faunas of the Ethiopian Province (Feldman, 1987; Feldman and Owen, 1988; Feldman et al., 1991; Feldman et al., 2001).We are now in a position to study the paleoecology of various Jurassic marine communities in southern Israel (e.g. Feldman and Brett, 1998).

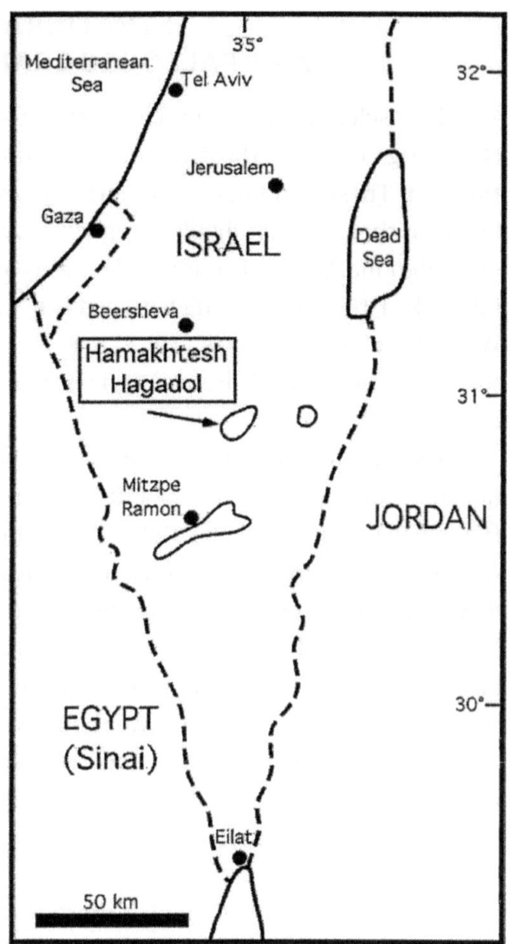

FIG. 1: Location of Hamakhtesh Hagadol, one of three major makhteshim in the Negev Desert, southern Israel. It is an erosionally-breached anticline in which are the exposures of the Middle Jurassic Matmor Formation studied here.

LOCATION

The sclerozoan fauna described here was collected on the northern interior wall of Hamakhtesh Hagadol, a breached anticlinal form approximately 15 km long and 6 km wide, in the northeastern Negev Desert of Israel (Locality number C/W-226; N30° 56.083′, E34° 58.537′, 420m elevation; Fig. 1). This outcrop is in the same area and within the same stratigraphic interval that Feldman and Brett (1998) did their work, but with a set of subunits they did not include in their study.

STRATIGRAPHIC AND PALEOGEOGRAPHIC CONTEXT

The sclerozoan assemblage was collected from subunits 53 and 54 (Goldberg, 1963) of the Matmor Formation (Late Callovian) in Hamakhtesh Hagadol. These subunits at this locality are indistinguishable. The boundary between them was originally designated by Goldberg (1963) as the base of an indurated marl, but it is not visible in this locality. Subunits 53 and 54 are easily marked, though, as being between the "Middle White Unit" and "First Upper White Unit" in Goldberg's (1963) scheme. Together they are here about 8 m thick, consisting of calcareous marl with local patch reefs of corals and calcareous sponges.

The Matmor Formation consists of 100 m of alternating marls and fossiliferous limestones (subunits 43–74 of Goldberg, 1963) above the Zohar Formation (Callovian) and below the Kidod Formation (Oxfordian) in southern Israel. Goldberg (1963) originally considered these subunits as the Matmor Member of the Zohar Formation and an unnamed member of the Kidod Formation. Hirsch and Roded (1996) revised this stratigraphy, noting that the marly units at the top of the Zohar and bottom of the Kidod were Late Callovian (*athleta* Zone) and distinct enough to have their own lithostratigraphic designation: the Matmor Formation. Hirsch and Roded (1996) cited co-occurrences of ostracods and foraminifers in the Matmor Formation to correlate them with the top beds of the Hermon Formation at Majdal Shams in the Golan and the top of the Zohar Formation in Sinai.

During the Middle Callovian there was a maximum marine transgression in southern Israel that inundated most of the Arabian platform with typical "Zohar/Matmor" type carbonates yielding nerinacean gastropods, foraminiferans (*Kurnubia* lineage), brachiopods, and abundant bivalves (e.g. *Eligmus*), mostly related to Ethiopian–Somalian taxa. The Ethiopian Faunal Province appears to have been colonized by brachiopods migrating from the north in the Early Jurassic that were then isolated for the remainder of the Jurassic. Subsequently these faunas evolved special morphological characters that distinguish them from their ancestors. In general, the faunas of Israel and Sinai are related to southern Tethyan shelf faunas and are strongly endemic in character.

During the Late Callovian this location would have been on or very close to the equator on the western coast of the Neotethys (Guiraud and Bosworth, 1999, Fig. 8A; Golonka, 2004, Fig. 8).

PALEOECOLOGICAL CONTEXT

Coral platters and microatolls

The most common hard substrate in the Matmor Formation assemblage is an unidentified species of Microsolena (Family Microsolenidae). These corals usually have a platter-like corallum 2–20 cm in diameter and up to 5 cm thick (Fig. 2A). Most are slightly dish-shaped with their upper surface concave. A few have distinctly upturned edges forming distinct "microatolls" (Kobluk and Noor, 1990) which are excellent indicators for very shallow water. Some juvenile coralla are horn-shaped with a narrow attachment expanding upward as a cone. These are described in more detail below as encrusting sclerozoans. The microsolenid coral platters and to a lesser extent the conical coralla have numerous encrusters and borings, primarily on their undersurfaces. These corals were loosely distributed on the sediment surface, only rarely forming small aggregates.

Calcareous sponge platters and mounds

The second most common hard substrate in this Matmor assemblage are platters and mounds made by various calcareous sponges traditionally considered "stromatoporoids" by most previous authors (e.g., Hudson, 1958, 1959, 1960). The taxonomy of these Jurassic sponges is very confused, but we can at least be certain that they are not "stromatoporoids," at least not in the sense of Paleozoic stromatoporoids (see Stock, 2001 for a brief review of post-Devonian "stromatoporoids" and references). The Matmor specimens here, both the platters and mounds, generally match the description for *Actostroma* Hudson, 1958. Wood (1987, p. 72) classified *Actostroma* as a "calcified demosponge" within the Order Axinellida. We are referring to this sponge as *Actostroma*?, with the query because some internal features and its variable external shape are not in the original generic description.

These calcareous sponges may not be true stromatoporoids, but their form and paleobiology may be most usefully described using stromatoporoids as ecological analogues. Using the analytical framework erected by Kershaw (1998, p. 520) for stromatoporoids, the Matmor sponges are primarily "tabular," "smooth," and with "mamelons." A few are "bulbous" and some smaller specimens are "expanding conical" or horn-shaped, like the *Microsolena* colonies described above and below. The platters grew from an initial hard substrate attachment (usually a bioclast) onto the surrounding soft sediment. The mounds and cones remained attached to larger hard substrates such as other sponges or microsolenid corals.

Soft-sediment fauna

The Matmor sclerozoan assemblage is found with a diverse assemblage of organisms which lived on or in the surrounding marly sediments. These include rhynchonellid brachiopods (*Burmirhynchia jirbaensis* being the most common species), terebratulid brachiopods, gastropods (preserved only as external molds underneath calcareous sponges; Fig. 2E), numerous regular echinoids (including representatives of the families Arbaciidae and Cidaridae, and Hessotiara of the Family Hemicidaridae), oysters, and rare articulate crinoid columnals. Other brachiopods found in subunits 53 and 54 include what appear to be depauperate specimens of two distinct species of *Somalirhynchia* as well as *Digonella* sp. The brachiopods seem to be mature (ephebic) forms that may have been geographically isolated from the main populations and lacked adequate nutrients, possibly due to inadequate current circulation, as would be the case in a lagoonal environment.

ENCRUSTING SCLEROZOANS

The Matmor sclerozoans are almost all confined to the upper and lower surfaces of platter-shaped *Microsolena* and *Actostroma*?. Many of the encrusters show signs of post-mortem abrasion, and some encrusting corals and sponges are known only from small attachments from which the erect portion of the skeleton has broken free. There is sufficient taphonomic loss in this assemblage to prevent us from attempting a quantitative study of the fauna. We can, though, describe the major sclerozoans and reach some conclusions about their paleoecological relationships.

Serpulid worms

There are two serpulid worm tubes common on the Matmor hard substrates (Fig. 2C). The most common is a thin, smooth, convoluted tube with a circular cross-section referred to as *Glomerula gordialis*. The other tube is thicker, less convoluted, slightly corrugated, and keeled along its dorsal surface. This serpulid is placed in the genus *Mucroserpula* as recently illustrated by Radwanska (2004). Whereas *G. gordialis* is found mostly on the undersides of corals and sponges, *Mucroserpula* sp. is present on both the under and upper surfaces.

Sponges

The small, attaching sponges fall into two categories: calcified demosponges and true members of the Class Calcarea (Fig. 2D). They are too small and underdeveloped to be distinguished further. The calcified demosponges are found primarily on the upper surfaces of the skeletal platters. They are either small mounds or upward-growing cones with small attachments. The calcareans are always found on the undersurfaces of the platters and are small mounds or encrusting sheets.

Corals

Like the sponges, the encrusting corals are also too abraded and juvenile to be unequivocally identified at genus level, but they are probably *Microsolena*. Their microscopic morphology is consistent with the larger microsolenids forming the platter substrates. Many of these corals are small mounds, but a few are upward-growing cones similar to the calcified demosponges (Fig. 2F). The corals are all found on the upward-facing surfaces of the skeletal platters.

Cyclostome bryozoans

The cyclostome bryozoans in this sclerozoan fauna are sheet-like encrusters. The colonies are small, often just a few zooids in extent, and fan-shaped. They do not possess gonozooids or other features that would allow us to identify them further. The bryozoans are found entirely on the undersurfaces of the platters.

Thecideide brachiopods

The cemented ventral valves of these small brachiopods occur ubiquitously but only on the undersurfaces of the platters, and their separated dorsal valves (Fig. 2B) are common in the surrounding sediment. These brachiopods represent an undescribed species of the genus *Moorelina* (G. Jaecks, pers. comm.). These are the first thecideide brachiopods reported from the Jurassic of the Middle East.

Oysters

Numerous oysters are present on the undersurfaces of the skeletal platters (Fig. 2A). They appear to be the cosmopolitan Jurassic species *Exogyra nana*, although there are few taxonomic features present on these attached valves.

Plicatulid bivalves

One specimen of *Plicatula* sp. A was found on the underside of a calcareous sponge platter. There are other patches of thin encrusting calcite which may be plicatulid in origin, but are too abraded or dissolved for identification.

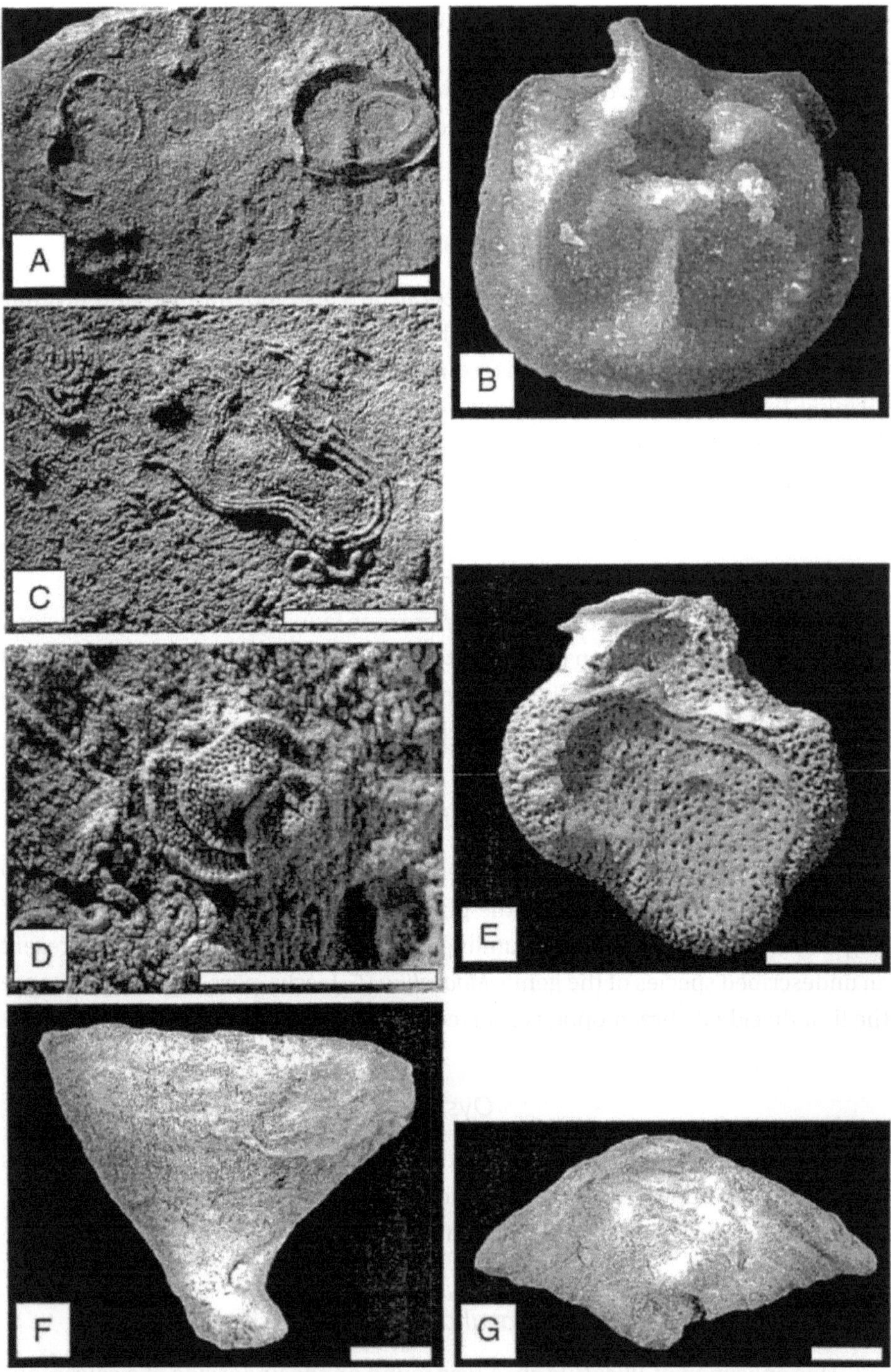

FIG. 2: The Matmor sclerozoan fauna. All scale bars are 5 ml except for Fig. 2B which is 0.5 mm. A. two oysters *(Exogyra nana)* encrusting the underside of the coral *Microsolena* sp. B, detached

BORINGS AND NESTLING BIVALVES

The boring Gastrochaenolites

The bivalve boring ichnospecies *Gastrochaenolites torpedo* is common on both the upper and undersurfaces of the coral and calcareous sponge platters. These borings are always perpendicular to the hard substrate and range in depth from 2 to 10 mm and greatest width from 3 to 5 mm. They show no preference for either substrate type, or for upper or lower surfaces.

Nestling bivalves

A mytilid bivalve is sometimes found nestling inside the *G. torpedo* borings with its posterior at the opening of the boring. These appear to be true nestlers because they are sometimes found inside the valves of previous bivalve occupants (thus they were not the makers of the borings; see Wilson, 1986) and they are always larger in diameter than the opening of the boring (thus they lived within the cavity and were not swept in from outside). These nestling bivalves are found in borings on the lower surfaces of the platters only.

CRYPTIC VS. EXPOSED SCLEROZOANS

The Matmor sclerozoans show a distinct polarity between cryptic and exposed surfaces of the skeletal platters (Table 1), and they show no preference for either coral or sponge substrates. Of the 12 encrusting and boring species, seven are exclusively on the undersurfaces and three are found only on the upper surfaces. Only the boring *G. torpedo* and the smooth serpulid *Glomerula gordialis* is found on both surfaces. The cryptic fauna here dominated by serpulids, oysters, plicatulids, bryozoans, and thecideide brachiopods is very similar to that seen in other Jurassic sclerozoan communities on calcareous sponge substrates (Palmer and Fürsich, 1981) and on corals (Mancenido and Damborenea, 1990; Bertling, 1994; Leinfelder et al., 1996). It is also very much like the cryptic fauna in Jurassic caves (Taylor and Palmer, 1994), mollusk shell interiors (Gaillard and Pajaud, 1971), and in the cavities produced underneath hardground slabs (Wilson, 1998; Baker and Wilson, 1999).

---

dorsal valve of *Moorelina* sp., a thecideide brachiopod; photograph courtesy of G. Jaecks. C, the keeled serpulid *Mucroserpula* sp. with a small specimen of the smooth serpulid *Glomerula gordialis* next to it on the upward-facing surface of the calcareous sponge *Actostroma?* D, calcareous sponge encruster and G, *gordialis* on the underside of *Microsolena* sp. E, gastropod biommured by a calcareous sponge. F, *Microsolena* sp. with a conical form. G. *Actostroma?* with a conical form.

TABLE 1: Distribution of Matmor Formation sclerozoans on coral and calcareous sponge skeletal platters

Sclerozoans on upward-facing surfaces	Sclerozoans on undersurfaces
Mucroserpula sp. (keeled serpulid)	*Glomerula gordialis* (smooth serpulid)
Glomerula gordialis (smooth serpulid)	Nestling mytilid bivalve sp. A
Actostroma? (calcified demosponge)	*Moorelina* sp. (thecideide brachiopod)
Microsolena sp. A	*Exogyra nana*
Gastrochaenolites torpedo (bivalve boring)	Cyclostome bryozoan sp. A
	Class Calcarea sp. A
	Class Calcarea sp. B
	Plicatula sp. (encrusting bivalve)
	Gastrochaenolites torpedo (bivalve boring)

The fossil record of cryptic marine communities is critical evidence for answering the question whether modern cryptic communities serve as ecological refuges for taxa which were excluded by competition from exposed habitats (Taylor and Wilson, 2003). Thus far the distribution of clades known in modern marine cryptic environments (e.g., Harmelin et al., 1985; Harmelin, 1997; Marti et al., 2004) is remarkably similar to those in the Jurassic equivalents, supporting the hypothesis that today's cryptic fauna has a very long history of adaptation to this environment. The Matmor sclerozoans show that Jurassic cryptic faunas did not change appreciably in general composition with latitude.

We can also conclude from the significant differences in the occurrences of sclerozoans on the upper and undersurfaces of the Matmor skeletal platters that they did not turn over very often in their depositional environment. Cryptic forms are rarely found on upper surfaces and vice versa, so most of these surfaces must have remained in their relative positions until burial. This conclusion is supported as well by the following observations of adaptations to sediment accumulation by some corals and sponges growing on the upper surfaces.

ADAPTATIONS TO SEDIMENT ACCUMULATION

One of the curious features of the Matmor sclerozoan fauna is that two unrelated taxa, the calcified demosponge *Actostroma*? (Fig. 2G) and the microsolenid coral (Fig. 2F), sometimes produced skeletons with small attachments and upwardly-expanding cones much like the classic rugosan "horn corals." We believe that these organisms grew upward from a hard substrate and were episodically covered with sediment. Their skeletons took the form of upwardly-expanding cones as they grew with sediment accumulating around their bases.

The same pattern has been observed in *Microsolena* from Upper Jurassic reef settings described by Leinfelder et al. (1996). The Matmor *Microsolena* and *Actostroma*? are both "umbrella-shaped" and "pseudobranched" using the Leinfelder et al. (1996, Fig. 2) terminology. This pattern is predicted in models of stromatoporoid growth produced by Swan and Kershaw (1994). Their computer-modeled stromatoporoids grew hemispherical forms when there was little or no sediment accumulation, and increasingly steeper cones with increasing sedimentation. Taylor and Wilson (1999) suggested that the Ordovician bryozoan *Dianulites fastigiatus*, which has a similar conical shape, also grew upward with sedimentation from a hard substrate attachment. The Matmor corals and sponges, then, likely lived in a quiet muddy lagoonal environment steadily filling with marly sediment. Most sclerozoans would have been smothered in the gathering mud, but the conical corals and sponges survived because they could keep above the sediment-water interface.

CONCLUSIONS

The sclerozoan fauna described here from the Matmor Formation of southern Israel represents one of few known from the tropics of the Jurassic. With at least 12 sclerozoan species, it shows a similar range of diversity as the better known sclerozoan faunas from northern Europe, North America, South America, and India. The fauna encrusted the upward-facing and undersurfaces of microsolenid coral platters and *Actostroma*?, a flattened calcified demosponge formerly referred to the stromatoporoids. This fauna was preserved in very shallow water as shown by the occurrence of microatoll corals. The depositional environment was probably an inter-reef lagoon since fine marly sediment accumulated at such a rate that some sponges and corals grew in upwardly-expanding cones to keep from being entirely buried. Additional work is now necessary to better classify the sclerozoans, some of which are new species, and further describe the complex facies mosaic in which these lagoonal sediments are found.

ACKNOWLEDGMENTS

We thank the Donors of the Petroleum Research Fund, administered by the American Chemical Society, for the partial support of this research. We also are grateful for the support from the Wengerd, Luce, Wilson and Faculty Development Funds at The College of Wooster. We especially thank Amihai Sneh of the Geological Survey of Israel for his assistance, and Paul Taylor, Tim Palmer, and James Nebelsick for their thoughtful reviews of the manuscript.

REFERENCES

Arkell, W. J. 1956. *Jurassic Geology of the World*. Edinburgh: Oliver and Boyd, Ltd.

Baker, P. G., and M. A. Wilson. 1999. The first thecideide brachiopod from the Jurassic of North America. *Palaeontology* 42: 887-895.

Bertling, M. 1994. Ökologie und Taxonomie koralleninkrustierender Bryozoen des norddeutschen Malm. *Paläontologische Zeitschrift* 68: 419-435.

Dommergues, J.-L. 1987. L'evolution chez les Ammonitina du Lias moyen (Carixien, Domerien Basal) en Europe occidentale. *Documents des Laboratoires de Géologie de la Faculté des Sciences de Lyon* 98: 1-297.

Feldman, H. R. 1987. A new species of the Jurassic (Callovian) brachiopod *Septirhynchia* from northern Sinai. *Journal of Paleontology* 61: 1156-1172.

Feldman, H. R., and C. E. Brett. 1998. Epi- and endobiontic organisms on Late Jurassic crinoid columns from the Negev Desert, Israel: implications for co-evolution. *Lethaia* 31: 57-71.

Feldman, H. R., and E. F. Owen. 1988. *Goliathyris lewyi*, new species (Brachiopoda, Terebratellacea) from the Jurassic of Gebel El-Minshera, northern Sinai. *American Museum Novitates* 2908: 1-12.

Feldman, H. R., E. F. Owen, and F. Hirsch. 1991. Brachiopods from the Jurassic of Gebel El-Maghara, northern Sinai. *American Museum Novitates* 3006: 1-28.

------. 2001. Brachiopods from the Jurassic (Callovian) of Hamakhtesh Hagdol (Kurnub Anticline), southern Israel. *Palaeontology* 44: 637-658.

Gaillard, C., Pajaud, D., 1971. *Rioultina virdurensis* (Buv.) cf. *ornata* (Moore) brachiopode thecideen de l'epifaune de l'Oxfordien superieur du Jura Meridional. *Geobios* 4, 227-242.

Goldberg, M. 1963. Reference section of Jurassic sequence in Hamakhtesh Hagadol (Kurnub Anticline). Detailed binocular sample description, including field observations. Israel Geological Survey, Unpublished Report, 50 pp.

Golonka, J. 2004. Plate tectonic evolution of the southern margin of Eurasia in the Mesozoic and Cenozoic. *Tectonophysics* 381: 235-273.

Guiraud, R., and W. Bosworth. 1999. Phanerozoic geodynamic evolution of northeastern Africa and the northwestern Arabian Platform. *Tectonophysics* 315: 73-108.

Hallam, A. 1975. *Jurassic Environments*. Cambridge: Cambridge University Press.

Harmelin, J.-G., J. Vacelet, and P. Vasseur. 1985. Les grottes sous-marines obscures: un mileu extrème et remarquable biotope refuge. *Tethys* 11: 214-219.

Harmelin, J.-G. 1997. Diversity of bryozoans in a Mediterranean sublittoral cave with bathyal like conditions: Role of dispersal processes and local factors. *Marine Ecology Progress Series* 153: 139-152.

Hirsch, F., and A. Roded. 1996. The Jurassic stratigraphic nomenclature in Hamakhtesh Hagadol, northern Negev. Geological Survey of Israel, Current Research 10: 10-14.

Hudson, R. G. S. 1958. The Upper Jurassic faunas of southern Israel. *Geological Magazine* 95: 415-425.

------. 1959. A revision of the Jurassic stromatoporoids *Actinostromina, strostylopsis,* and *Trupetostromaria Germovsek. Palaeontology* 2: 28-38.

------. 1960. The Tethyan Jurassic stromatoporoids *Stromatoporina, Dehornella,* and *Astroporina. Palaeontology* 2: 180-199.

Johnson, M. E., and B. G. Baarli. 1999. Diversification of rocky-shore biotas through geologic time. *Geobios* 32: 257-273.

Kershaw, S. 1998. The applications of stromatoporoid palaeobiology in palaeoenvironmental analysis. *Palaeontology* 41: 509-544.

Kobluk, D. R., and I. Noor. 1990. Coral microatolls and a probable Middle Ordovician example. *Journal of Palaeontology* 64: 39-43.

Leinfelder, R. R., W. Werner, M. Nose, D. U. Schmid, M. Krautter, R. Laternser, M. Takacs, and D. Hartmann. 1996. Paleoecology, growth parameters and dynamics of coral, sponge and microbolite reefs from the Late Jurassic. *In* J. Reitner, F. Neuweiler, and F. Gunkel (eds.), Global and Regional Controls on Biogenic Sedimentation. I. Reef Evolution. Research Reports. *Göttinger Arbeiten zur Geologie und Paläontologie*: 227-248. Sb2.

Manceñido, M. O., and S. E. Damborenea. 1990. Corallophilous micromorphic brachiopods from the Lower Jurassic of west central Argentina. *In* D. MacKinnon, D. Lee, and D. Campbell, (eds.), *Brachiopods Through Time.* Rotterdam: A.A. Balkema, 89-96.

Marti, R., M. J. Uriz, E. Ballesteros, and X. Turon. 2004. Benthic assemblages in two Mediterranean caves: Species diversity and coverage as a function of abiotic parameters and geographic distance. *Journal of the Marine Biological Association of the United Kingdom* 84: 557-572.

McKinney, F. K. 1995. One hundred million years of competitive interactions between bryozoan clades: Asymmetrical but not escalating. *Biological Journal of the Linnaean Society* 56: 465-481.

Palmer, T. J. 1982. Cambrian to Cretaceous changes in hardground communities. *Lethaia* 15: 309-323.

Palmer, T. J., and F. T. Fürsich. 1981. Ecology of sponge reefs from the Middle Jurassic of Normandy. *Palaeontology* 24: 1-23.

Radwanska, U. 2004. Tube-dwelling polychaetes from the Upper Oxfordian of Wapienno/Bielawy, Couiavia region, north-central Poland. *Acta Geologica Polonica* 54: 35-52.

Stanley, S. M., and L. A. Hardie. 1998. Secular oscillations in the carbonate mineralogy of reefbuilding and sediment-producing organisms driven by tectonically forced shifts in seawater chemistry. *Palaeogeography, Palaeoclimatology, Palaeoecology* 144: 3-19.

Stock, C. W. 2001. Stromatoporoidea, 1926–2000. *Journal of Paleontology* 75: 1079-1089.

Swan, A. R. H., and S. Kershaw. 1994. Computer model for skeletal growth of stromatoporoids. *Palaeontology* 37: 409-423.

Taylor, P. D., and T. J. Palmer. 1994. Submarine caves in a Jurassic reef (La Rochelle, France) and the evolution of cave biotas. *Naturwissenschaften* 81: 357-360.

Taylor, P. D., and M. A. Wilson. 1999. *Dianulites* Eichwald, 1829: an unusual Ordovician bryozoan with a high-magnesium calcite skeleton. *Journal of Paleontology* 73: 38-48.

------. 2002. A new terminology for marine organisms inhabiting hard substrates. *Palaios* 17: 522-525.

------. 2003. Paleoecology and evolution of marine hard substrate communities. *Earth-Science Reviews* 62: 1-103.

Westermann, G. E. D. 2000. Biochore classification and nomenclature in paleobiogeography: an attempt at order. *Palaeogeography, Palaeoclimatology, Palaeoecology* 158: 1-13.

Wilson, M. A. 1986. Coelobites and spatial refuges in a Lower Cretaceous cobble-dwelling hardground fauna. *Palaeontology* 29: 691-703.

------. 1998. Succession in a Jurassic marine cavity community and the evolution of cryptic marine faunas. *Geology* 26: 379-381.

Wilson, M. A., and T. J. Palmer. 1992. *Hardgrounds and hardground faunas.* University of Wales, Aberystwyth, Institute of Earth Studies Publications 9: 1-131.

Wood, R. 1987. Biology and revised systematics of some Late Mesozoic stromatoporoids. *Special Papers in Palaeontology* 37: 1-89.

Chapter Eleven

Bioerosion in an equatorial Middle Jurassic coral–sponge reef community (Callovian, Matmor Formation, southern Israel)

ABSTRACT

The Matmor Formation (Middle Jurassic, Callovian, 155-151 mya) in southern Israel contains abundant coral-sponge patch reefs and large crinoids which have been extensively bioeroded by bivalves, worms, barnacles, phoronids, and others producing eight ichnospecies. It is significant for the evolutionary history of bioerosion because this is the first equatorial Middle Jurassic boring ichnofauna to be documented. When compared to contemporaneous ichnofaunas, this assemblage is of average diversity and abundance but has only rare sponge borings and contains abundant specimens of *Oichnus paraboloides* as shallow pits on crinoid stems. The Matmor Formation surprisingly lacks carbonate hardgrounds, which are otherwise abundant in subtropical and temperate equivalents.

INTRODUCTION

This is one of the first detailed studies of bioerosion (the destruction of hard substrates by biological processes) in an equatorial Jurassic ecosystem. The Matmor Formation (Jurassic, Callovian. 165-161 mya) is a shallow-water carbonate and siliciclastic unit deposited in shallow water along the western shore of the Neotethys on or very near the paleoequator (Guiraud and Bosworth, 1999, Fig. 8A; Golonka, 2004, Fig. 8). It is approximately 120 m thick and composed of poorly consolidated marls and limestones with very fossiliferous horizons dominated by calcareous sponges, scleractinian corals, mollusks and echinoderms, with minor brachiopods, sabellids, serpulids and bryozoans. It is completely and exclusively exposed in an erosional cirque in southern Israel (Hamakhtesh Hagadol = Kurnub Anticline; see Goldberg, 1963). Most of the corals and sponges constructed small patch reefs with diverse associated echinoderm, molluscan and sclerobiont (hard substrate dwelling organisms; see Taylor and Wilson, 2003) communities (Feldman and Brett, 1998; Wilson et al., 2008).

Many of the coral, sponge and crinoid skeletons in the Matmor Formation are bored by worms of some type (producing the ichnogenus *Trypanites*), barnacles (*Rogerella*) and bivalves (*Gastrochaenolites*). These borings are abundant and of average ichnodiversity when compared to other Middle Jurassic bioerosion assemblages (see Taylor and Wilson, (2003) and Bromley (2004) for a review of bioerosion history). This is also true for the encrusting taxa on these equatorial skeletal substrates, which are as diverse as most other Jurassic sclerobiont assemblages (Feldman and Brett, 1998; Wilson et al., 2008).

Feldman and Brett (1998) were the first to describe bioerosion in the Matmor Formation when they examined sclerobionts on crinoid stems (*Apiocrinites* and *Millericrinus*) in Hamakhtesh Hagadol. They found numerous borings, many of which produced embedment structures ("*Tremichnus*") in living crinoid skeletal tissue. This study expands the initial work of Feldman and Brett (1998) to all bioeroded substrates in the Matmor Formation and updates the ichnotaxonomy and interpretations.

On a global scale, bioerosion began a significant increase in abundance and diversity during the Jurassic (Fürsich et al., 1994; Bertling, 1997, 1999; Wood, 1999; Perry and Bertling, 2000; Taylor and Wilson, 2003; Bromley, 2004; Wilson, 2007). From very low levels in the Triassic, macroborings become far more numerous in Jurassic corals and calcareous sponges, especially in reef settings (Pisera, 1987; Bertling, 1999; Wood, 1999), on carbonate clasts (Schlagintweit, 2008), and on mollusk shells (Hillmer and Schulz, 1973; Mayoral and Sequeiros, 1979). Microborings are also common (Gatrall and Golubic, 1970; Glaub, 1988, 1994; Glaub and Schmidt, 1994; Glaub and Bundschuh, 1997; Kolodziej, 1997; Vogel et al., 1999; Vogel and Glaub, 2004). The questions now are: (a) what is the timing of this bioerosion diversification and intensification?; (b) how is this increase related to diversifications of both bioeroders and the organisms which produce the organic substrates that they bore?; and (c) is there a latitudinal gradient for these diversity and abundance changes? Several projects are currently addressing the first two questions (see Perry and Bertling, 2000, for a review of these questions and others). This study is the start of an answer to the third question with the establishment of one bioerosion data set in the Middle Jurassic equatorial shallow-water tropics.

This paper is part of a continuing study on the marine faunas of the Ethiopian Province of the Jurassic Tethyan Realm (the Tethys-Panthalassa Realm *sensu* Dommergues, 1987; see also Westermann, 2000). The Ethiopian Province is characterized by the presence of endemic taxa at the family and lower levels (Feldman et al., 2001). It appeared in the Early Jurassic and lasted until the middle and possibly the end of the Cretaceous in India, Madagascar, East Africa and, at the end of the Jurassic, in South America. This long-term study began with a taxonomic revision of the brachiopod faunas of the Ethiopian Province (Feldman, 1987; Feldman and Owen, 1988; Feldman et al., 1991: Feldman et al., 2001) and has continued with paleoecological investigations of Jurassic faunas in Israel (Feldman and Brett, 1998; Wilson et al., 2005, 2008, 2009).

LOCATIONS

The specimens for this study were collected from 11 exposures of the Matmor Formation in Hamakhtesh Hagadol, Israel (Fig. 1; Table 1). They range from 12 to 90 m above the base of the Matmor where it overlies the Madsus Member of the Zohar Formation (Hirsch and Roded, 1996). Bored corals, sponges and crinoids are abundant in these horizons. Examples of bioerosion are found elsewhere throughout the Matmor Formation at Hamakhtesh Hagadol, but these locations contain all the ichnospecies known from the unit. They include the same primary location for the sclerobiont study by Wilson et al. (2008) and the same horizons used in the crinoid sclerobiont study by Feldman and Brett (1998).

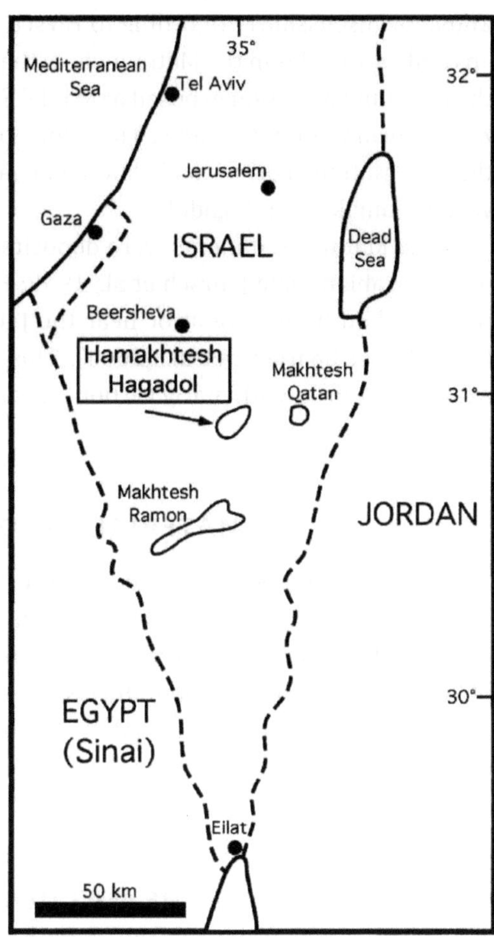

FIGURE 1: Location of Hamakhtesh Hagadol, one of three large makhteshim in the Negev desert, southern Israel. The Middle Jurassic Matmor Formation is entirely and exclusively exposed in this structural and geomorphological structure.

The makhtesh is an erosional cirque with an elliptical shape approximately 15 km long and 6 km wide and striking to the northeast. The floor of the makhtesh and low hills in the center are formed from the relatively soft Jurassic Zohar and Matmor Formations, while the resistant walls of the structure are composed mostly of Cretaceous and Eocene limestones and sandstones. The makteshim of the

Negev Desert (Hamakhtesh Hagadol, Makhtesh Qatan and Makhtesh Ramon) developed through a complex process of folding and deep desert erosion; they are essentially breached anticlines (Avni, 2001).

STRATIGRAPHY

The Matmor Formation was defined by Hirsch and Roded (1996) as a distinctive series of marls and limestone, with its type section in the Matmor Hills near the center of Hamakhtesh Hagadol. There is debate over the name of this unit, with some referring to it as the Be'er Sheva Formation (see Wilson et al., 2009, for discussion). According to Hirsch and Roded (1996), the foraminiferans and ostracodes in the Matmor show that it is correlated with the top beds of the Hermon Formation exposed at Majdal Shams in the Golan and the top of the Zohar Formation in the Sinai. The fauna, particularly the ammonites, indicate that the Matmor Formation is late Callovian (*athleta* Zone). It is exposed only within Hamakhtesh Hagadol.

The Matmor sediments were deposited during a significant transgression on the Arabian Plate (Hirsch et al., 1998; Haq and Al-Qahtani, 2005). The sections studied here were at or near the J40 major maximum flooding surface (163 Ma) of Sharland et al. (2004). This transgression covered most of the Arabian Platform with marly carbonates and patch reefs of corals and calcareous sponges.

MATERIALS AND METHODS

All specimens were collected from sections of the Matmor Formation measured and described during field seasons from 2004 to 2009 in Hamakhtesh Hagadol. The specimens were cleaned, prepared and identified in the College of Wooster paleontology laboratories, where they are curated under the accession and locality numbers in Table 1. The borings were studied by thin-section, epoxy casts (using the methods outlined by Nielsen and Maiboe, 2000), and acetate peels (following the procedures of Wilson and Palmer, 1989).

THE MATMOR CORAL AND SPONGE REEF COMMUNITY

This coral and sponge reef community has been described in more detail in Wilson et al. (2008). Here we have updated the identifications and added a few more minor taxa to the record (Table 2). In summary, the most common fossils in the community are large (up to 20 cm in diameter) platters of the microsolenid coral *Microsolena* aff. *M. sadeki.* The upper surfaces of these coralla are often slightly concave with raised edges, indicating that they were "microatolls" (Kobluk and Noor, 1990) formed in very shallow water. The next most common fossils are calcified axinellid demosponges (*sensu* Wood, 1987, p. 72; sometimes termed "stromatoporoids") that take the form of pillars, platters and mounds up to 15 cm in diameter. The most abundant sponges are the milleporellid *Dehornella harrarensis* and the actinostromarid *Actostroma damesini.* These corals and sponges form the growth fabric of the Matmor Formation patch reefs, which would be termed "stromatoporoid-rich" reefs by Leinfelder et al. (2005, p. 287). These Matmor reefs were briefly described in Wood (1999) as living in a shallow shoal environment with frequent wave disturbances. These disturbances are easily seen in the number of corals and sponges that were toppled onto their sides and then regrew. Crinoids are abundant in some horizons as columnals and calyx ossicles from *Apiocrinites* and hold fasts of *Millericrinus.*

TABLE 1: Localities where bioeroded substrates were collected from the Matmor Formation in Hamakhtesh Hagadol. Israel.

Location #	Coordinates	Meters above base	Bored Substrate
C/W-170	N30.93299° E34.97572°	30 m	Corals
C/W-171	N30.93430° E34.97607°	28 m	Crinoids, sponges and isolated corals
C/W-172	N30.93375° E34.97582°	28 m	Crinoids, sponges and isolated corals
C/W-173	N30.92907° E34.97295°	28 m	Crinoids, sponges and isolated corals
C/W-226	N30.93374° E34.97533°	33 m	Sponges and isolated corals
C/W-242	N30.95038° E35.00531°	12 m	Sponges
C/W-244	N30.95038° E35.00531°	18 m	Sponges
C/W-245	N30.95131° E35.00550°	21 m	Crinoids
C/W-246	N30.95131° E35.00550°	25 m	Corals
C/W-247	N30.95208° E35.00475°	25 m	Crinoids and sponges
C/W-252	N30.95834° E35.01089°	90 m	Isolated corals

Table 2: Fossil taxa in the Matmor Formation coral and sponge reef complexes associated with bioerosion in Hamakhtesh Hagadol, southern Israel.

Phylum	Class	Order	Family	Species
Porifera	Demospongea	Axinellida	Milleporellidae	*Dehornella crustans* *Steineria somaliensis*
				Promillepora pervinquieri
				Shuqraia zuffardi
				Steineria somaliensis
			Actinostromaridae	*Actostroma damesini*
Cnidaria	Anthozoa	Scleratinia	Microsolenidae	*Microsolena* aff. *M. Sadeki*
				Microsolena aff. *arishensi*
			Agariciidae	*Craterastraea* cf. *craterformis*
			Thamnasteriidae	*Thamnasteria* sp.
			Amphiastraeidae	*Amphiastraea* sp.
Mollusca	Gastropoda			
	Bivalvia	Ostreoida	Gryphaeidae	*Nanogyra nana*
		Pectinoida	Plicatulidae	*Plicatula* sp.
Brachiopoda	Rhynchonellata	Thecideida	Thecidellinidae	*Moorellina* sp.
		Rhynchonellida	Tetrarhynchiidae	*Burmirhynchia jirbaensis*
				Somalirhynchia sp.
		Terabratulida	Zeilleriidae	*Digonella* sp.
Bryozoa	Gymnolaemata	Cyclostomata	Plagioeciidae	*Microeciella* sp.
Echinodermata	Crinoidea	Millericrinida	Millericridinae	*Millericrinus* sp.
			Apiocrinidae	*Apiocrinites* sp.
	Echinoidea	Cidaroida	Rhabdocidariae	
			Cidariae	
		Hemicidaroida	Hemicidaridae	*Hessotaria* sp.
		Arbacoida	Arbaciidae	
Annelida	Polychaeta	Sedentaria	Serpulidae	*Vermiliopsis negevensis*
				Nogrobs cf. *N. quadrangularis*
				Propomatoceros aff. *P. trigona*
				Serpula (Cementula?) Sp.
			Sabellidae	*Glomerula gordialis*

The corals, sponges and crinoids were encrusted by a variety of sclerozoans. The encrusters include four species of serpulids, one species of sabellids, small attached corals and sponges, thecideide brachiopods, rare plicatulid bivalves, and rare sheet-like cyclostome bryozoans (Table 2). The distribution patterns of these sclerozoans on the corals and sponges are described in detail by Wilson et al. (2008) and on the crinoids by Feldman and Brett (1998). The numerous borings are described below in a separate section.

TABLE 3: Boring ichnotaxa and their host substrates in the Matmor Formation. See Table 2 for the higher classification of the substrate taxa.

Boring Ichnotaxa	Bored substrate taxa
Clionolithes ichnosp. (rare)	*Apiocrinites*
Gastrochaenalites torpedo (common)	*Microsolena* aff. *M. sadeki, Thamnasteria* sp., *Dehornella crustans, Shuqraia zuffardi, Actostroma damesini*
Gastrochaenolites lapidicus (common)	*Microsolena* aff. *M. sadeki, Thamnasteria* sp., *Dehornella crustans, Shuqraia zuffardi, Actostroma damesini*
Gastrochaenolites dijugus (rare)	*Apiocrinites*
Oichnus paraboloides (common)	*Apiocrinites, Millericrinus*
Rogerella elliptica (rare)	*Apiocrinites*
Talpina ichnosp. (rare)	*Apiocrinites*
Trypanites weisi (common)	*Microsolena* aff. *M. sadeki, Thamnasteria* sp., *Dehornella crustans, Shuqraia zuffardi, Actostroma damesini, Apiocrinites, Millericrinus*

Diverse soft-sediment organisms lived between the coral and sponge skeletons. These include common rhynchonellid brachiopods, less common terebratulid brachiopods, abundant regular echinoids of three families, common small oysters, rare millericrinid crinoids, and surprisingly rare gastropods preserved only as unidentifiable internal molds (Table 2).

The Matmor Formation coral and sponge patch reefs represent very shallow, warm water conditions. The deposition rates of marl sediments were probably high because the corals and sponges are often found in inverted cone or mushroom shapes, suggesting that they grew upwards to pace sediment accumulation below (Wilson et al., 2008), The bioeroders described in this paper thus lived in an environment much like that of a modern coral reef lagoon growing off a continental margin that is shedding fine-grained sediment such as the coral reefs of the Red Sea.

THE BIOEROSION ASSEMBLAGE

Eight boring ichnotaxa have been identified in the coral, sponge and crinoid substrates of the Matmor Formation (Table 3).

Clionolithes

Only one specimen of this ichnogenus was found in the Matmor material. It is a radiating boring cut into a crinoid stem (Fig. 2A). At its greatest dimension it is approximately 16 mm long. It has a wider central portion that develops on one side into thin channels down to about 0.20 mm wide, Its form is clearly constricted by the size and curvature of the crinoid stem substrate. There is not enough detail to place this single specimen into an ichnospecies of *Clionolithes.*

Clionolithes is most commonly considered a sponge boring in the literature (Clarke, 1908, 1921; Fenton and Fenton, 1932; Bromley, 2004; Tapanila et al., 2004; Tapanila, 2006) but it has occasionally been attributed to algae (e.g,. Parras and Casadio, 2006), Plewes (1996. p. 175), as part of a very

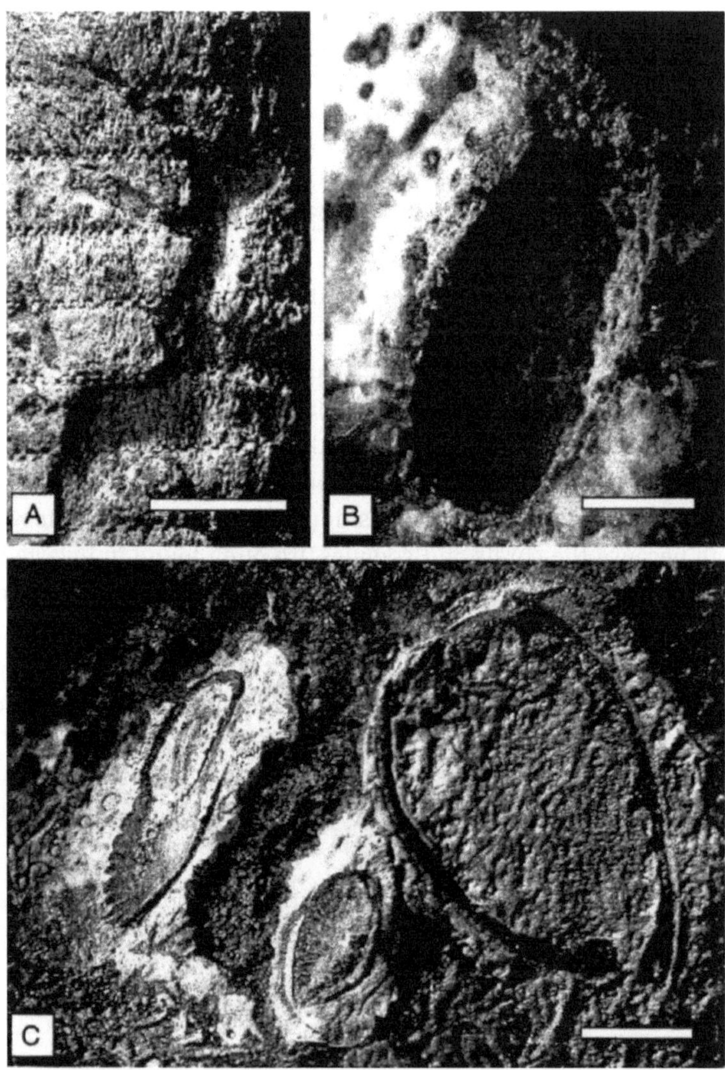

FIGURE 2: Matmor Formation bioerosion features. (A) *Clionolithes* bored in a crinoid stem (scale bar=2.0 mm; C/W-247-1). (B) *Gastrochaenolites lapidicus* in a microsolenid coral (scale bar=2.0 mm; C/W-226-1). (C) Lithophagid bivalve (left) and a nestling bivalve (right) in *Gastrochaenolites* crypts within a microsolenid coral (scale bar=5.0 mm; C/W-170-1).

detailed study of borings, concluded that a "compelling" case could be made for *Clionolithes* as the product of sponge excavation. We consider it a sponge boring here. This single specimen is the only evidence of endolithic sponges thus far found in the Matmor.

Gastrochaenolites

The most common borings in the corals and sponges of the Matmor Formation are those of bivalves that produced the ichnogenus *Gastrochaenolites.* Where adequate cross-sections and apertures can be observed, three ichnospecies can be identified: *G. lapidicus* Kelly and Bromley 1984 (a smooth, clavate boring with a relatively blunt parabolic base; Fig. 2B), *G. torpedo* Kelly and Bromley 1984 (with a more pointed parabolic base), and *G. dijugus* Kelly and Bromley 1984 (smooth, clavate boring with a longitudinally-restricted neck forming a figure-8 apertural opening; Fig. 3B). *Gastrochaenolites* is excavated today by the mytiloid *Lithophaga* and the myoids *Gastrochaena* and *Hiatella.* The Matmor specimens often have articulated lithophagid bivalves preserved inside them (Fig. 2C). In a few cases there are multiple sets of bivalve shells visible in cross-sections of the borings, demonstrating that the cavities were sometimes occupied by nestling bivalves after the borers had died.

Gastrochaenolites lapidicus and *G. torpedo* in the Matmor range from 5 to 15 mm in diameter at their widest cross-sections. Almost all are found in corals and sponges, showing no preference for exposed or cryptic surfaces. Some clearly bored into living corals because the host produced a blister of skeletal material around the apertures (Fig. 3A). A few are found in crinoid columns, in one case having bored almost entirely through the column. *G. dijugus* is found only in crinoid stem fragments, and it ranges from 2 to 5 mm in diameter at greatest width. *Gastrochaenolites* boring interiors are occasionally encrusted by serpulid and sabellid worm tubes as well as calcareous sponge recruits and thecideide brachiopods.

Oichnus

The stem fragments of *Apiocrinites* in the Matmor often have multiple circular parabolic shallow pits drilled into them normal to their exteriors and usually cutting two or more contiguous columnals (Fig. 3C and D). These match the description of *Oichnus paraboloides* Bromley 1981. Some cut into the articulating faces of columnals, demonstrating thar they were drilled after the death and disarticulation of the crinoid. Others were formed while the crinoid was alive as shown by impressive skeletal swellings around the pits (Fig. 3E). They range in size from 0.5 to 4.0 mm in width and (in unswollen specimens) 0.2 to 3.0 mm in depth. In no individual boring does depth exceed diameter of the opening.

Feldman and Brett (1998) referred to these borings as *Trypanites* (when there was no skeletal swelling) and *Tremichnus* (with skeletal swelling). These are not *Trypanites,* though, because the diameter of each equals or exceeds its depth, and they have parabolic rather than cylindrical cross-sections.

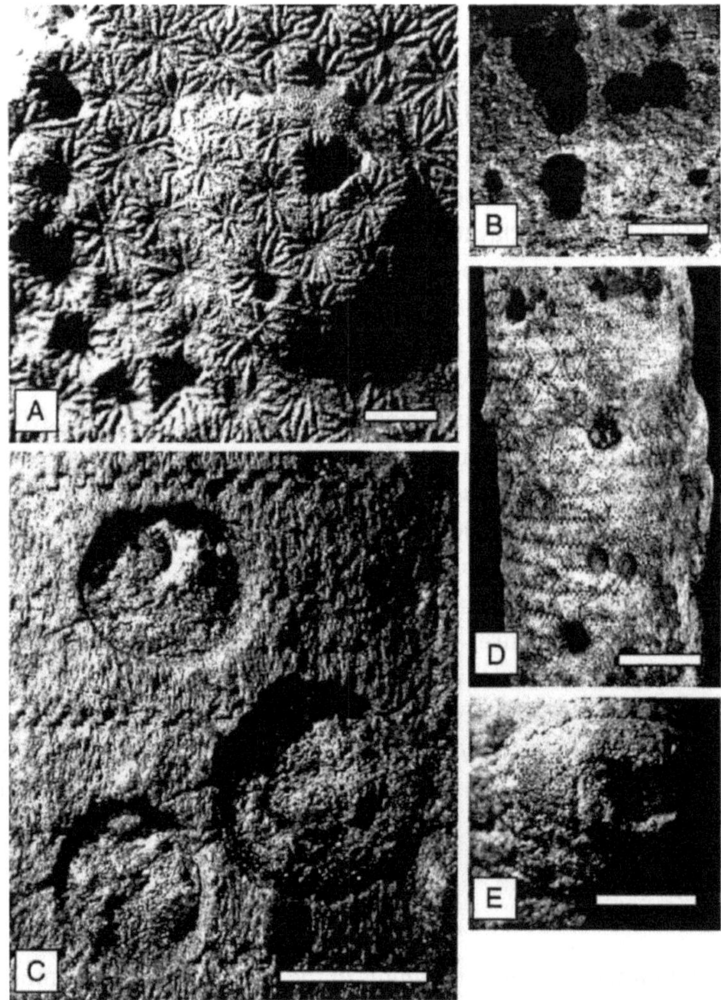

FIGURE 3: Additional Matmor Formation bioerosion features. (A) *Gastrochaenolites* in a coral which developed a blister in response to the boring (scale bar=2.0 mm; C/W-247-2). (B) *Gastrochaenolites dijugus* in a sponge attached to a crinoid stem (scale bar=3.0 mm; C/W-247-3). (C) Three *Oichnus paraboloides* in a crinoid stem (scale bar=2.0 mm; C/W-247-3). (D) Multiple *Oichnus paraboloides* in a crinoid stem (scale bar=5.0 mm; C/W-247-4). (E) *Oichnus paraboloides* in a crinoid stem which has swollen in response to the boring (scale bar=2.0 mm; C/W-247-5).

Pickerill and Donovan (1998) synonymized *Tremichnus* Brett 1985 with *Oichnus*, as did Nielsen and Nielsen (2001). The definition of *Oichnus* was further refined by Donovan and Jagt (2002). These are complicated structures when there was concurrent growth of the crinoid skeleton around them, placing them on a spectrum from embedment structures (see Tapanila, 2005, and references therein) to true borings. The organism that produced *Oichnus paraboloides* on the Matmor crinoids is unknown. It was likely a soft-bodied sclerobiont of some type that settled on the stem for an extended period. A. Radwański (University of Warsaw, pers. com.) suggests these may be myzostomid infections.

Rogerella

Acrothoracican barnacle borings of the ichnogenus *Rogerella* are relatively rare in the Matmor fauna. They are found only on crinoid columns (Fig. 4A). When present on a particular column they are abundant, but we recorded them on less than 5% of the columns we examined. The Matmor barnacle borings are consistent with descriptions of *Rogerella elliptica* Codez and de Saint-Seine, 1958. They range from 1.10 to 1.85 mm in length, 0.50 to 0.95 in width, and 1.35 to 2.05 mm in depth. The crinoid skeletal host does not show swellings associated with these borings, so they were most likely produced post-mortem on the seafloor.

Talpina

Phoronid borings are found only on three crinoid stem specimens in the Matmor Formation material. In one of these cases the boring is within an encrusting calcareous sponge on a crinoid stem fragment (Fig. 4B). These tubular, gently curved, branching borings resemble the ichnospecies *Talpina ramosa* von Hagenow 1840, but they are too incompletely preserved and too often filled with sediment to adequately cast. We thus cannot determine the ichnospecies and have left this boring in open nomenclature as *Talpina* ichnosp. Eroded specimens of *Talpina* were identified as "channel-like borings" in Feldman and Brett (1998. p. 64).

Trypanites

The second most common boring ichnotaxon in the Matmor Formation corals and sponges is the long cylindrical *Trypanites weisi* Mägdefrau 1932 (Figs. 4C and 5). It is also found in a few of the larger crinoid columns. It ranges in size from 2.0 to 5.0 mm in diameter and 3.0 to an estimated 35 mm in length.

It is distinguished here from *Oichnus* by having parallel sides and a depth greater than the diameter of the opening. This may be an artificial distinction in some cases because an early unfinished *Trypanites* would naturally be indistinguishable from a small *Oichnus* shallow pit. Like *Oichnus*, the skeletal host of *Trypanites* sometimes responded with living tissue, developing a blister around the opening of the boring.

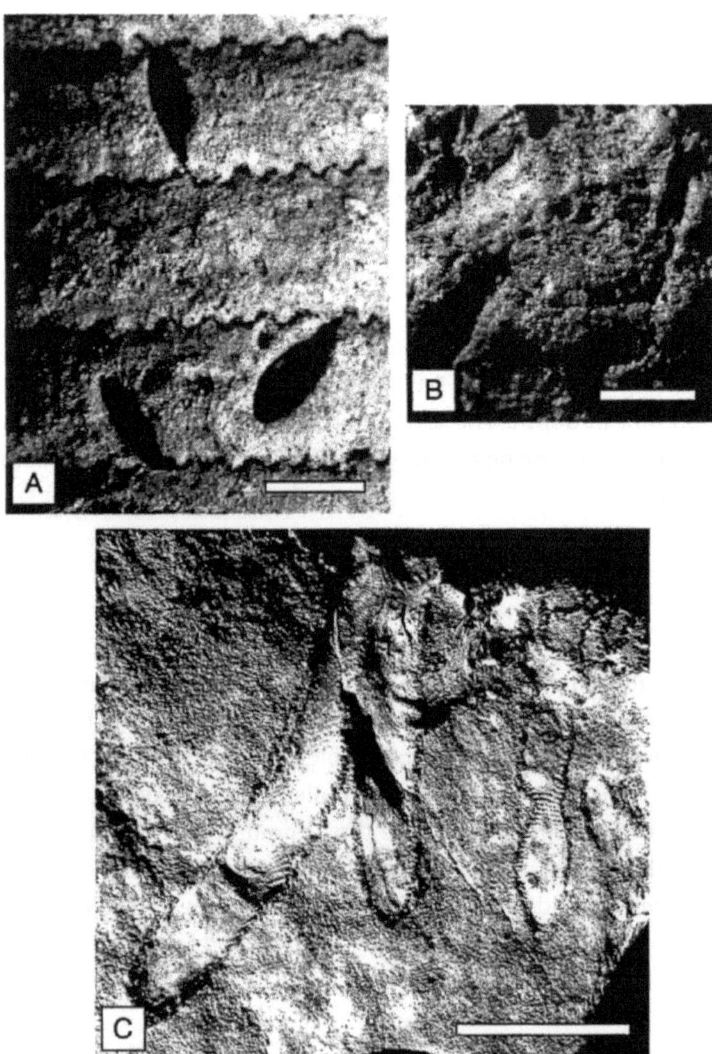

FIGURE 4: Additional Matmor Formation bioerosion features. (A) *Rogerella elliptica* in a crinoid stem (scale bar=1.0 mm; C/W-247-6). (B) *Talpina* ichnosp. in a sponge attached to a crinoid stem (scale bar=1.0 mm; C/W-247-7). (C) Three *Trypanites weisi* in a microsolenid coral in the field (scale bar=2.0 mm).

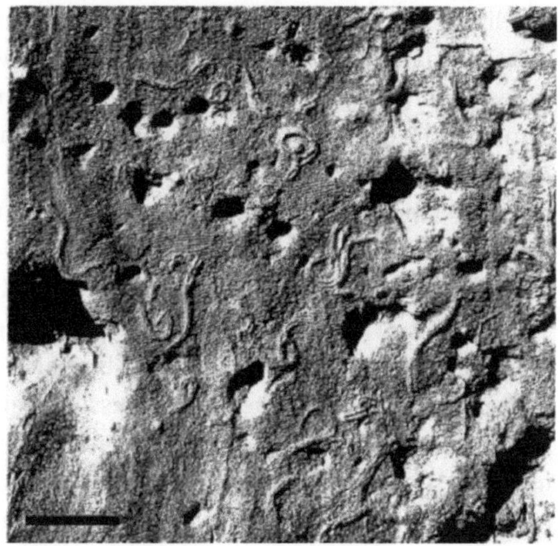

FIGURE 5: Underside of a microsolenid coral in the field showing the openings of *Gastrochaenolites* and *Trypanites* borings as well as numerous serpulid worm tubes (scale bar=5.0 mm).

Trypanites is abundant on the undersides and in cavities of the Matmor corals, usually sharing the space with thecideide brachiopods, serpulids and sabellids (Fig. 5). They were thus part of a classic Jurassic cryptic community (Taylor and Palmer, 1994; Wilson, 1998). Those found on the sponges and corals may have been formed post-mortem as these skeletal parts lay on the seafloor and created cryptic spaces.

DISCUSSION

The Middle Jurassic was a pivotal time for both scleractinian coral reef evolution (Leinfelder et al., 2002) and the development of modern bioerosion regimes on coral and sponge reefs (Bertling, 1999; Wood, 1999). Late Jurassic reef systems and their bioeroders are relatively well known, but there are still few studies of Middle Jurassic cases. Palmer and Fürsich (1981) described sponge reef frameworks and their bioeroders in the Middle Jurassic of Normandy, France. Wilson et al. (1998) detailed borings in Middle Jurassic oyster accumulations ("ostreoliths") from the Middle Jurassic of Utah that are at times reeflike with their solid frameworks and significant relief. Perry and Bertling (2000) have the most comprehensive analysis of Jurassic reef bioerosion patterns in space and time. Prior to their work it appeared that Middle Jurassic reef bioerosion was entirely the work of bivalves producing *Gastrochaenolites* crypts. This bivalve-dominated pattern was recorded for Middle Jurassic reefs in Switzerland (Wullschleger, 1966, 1971), France (Hallam, 1975; Lathuilière, 1982), and northern Chile (Prinz, 1986). The additional work by Perry and

Bertling (2000) diversified this record with additional data from the Bajocian and Callovian of Chile, Iran and India. They showed that bivalve and "worm" bioeroders were roughly equal in abundance and more taxonomically diverse, with significantly less bioerosion from barnacles, phoronids and sponges (Perry and Bertling, 2000, Fig. 2).

The ichnodiversity and composition of the community in the Matmor Formation is roughly equal to the other recorded Middle Jurassic reef-bioeroding assemblages. This is significant because observations suggest that bioerosion in modern low and high latitude assemblages (e.g. Cutler and Flessa, 1995) is also generally of the same intensity, but this has not been demonstrated before in the Jurassic. As with the other Jurassic assemblages, *Gastrochaenolites* is the most abundant and diverse set of borings and the "worm" borings (*Trypanites* in this case) are a close second; both borings are found in corals, sponges and crinoids. Phoronid borings (*Talpina*) and barnacle borings (*Rogerella*) are minor and restricted to just one substrate type (crinoid ossicles).

There are two major differences between the Matmor Formation bioerosion assemblage and those elsewhere in the Middle Jurassic. First, sponge borings (represented here by *Clionolithes*) are very rare in the Matmor. The unequivocal sponge boring *Entobia* is present albeit rare in the Bajocian reefs of Chile and the Bajocian-Callovian reefs of Iran. Perry and Bertling (2000, p. 40) note that these sites "represent coastal high-energy settings with strong siliciclastic influx" and suggest that these conditions were not conducive to sponge bioerosion. In contrast, the Matmor Formation was deposited under lower energy conditions with only moderate siliciclastic influx, yet there was even less sponge bioerosion. That the environment was suitable for sponges in general is apparent, of course, from the diversity and abundance of the skeletal demosponges ("stromatoporoids"). The rarity of boring sponges in this fauna is thus a mystery. Leinfelder et al. (2005, p. 288) suggested that the stromatoporoids in this region "might have been adapted to overheated and slightly hypersaline waters" to explain compositional differences with other Jurassic assemblages. This may be a clue that environmental factors discouraged boring sponges.

The second unique feature of the Matmor bioerosion community is the abundance of the shallow pits on crinoid stems here referred to *Oichnus paraboloides.* These have not been recorded in other Middle Jurassic reef systems. *O. paraboloides* is common throughout the Phanerozoic and was no doubt formed by a variety of different organisms. It is likely just a matter of recognition in other Middle Jurassic reef faunas. Curiously, there is one other case of crinoid skeletons with numerous *O. paraboloides* borings: Donovan et al. (2006) describe a Lower Carboniferous crinoid theca *(Amphoracrinus gilbertsoni)* from England "infested" with *O. paraboloides.* Except for the fact that the *O.*

paraboloides borings in the Matmor are on the stems and not the thecal plates of crinoids, the occurrences are remarkably similar.

Another curious difference between the Matmor bioerosion suite and its Jurassic equivalents elsewhere is the complete lack of a common bioerosion substrate: carbonate hardgrounds. Hardgrounds are very common in subtropical and temperate Jurassic carbonate sequences, and they host abundant and diverse borings (Wilson and Palmer, 1992, 1994; Taylor and Wilson, 2003; Bromley, 2004). No hardgrounds have been reported from the Matmor Formation, despite the presence of suitable lithologies (especially ooid-rich units) and an intensive search by the authors. Rockgrounds (rock exposed on the seafloor which was encrusted and bored) are known from the Zohar Formation below the Matmor in nearby Makhtesh Qatan (Wilson et al., 2005), but these are not the depositional equivalent to contemporaneously-cemented hardgrounds. It is possible that the Matmor sedimentation rate was higher than at the hardground-bearing locations, inhibiting intragranular calcite cementation, or that there was some undetermined difference in seawater chemistry between this equatorial system and the temperate sites.

CONCLUSIONS

1. Eight ichnotaxa of borings are found in the coral and sponge patch reefs of the Matmor Formation (Callovian) in southern Israel. The environment was a shallow carbonate shelf just above the storm wavebase at or near the paleoequator.

2. The bored substrates are microsolenid corals, calcareous axinellid demosponges, and crinoid columns. They are the most common hard substrates in this depositional system.

3. The bioerosion assemblage is dominated by bivalve borings (*Gastrochaenolites*) that also served as secondary living spaces for later cryptic bivalves. Worm borings (*Trypanites*) and shallow pits (*Oichnus paraboloides*) in crinoid stems are also common. Phoronid borings (*Talpina*) and acrothoracican barnacle borings (*Rogerella*) are relatively rare and confined to crinoid stems. Only one specimen of a putative sponge boring (*Clionolithes*) was found in this study.

4. *Gastrochaenolites* and *Oichnus* were sometimes excavated into living coral and crinoid substrates as shown by reactive host skeletal tissue. *Rogerella, Talpina* and *Trypanites* were apparently only formed after the death of the host.

5. There are few studies of bioerosion in Middle Jurassic reef systems. The Matmor bioerosion assemblage is similar to contemporary reef assemblages with the exception that sponge borings are very rare and *Oichnus paraboloides*

is not yet recorded elsewhere. Carbonate hardgrounds have not been found in the Matmor Formation even though they are very common in contemporaneous subtropical and temperate depositional sequences elsewhere.

ACKNOWLEDGMENTS

We thank the Donors of the Petroleum Research Fund, administered by the American Chemical Society, for partial support of this research. We also thank the Luce, Copeland and Wengerd Funds at The College of Wooster for supporting the field and laboratory work, and the Geological Survey of Israel (GSI) for logistics, especially vehicles, library access and specimens. Yoav Avni and Amihai Sneh of the GSI were especially helpful colleagues. Rowan Martindale provided a thorough review which improved the value of this paper. An anonymous second reviewer also had helpful comments.

REFERENCES

Avni, Y. 2001. Structure and landscape evolution of the Makhteshim country—interrelations between monoclines, truncation surfaces and rhe evolution of the Makhteshim. In B. Krasnov and E. Mazor (eds.), *Makhteshim Country: A Laboratory ot Nature: Geological and Ecological Studies in the Desert Region of Israel*, 33-58. Sofia: Pensoft Publishers.

Bertling, M. 1997. Bioerosion of Late Jurassic reef corals—implications for reef evolution. Proceedings 8th International Coral Reef Symposium, 2: 1663-1668.

------. 1999. Late Jurassic reef bioerosion—the dawning of a new era. *Geological Society of Denmark Bulletin* 45: 173-176.

Brett, C. E. 1985. *Tremichnus:* A new ichnogenus of circular-parabolic pits in fossil echinoderms. *Journal of Paleontology* 59: 625-635.

Bromley, R. G. 1981. Concepts in ichnotaxonomy illustrated by small round holes in shells. *Acta Geológica Hispànica* 16: 55-64.

------. 2004. A stratigraphy of marine bioerosion. In D. Mcilroy (ed.), *The Application of Ichnology to Palaeoenvironmental and Stratigraphic Analysis: Geological Society of London, Special Publication*, 228: 455-481.

Clarke, J. M. 1908. The beginnings of dependent life. *New York State Museum Bulletin* 121: 146-196.

------. 1921. Organic dependence and disease: their origin and significance. *New York State Museum Bulletin* 221-222: 1-113.

Codez, J., and R. de Saint-Seine. 1958. Revision des Cirripedes Acrothoraciques fossiles. Bulletin. *Société Géologique de France* 7: 699-719.

Cutler, A. H., and K. W. Flessa. 1995. Bioerosion, dissolution and precipitation as taphonomic agents at high and low latitudes. *Senckenbergiana Maritima* 25: 115-121.

Dommergues. J.-L. 1987. L'evolution chez les Ammonitina du Lias moyen (Carixien, Domerien Basal) en Europe occidentale. *Documents des Laboratoires de Géologie de la Faculté des Sciences de Lyon* 98: 1-297.

Donovan, S. K., and J. W. M. Jagt. 2002. *Oichnus* Bromley borings in the irregular echinoid *Hemipneustes agassiz* from the Type Maastrichtian (Upper Cretaceous, The Netherlands and Belgium). *Ichnos* 9: 67-74.

Donovan. S., D. Lewis, and P. Kabrna. 2006. A dense epizoobiontic infestation of a lower Carboniferous crinoid (*Amphoracrinus gilbertsoni* (Phillips)) by *Oichnus paraboloides* Bromley. *Ichnos* 13: 43-45.

Feldman, H. R. 1987. A new species of the Jurassic (Callovian) brachiopod *Septirhynchia* from northern Sinai. *Journal of Paleontology* 61: 1156-1172.

Feldman, H.R., and C. E. Brett. 1998. Epi- and endobiontic organisms on Late Jurassic crinoid columns from the Negev Desert Israel: Implications for Co-evolution: *Lethaia* 31: 57-71.

Feldman, H.R., and E. F. Owen. 1988. *Goliathyris lewyi*, new species (Brachiopoda, Terebratellacea) from the Jurassic of Gebel El-Minshera, northern Sinai. *American Museum Novitates* 2908: 1-12.

Feldman, H.R., E. F. Owen, and F. Hirsch. 1991. Brachiopods from the Jurassic of Gebel El-Maghara, northern Sinai. *American Museum Novitates* 3006: 1-28.

Feldman, H.R., E. F. Owen, and F. Hirsch. 2001. Brachiopods from the Jurassic (Callovian) of Hamakhtesh Hagadol (Kurnub Anticline), southern Israel. *Palaeontology* 44: 637-658.

Fenton, C.L., and M. A. Fenton. 1932. Boring sponges in the Devonian of Iowa. *American Midland Naturalist* 13: 42-54.

Fürsich, F.T., T. J. Palmer, and K. L. Goodyear. 1994. Growth and disintegration of bivalve-dominated patch reefs in the Upper Jurassic of southern England. *Palaeontology* 37: 131-171.

Gatrall, M., and S. Golubic. 1970. Comparative study on some Jurassic and Recent endolithic fungi using scanning electron microscope. In T. P. Crimes and J. C. Harper (eds.), *Trace Fossils: Geological Journal Special Issue* 3: 167-178.

Glaub, I. 1988. Mikrobohrspuren in verschiedenen Faciesbereichen *des* Oberjura Westeuropas (vorläufige Mitteilungen). *Neues Jahrbuch Geologie Paläontologie,* Abhandlungen 177: 135-164.

------. 1994. Mikrobohrspuren in ausgewählten Ablagerungsräumen des europäischen Jura und der Unterkreide (Klassifikation und Palökologie). *Courier Forschungsinstitut Senckenberg* 174: 1-324.

Glaub, I., and M. Bundschuh. 1997. Comparative studies on Silurian and Jurassic/ Lower Cretaceous microborings. *Courier Forschunginstitut Senckenberg* 201: 123-135.

Glaub, I., and H. Schmidt. 1994. Traces or endolithic microboring organisms in Triassic and Jurassic bioherms. *Kaupia. Darmstädter Beiträge zur Naturgeschichte* 4: 103-112.

Goldberg, M. 1963. Reference section of Jurassic sequence in Hamakhtesh Hagadol (Kurnub Anticline). Detailed binocular sample description, including field observations. *Geological Survey of Israel*, Unpublished Report: 1-50.

Golonka, J. 2004. Plate tectonic evolution of the southern margin of Eurasia in the Mesozoic and Cenozoic. *Tectonophysics* 381: 235-273.

Guiraud, R., and W. Bosworth. 1999. Phanerozoic geodynamic evolution of northeastern Africa and the northwestern Arabian Platform. *Tectonophysics* 315: 73-108.

Hallam, A. 1975. Coral patch reefs in the Bajocian (Middle Jurassic) of Lorraine. *Geological Magazine* 112: 383-392.

Haq, B. U., and A. M. Al-Qahtani. 2005. Phanerozoic cycles of sea-level change on the Arabian Platform. *GeoArabia* 10: 127-160.

Hillmer, G., and M.-G. Schulz. 1973. Ableirung der biologie und ökologie eines polychaeten der bohrganges *Ramosulcichnus biforans* (Gripp) nov. ichnogen. *Geologisch-Paläontologisches Institut der Universitat Hamburg* 42: 5-24.

Hirsch, F., and A. Roded. 1996. The Jurassic stratigraphic nomenclature in Hamakhtesh Hagadol, northern Negev. *Geological Survey of Israel, Current Research* 10: 10-14.

Hirsch. F., J.-P. Bassoulett, É Cariou, B. Conway, H. R. Feldman, L. Grossowicz, A. Honigstein, E. F. Owen, and A. Rosenfeld. 1998. The Jurassic of the southern Levant. Biostratigraphy, palaeogeography and cyclic events. In S. Carasquin-Soleau and E. Barrier (eds.), Peri-Tethys Memoir 4: Epicraronic Basins of Peri-Tethyan Platforms. Mémoires du Muséum national d'Histoire Naturelle 179: 213-235.

Kelly, S. R. A., and R. G. Bromley. 1984. Ichnological nomenclature of clavate borings. *Palaeontology* 27: 793-807.

Kobluk, D. R., and I. Noor. 1990. Coral microatolls and a probable Middle Ordovician example. *Journal of Palaeontology* 64: 39-43.

Kolodziej, B. 1997. Boring foraminifera from exotics of the Stramberk-type limestones (Tithonian-lower Berriasian, Polish Carpathians). *Annales, Societatis Geologorum Poloniae* 67: 249-256.

Lathuilière, B. 1982. Bioconstructions bajociennes a madréporaires et faciès associés dans l'ile Crémieu (Jura du Sud: France). *Geobios* 15: 491-504.

Leinfelder, R. R., D. U. Schmid, M. Nose, and W. Werner. 2002. Jurassic reef patterns—the expression of a changing globe. In E. Flugel, W. Kiessling, and J. Golonka (eds.), *Phanerozoic Reef Patterns: SEPM Special Publication* 72: 465-520.

Leinfelder, R. R., F. Schlagintweit, W. Werner, O. Ebli, M. Nose, D. U. Schmid, and G. W. Hughes. 2005. Significance of stromatoporoids in Jurassic reefs and carbonate platforms—concepts and implications. *Facies* 51: 288-326.

Mayoral, E., and L. Sequeiros. 1979. Significado paleoecologico de algunos epizoos y "borers" del Jurásico Inferior y media de Belchite (Zaragoza, Cordillera Ibérica). *Caudemos de Geología* 10: 121-135.

Nielsen, J. K., J. Maiboe. 2000. Epofix and vacuum: An easy method to make casts of hard substrates. *Palaeontologia Electronica* 3(1): article 2, 10 pp., 1.3 MB.

Nielsen, K. S. S., and J. K. Nielsen. 2001. Bioerosion in Pliocene to late Holocene tests of benthic and planktonic foraminiferans, with a revision of the ichnogenera *Oichnus* and *Tremichnus. Ichnos* 8: 99-116.

Palmer, T.J., and F. T. Fürsich. 1981. Ecology of sponge reefs from the Middle Jurassic of Normandy. *Palaeontology* 24: 1-23.

Parras, A., and S. Casadio, 2006. The oyster *Crassostrea? hatcheri* (Ortmann, 1897), a physical ecosystem engineer from the Upper Oligocene-Lower Miocene of Patagonia, southern Argentina. *Palaios* 21: 168-186.

Perry, C.T., and M. Bertling. 2000. Spatial and temporal patterns of macroboring within Mesozoic and Cenozoic coral reef systems. In E. Insalaco, P. W. Skelton, and T. J. Palmer (eds.), *Carbonate Platform Systems: Components and Interactions: Geological Society, London, Special Publications,* 178: 33-50.

Pickerill, R. K., and S. K. Donovan. 1998. Ichnology of the Pliocene Bowden shell bed, southeast Jamaica. In S. K. Donovan (ed.), *The Pliocene Bowden shell bed, southeast Jamaica:* Contributions to Tertiary and Quaternary Geology, 35: 161-175.

Pisera, A. 1987. Boring and nestling organisms from Upper Jurassic coral colonies from northern Poland. *Palaeontologica* 32: 83-104.

Plewes, C. R. 1996. Ichnotaxonomic studies of Jurassic endoliths. Unpublished Ph.D. dissertation, University of Wales, Aberystwyth, 313 p.

Prinz, P. 1986. Mitteljurassische Korallen aus Nordchile. *Neues Jahrbuch für Geologie und Paläontologie,* Monatsshefte 1986: 736-750.

Schlagintweit, F. 2008. Bioerosional structures and pseudoborings from Late Jurassic and Late Cretaceous-Paleocene shallow-water carbonates

(Northern Calcareous Alps, Austria and SE France) with special reference to cryptobiotic foraminifera. *Facies* 54: 377-402.

Sharland, P. R., D. M. Casey, R. B. Davies, M. D. Simmons, and O. E. Sutcliffe. 2004. Arabian Plate sequence stratigraphy. *GeoArabia* 9: 199-214.

Tapanila, L. 2005. Palaeoecology and diversity of endosymbionts in Palaeozoic marine invertebrates: trace fossil evidence. *Lethaia* 38: 89-99.

------. 2006. Devonian *Entobia* borings from Nevada, with a revision of *Topsentopsis*. *Journal of Paleontology* 80: 760-767.

Tapanila, L., P. Copper, and E. Edinger. 2004. Environmental and substrate control on Paleozoic bioerosion in corals and stromatoporoids, Anticosti Island, eastern Canada. *Palaios* 19: 292-306.

Taylor, P. D., and T. J. Palmer. 1994. Submarine caves in a Jurassic reef (La Rochelle, France) and the evolution of cave biotas. *Naturwissenschaften* 81: 357-360.

Taylor, P. D., and M. A. Wilson. 2003. Palaeoecology and evolution of marine hard substrate communities. *Earth Science Reviews* 62: 1-103.

Vogel, K., and I. Glaub. 2004. 450 Millionen Jahre Beständigkeit in der Evolution Endolithischer Mikroorganismen. *Sitzungsberichte der Wissenschaftlichen Gesellschaft an der Johann Wolfgang Goethe-Universitat Frankfurt am Main* 42: 1-42.

Vogel, K., S.-J. Balog, M. Bundschuh, M. Gektidis, I. Glaub, J. Krutschinna, and G. Radtke. 1999. Bathymetrical studies in fossil reefs, with microendoliths as paleoecological indicators. *Profil* 16: 181-191.

von Hagenow, K. F. 1840. Monographie der Rügenschen Kreideversteiner ungen II Abth. Radiarien und Annulaten. *Neues Jahrbuch Mineralogia, Geognosie, Geologie. Petrefaktenkunde* 1840: 631-672.

Westermann, G. E. D. 2000. Biochore classification and nomenclature in paleobiogeography: An attempt at order. *Palaeogeography, Palaeoclimatology, Palaeoecology* 158: 1-13.

Wilson, M. A. 1998. Succession in a Jurassic marine cavity community and the evolution of cryptic marine faunas. *Geology* 26: 379-381.

------. 2007. Macroborings and the evolution of bioerosion. In W. Miller III (ed.), *Trace Fossils: Concepts, Problems, Prospects*, 356-367. Amsterdam: Elsevier.

Wilson, M. A., and T. J. Palmer. 1989. Preparation of acetate peels. In R. M. Feldmann, R. E. Chapman, and J. T. Hannibal, (eds.), *Paleotechniques: The Paleontological Society Special Publication*, 4: 142-145.

Wilson, M. A., and T. J. Palmer. 1992. Hardgrounds and hardground faunas. *University of Wales, Aberystwyth, Institute of Earth Studies Publications* 9: 1-131.

Wilson, M. A., and T. J. Palmer. 1994. A carbonate hardground in the Carmel Formation (Middle Jurassic, SW Utah, USA) and its associated encrusters, borers and nestlers. *Ichnos* 3: 79-87.

Wilson, M. A., C. R. Ozanne, and T. J. Palmer. 1998. Origin and paleoecology of free-rolling oyster accumulations (ostreoliths) in the Middle Jurassic of southwestern Utah, USA. *Palaios* 13: 70-78.

Wilson, M. A., K. R. Wolfe, and Y. Avni. 2005. Development of a Jurassic rocky shore complex (Zohar Formation, Makhtesh Qatan, southern Israel). *Israel Journal of Earth Sciences* 54: 171-178.

Wilson, M. A., H. R. Feldman, J. C. Bowen, and Y. Avni. 2008. A new equatorial, very shallow marine sclerozoan fauna from the Middle Jurassic (late Callovian) of southern Israel. *Palaeogeography, Palaeoclimatology, Palaeoecology* 263: 24-29.

Wilson, M. A., E. B. Krivicich, Y. Avni, and M. Goldberg. 2009. Addition to the top of the stratigraphic column of the Be'er Sheva Formation (Jurassic, Callovian-Oxfordian) in Hamakhtesh Hagadol, Israel. *Israel Journal of Earth Sciences* 58: 81-85.

Wood, R. 1987. Biology and revised systematics of some Late Mesozoic stromatoporoids. *Special Papers in Palaeontology* 37: 1-89.

Wood, R. 1999. *Reef Evolution*. Oxford: Oxford University Press, 414 pp.

Wullschleger, E. 1966. Bemerkungen zum fossilen Korallenriff Gisliflue-Homberg. *Mitteilungen der Aargauischen Naturforschenden Gesellschaft* 27: 101-152.

Wullschleger, E. 1971. Bemerkungen zum fossilen Korallenvorkommen Tiersteinberg-Limperg-Kei. *Mitteilungen der Aargauischen Naturforschenden Gesellschaft* 28: 251-292.

CHAPTER TWELWE

JURASSIC RHYNCHONELLIDE BRACHIOPODS FROM THE JORDAN VALLEY

ABSTRACT

Jurassic rhynchonellide brachiopods from the Jordan Valley are herein revised and new taxa are added to the faunal list. In this study of Jurassic rhynchonellides from Wadi Zarqa, northwestern Jordan, we recognize the following taxa: *Eurysites rotundus*, *Cymatorhynchia quadriplicata*, *Daghanirhynchia triangulata*, *D. angulocostata*, *Pycnoria magna*, *Schizoria elongata*, and *Schizoria* cf. *intermedia*. The following new taxa are described: *Daghanirhynchia susanae* sp. nov. and *Amydroptychus markowitzi* sp. nov. The Middle Jurassic Mughanniyya Formation of northwest Jordan is dominated by limestone beds. The sedimentary environment is interpreted as neritic, light, and nutrient-rich, resulting in high faunal diversity. The high rhynchonellide endemism of this fauna is yet another confirmation of pronounced Middle Jurassic endemism along the southern Tethyan margin of the Ethiopian Province. Brachiopods of the Jordanian Mughanniyya Formation can be correlated with the fauna of the Aroussiah Formation in Sinai and the Zohar and Matmor formations in Southern Israel.

INTRODUCTION

During the Jurassic Period the Levant was a part of the Gondwanian Tethys platform shelf that extended from Morocco in the west to the Arabian Peninsula in the east and southward to the Horn of Africa. The collecting area for this study (Fig. 1) is a part of the elevated platform terrain of the Arabian Nubian Shield, which is covered by intermittent Palaeozoic to Cenozoic sedimentary successions consisting mainly of clastic units with marine carbonates increasing upward (Rybakov and Segev 2005). The Late Triassic in the Levant consisted of an area that was covered by broad sea-marginal flats, analogous to modern sabkhas, in which evaporites and dolomite rocks were deposited, whereas at the beginning of the Jurassic, emergence led to subaerial exposure that was accompanied by extensive freshwater runoff and subaerial weathering (Goldberg and Friedman 1974). Subsequent subsidence allowed the formation of shallow and marginal shelf environments interrupted by lagoons resulting in the deposition of thick, partly calcareous sandstone beds that were overlain by thick, partly gypsiferous carbonate, marl, and sandy marl rocks (Basha 1980).

Cox (1925) and Muir-Wood's (1925) descriptions of brachiopods and molluscs provided the first reliable data on the age of the Jurassic formations of Jordan. Their faunas were collected in Wadi Zarqa by Bryce Kerr Nairn Wyllie, Keir Arthur Campbell, and George Martin Lees on behalf of the Turkish Petroleum Company, later known as the Anglo-Persian Oil Company (Picard and Hirsch 1987).

This paper is one of a series that was begun with an analysis of the megafossils of northern Sinai (Feldman et al. 1982, 1991; Feldman 1986, 1987; Feldman and Owen 1988, 1993) and Israel (Feldman et al. 2001) in order to systematically study the brachiopod faunas. These data will help establish the history of brachiopod species and their evolution within the Jurassic Ethiopian Province, that is the part of the Tethyan Realm that laid along the southern Tethyan margin. The taxonomic information recorded as a result of these studies will elucidate the biogeographic history of the Ethiopian Province and help to interpret the structure and palaeoecology of its marine communities. For example, in the Jurassic sequence exposed at Hamakhtesh Hagadol, southern Israel, Feldman and Brett (1998) recognised sclerobionts (epi- and endobiontic) organisms on crinoids. Wilson et al. (2008, 2010) further developed the invertebrate community palaeoecological framework.

Previous studies of Ethiopian Province brachiopods from northern Sinai (Feldman et al. 1982, 1991a; Feldman 1987; Feldman and Owen 1988; Hegab 1988, 1989, 1991, 1992, 1993), along with Cooper's (1989) monograph and Almeras' (1987) work, complement the present paper on the brachiopods of the Jordan Valley. The Ethiopian Province has been recognised from the Early Jurassic until the mid- and possibly end of the Cretaceous by the presence of endemic taxa at the species, genus, and family level. These endemics (e.g., *Somalirhynchia africana, Daghanirhynchia daghaniensis, Somalithyris bihendulensis, Striithyris somaliensis, Bihenithyris barringtoni, B. weiri*) have been noted, especially in the brachiopods, by Weir (1925) and Muir-Wood (1935). Arkell (1952, 1956) recognised endemic faunas in the ammonoids and Kitchin (1912) in the trigoniacean and crassatellacean bivalves.

The Jurassic brachiopods of Saudi Arabia studied by Cooper (1989) were collected during the years 1933-1953 by field geologists of the Arabian-American Oil Company (Aramco). Cooper (1989) also used the Porter M. Kier-Earl G. Kauffman collections (1962) in his analyses (Gus Arthur Cooper, personal communication 1977). He noted that, whereas several rock types are recorded, some of the specimens cannot be assigned to specific parts of the column, but most have been referred to various ammonite zones. Consequently, it is not possible to determine the palaeoecological conditions under which they existed during the Jurassic. In Jordan, however, collecting was accomplished under strict stratigraphical control and an eventual palaeoecological study of the brachiopods and their communities and a comparison between the Negev communities can be accomplished.

GEOGRAPHICAL AND GEOLOGICAL SETTING

In northwestern Jordan Jurassic outcrops can be found (Fig. 1) along the western part of Wadi Zarqa beginning near the old Jerash Bridge and extending westward to Deir-Alla, a distance of about 20 km; toward the south the outcrop belt passes through Ain-Khuneizir, Subeihi, and Arda Road (Ahmad 2002). Two outcrops of dolomite and limestone can be found in the Baqa depression near the intersection of Wadi Mahis and Wadi Shueib. The Jurassic succession decreases in thickness from the Zarqa River and Wadi Huni eastward toward the Zarqa Bridge and from there southeast toward Suweileh-1 and Safra-1.

The Jurassic rocks of Jordan have been subdivided into seven formations: Hihi, Nimr, Silal, Dhahab, Ramla, Hamam, and Mughanniyya (Khalil and Muneizel 1992). The Mughanniyya Formation belongs to the Callovian, the Hamam and Ramla formations belong to the Bathonian and the Dhahab and

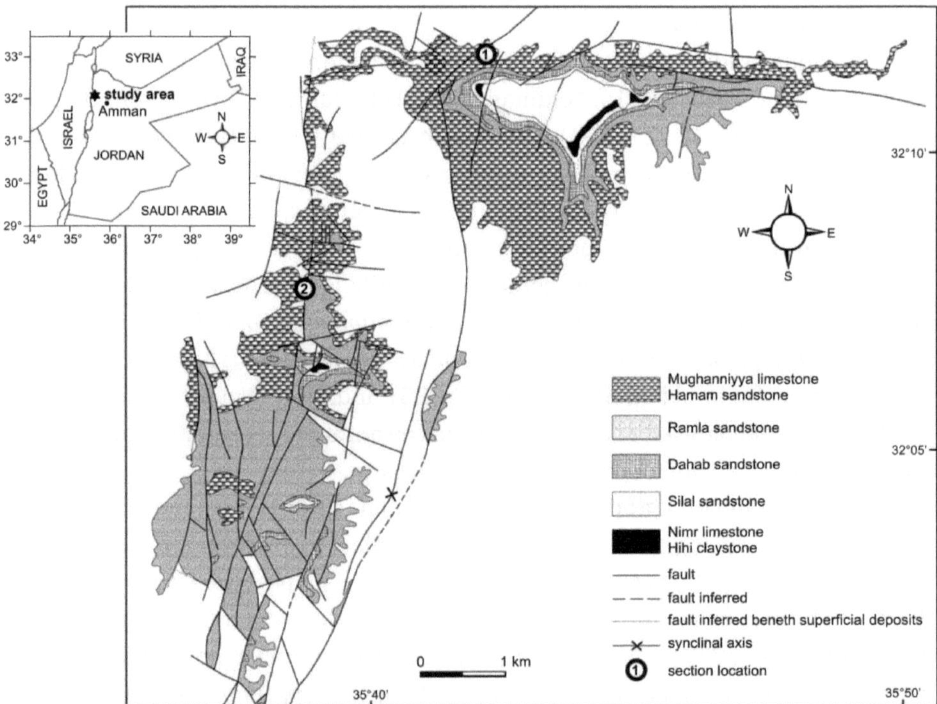

FIGURE 1: Geographical map of the study area in northwestern Jordan. The collecting localities are indicated by circles: Locality 1, Tel el Dhahab section; Locality 2, Arda section. Modified from Muneizel and Khalil (1993) and Swarieh and Barjous (1993).

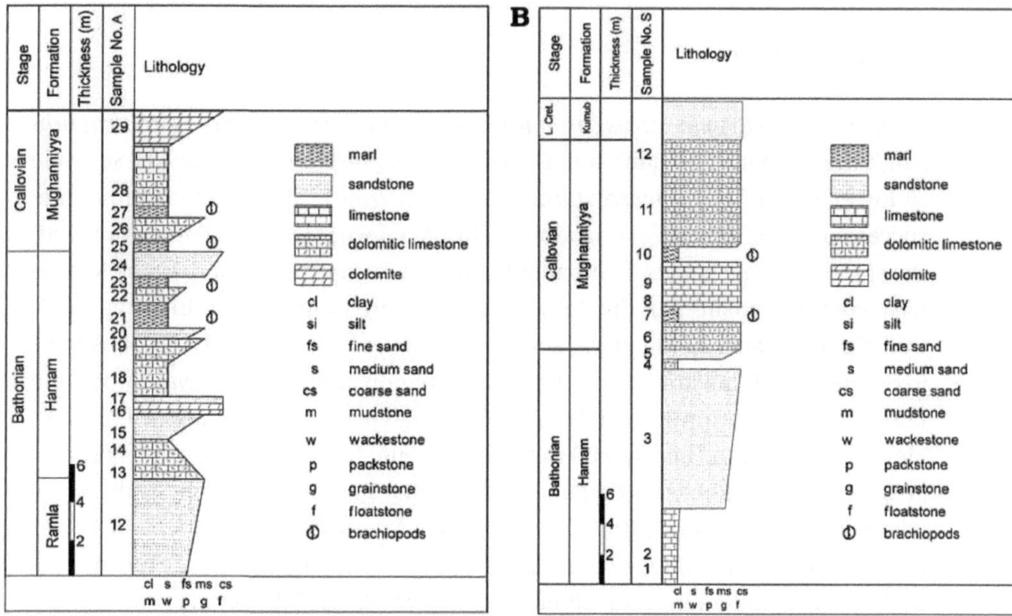

FIGURE 2: Stratigraphical columns of the sections sampled showing lithology and brachiopod yielding beds. A. Arda section. B. Tel el Dhahab section. Abbreviation: L. Cret., Lower Cretaceous.

Silal formations belong to the Bajocian Stage. Blake (1935, 1940) was the first to construct a geological cross section midway between the "Zerqamouth" and the Jerash Crossing; however, detailed stratigraphic sections were only provided in 1959 by Wetzel and Morton (Picard and Hirsch 1987). The Mughanniyya Formation (Fig. 2) represents the upper part of the Jurassic sequence in Jordan and outcrops consistently below the Kurnub Sandstone Group. The formation is composed of alternating claystone, siltstone, and marly limestone beds with minor dolomite, dolomitic limestone, and coquinas.

AMMONITE ZONATION

The paucity of ammonites in the Jurassic sediments of Jordan makes the formations difficult to date. Three species of ammonites have been described from northwestern Jordan: *Ermoceras* cf. *coronatoides* Douvillé, 1916, *Teloceras labrum* Buckman, 1922, and *Normannites* gr. *braikenridgii* (Buckman, 1927). These taxa, collected from the Dhahab Formation of early middle to late Bajocian age, were described by Basha and Aqrabawi (1994). The closest Jurassic rocks to those in Jordan occur in southern Israel.

The Callovian ammonite biozonation of the sequence at Hamakhtesh Hagadol, southern Israel, was refined by Gill and Tintant (1975), Lewy (1983), Gill et al. (1985), and more recently by Cariou et al. (1997). The same authors and Hirsch et al. (1998) considered the section at Hamakhtesh Hagadol to be a nearly continuous middle to upper Callovian sequence, unique in the Levant, comprising the *Erymnoceras coronatum* Zone (middle Callovian), *Peltoceras athleta* Zone (upper Callovian), and *Orionoides* (*Poculisphinctes*) *poculum* Subzone (= lower part of the *Quenstedtoceras lamberti* Zone). The Callovian ammonite succession in Hamakhtesh Hagadol (Arabic Province) facilitates correlation with the standard scale stages of Submediterranean Europe. Elsewhere in the Middle East, coeval exposures of the Zohar Formation are found at Abu El Darag (Suez Gulf, Egypt), Gebel El-Maghara, and Gebel El-Minshera (northern Sinai, Egypt), in the Hermon Limestone Formation (Antilebanon), and in the Tuwaiq Mountain Limestone Formation (Saudi Arabia). Coeval sequences of the upper Callovian Matmor Formation (nearly 100 m thick) are only found in the Negev boreholes at Qeren, Sherif, Rehkme, Boqer, and the Hallal borehole in Sinai, Egypt, all of which are part of the platform of the Higher Negev (Picard and Hirsch 1987). In the Gebel El-Maghara section (Sinai Deep) and in the Hermon exposures of the Judean Embayment (Hirsch et al. 1998), the later Callovian is reduced to a few dm of section covered by Oxfordian shales (Cariou et al. 1997). Goldberg (1963) subdivided the 206-m-thick section at Hamakhtesh Hagadol into 69 subunits (numbered 5-74). The sequence consists of the Ziyya and Madsus members of the Zohar Formation (Coates et al. 1963) and of the Matmor Formation (Goldberg 1963, emended Hirsch and Roded 1997).

MATERIAL AND METHODS

All specimens are preserved as articulated shell material. The specimens are from the Hamam and Mughanniyya formations (Bathonian-Callovian) in northwest Jordan. Serial sections were prepared with a WOKO 50P grinding machine with slice-spacing of 100 to 200 µm. Drawings were made with the help of a camera lucida. Measurements were taken with digital caliper and rounded to 0.1 mm. Specimens were coated with ammonium chloride prior to photographing. All measurements are in millimetres (mm); the width of the dorsal fold is measured at the anterior commissure. Est. indicates the specimen was incomplete and the measurement was estimated. The systematics follows the revised *Treatise on Invertebrate Palaeontology* (Manceñido et al. 2002).

ABBREVIATIONS

AMNH = Division of Palaeontology, American Museum of Natural History, New York, NY, USA

GSI = Geological Survey of Israel, Jerusalem, Israel

USNM = United States National Museum, Smithsonian Institution, Washington, D.C., USA

(L), length; (T), thickness; (W), width

SYSTEMATIC PALEONTOLOGY

Phylum Brachiopoda Dumeril, 1806
Subphylum Rhynchonelliformea Williams, Carlson, Brunton, Holmer, and Popov, 1996
Class Rhynchonellata Williams, Carlson, Brunton, Holmer, and Popov, 1996
Order Rhynchonellida Kuhn, 1949
Superfamily Hemithiridoidea Rzhonsnitskaya, 1956
Family Cyclothyrididae Makridin, 1955
Subfamily Cardinirhynchiinae Makridin, 1964
Genus *Eurysites* Cooper, 1989

Type species: *Eurysites transversus* Cooper, 1989; Bajocian/Callovian, Saudi Arabia.

Eurysites rotundus Cooper, 1989
Fig. 3A.

1989 *Eurysites rotundus* sp. nov.; Cooper 1989: 32-33, pl. 8: 27-49, pl. 17: 1-11, fig. 17.

Material.—Three articulated specimens: AMNH FI-72366, AMNH FI-72367, AMNH FI-72368, bed S-7, Mughanniyya Formation, Tel el Dhahab section, Jordan.

Description.—Shell medium size (Table 1), somewhat worn, outline roundly triangular and almost pyriform but slightly rounded at margins, length almost equal to width; strongly dorsibiconvex; anterior commissure uniplicate, lateral commissure straight but rises slightly posteriorly. Ventral beak straight, long, and pointed; costae high, subangular, numbering about 14.

Ventral valve broadly convex in anterior view, less convex in lateral view. Ventral interarea appears high but partially obscured as is the pedicle foramen. Umbo convex, swollen, becoming flatter anteriorly; sulcus low bearing 4 costae, flanks slightly curved with 10 costae.

Dorsal valve with weakly defined fold rising slightly above flanks begins just past midlength becoming more distinct near anterior commissure. The fold bears 5 costae, with 8 costae on each dorsal flank; flanks rounded when viewed anteriorly. In lateral view convexity of dorsal valve is at maximum at midvalve; sides steep.

Due to a paucity of material we did not section any specimens. However, Cooper (1989) illustrates transverse serial sections which show relatively thick dental plates that disappear before midvalve. His dorsal valves have a very short median septum or ridge with thick outer hinge plates; the crura are raduliform with well-developed bases.

FIGURE 3: Jurassic rhynchonellide brachiopods from Jordan. A. *Eurysites rotundus* Cooper, 1989. Articulated specimen (AMNH FI-72366), bed S-7, Mughanniyya Formation, Tel el Dhahab, Jordan, in ventral (A1), dorsal (A2), posterior (A3), anterior (A4), and lateral (A5) views. B. *Cymatorhynchia quadriplicata* (Zieten, 1830). Articulated specimen (GSI M 2975), bed 21a, Hamam Formation, Arda, Jordan, in ventral (B1), dorsal (B2), posterior (B3), anterior (B4), and lateral (B5) views. C. *Daghanirhynchia susanae* sp. nov. Articulated specimen, holotype (AMNH FI-72380), bed S-7, Mughanniyya Formation, Arda section, Jordan, in ventral (C1), dorsal (C2), posterior (C3), anterior (C4), and lateral (C5) views. D. *Daghanirhynchia angulocostata* Cooper, 1989. Articulated specimen (AMNH FI-72395), bed S-7, Mughanniyya Formation, Tel el Dhahab, Jordan, in ventral (D1),

TABLE 1: Measurements (in mm) of *Eurysites rotundus* Cooper, 1989.

Specimen	Length	Dorsal valve length	Width	Thickness	Fold Width
AMNH FI-72366	21.4	19.3	21.6	15.7	10.4
AMNH FI-72367	20.2	18.1	20.6	13.0	10.3
AMNH FI-72368	14.5	12.3	15.1	10.1	6.4

Discussion.—*Eurysites rotundus* differs from *E. transversus* (see Cooper 1989: fig. 18; pl. 8: 50-59; pl. 9: 1-16; pl. 17: 12) in its triangular shape and more swollen dorsal valves that are distinctly more convex in lateral view and less elliptical outline. On our specimen of *E. rotundus* there are no fila visible possibly due to the somewhat weathered shell surface. Cooper (1989) noted that on his specimens of *E. rotundus* from Saudi Arabia (housed in the USNM) there were fine concentric fila in the interspaces between the costae. *E. rotundus* is similar to *Flabellirhynchia* Buckman, 1918 from the Bajocian of Great Britain in its elongated beak and outline. It resembles *Flabellirhynchia delicata* (Buckman 1918: pl. 19: a-c) in general outline but differs in its lower fold, thicker and stubbier anterior commissure and longer beak.

Remarks.—The genera *Cymatorhynchia* and *Flabellirhynchia* were introduced by Buckman (1918) in a paper that has been dated originally as 1917, however, the footnote to Corrigenda sheet from the Geological Survey of India states: "Print off orders regarding this memoir were given to the press in September, 1917, but owing to shortage of paper, copies were not actually issued till July, 1918"; Shi and Grant (1993: 144). As not all following authors noticed the actual publication date *Cymatorhynchia* and *Flabellirhynchia* are cited in some papers as 1918 and elsewhere as 1917.

Stratigraphic and geographic range.—Bajocian to Callovian; Saudi Arabia, Jordan.

---

dorsal (D2), posterior (D3), anterior (D4), and lateral (D5) views. E. *Daghanirhynchia triangulata* Cooper, 1989. Articulated specimen (AMNH FI-72398), bed 21a, Hamam Formation, Arda, Jordan, in ventral (E1), dorsal (E2), posterior (E3), anterior (E4), and lateral (E5) views. F. *Pycnoria magna* Cooper, 1989. Articulated specimen (AMNH FI-72399), bed 21a, Hamam Formation, Arda, Jordan, in ventral (F1), dorsal (F2), posterior (F3), anterior (F4), and lateral (F5) views. G. *Amydroptychus markowitzi* sp. nov. AMNH FI-72405, holotype, bed S-7, Mugannhiyya Formation, Arda section, Jordan, in ventral (G1), dorsal (G2), posterior (G3), anterior (G4), and lateral (G5) views. H. *Schizoria elongata* Cooper, 1989. Articulated specimen (AMNH FI-72408), bed S-7, Mughanniyya Formation, Tel el Dhahab section, Jordan, in ventral (H1), dorsal (H2), posterior (H3), anterior (H4), and lateral (H5) views. I. *Schizoria* cf. *intermedia* Cooper, 1989. Articulated specimen (AMNH FI-72409), bed 21b, Hamam Formation, Arda section, Jordan, in ventral (I1), dorsal (I2), posterior (I3), anterior (I4), and lateral (I5) views.

Family Tetrarhynchiidae Ager, 1965
Subfamily Tetrarhynchiinae Ager, 1965
Genus *Cymatorhynchia* Buckman, 1918

Type species: *Rhynchonella cymatophorina* Buckman, 1910; Aalenian to upper Bajocian of Europe, Africa, Saudi Arabia, India, Burma, China, and Tibet.

Cymatorhynchia quadriplicata (Zieten, 1830)
Figs. 3B, 4.

1830 *Terebratula quadriplicata*; Zieten 1830: 55, pl. 41: 3.
1851 *Terebratula quadriplicata*; Quenstedt 1851: 453, pl. 36: 16; 1858: 423, pl. 58: 5-8; 1868-1871, pl. 38: 42.
1851-1852 *Rhynchonella quadriplicata* (Zieten); Davidson 1851-1852: appendix: 23, pl. A: 22; 201, pl. 29: 1-3.
1918 *Cymatorhynchia quadriplicata* (Zieten); Buckman 1918: 53-54.
1918 *Rhynchonella quadriplicata* (Zieten); Rollier 1918: 148.
1936 *Rhynchonella quadriplicata* (Zieten); Arcelin and Roche 1936: 59, pl. 2: 1-6, pl. 11: 1, 4, 5, pl. 12: 5, pl. 13: 3, 4.
1939 *Rhynchonella quadriplicata* (Zieten); Roche 1939: 268, pl. 4: 11.
1965 *Cymatorhynchia quadriplicata* (Zieten in Davidson); Rousselle 1965: 52, pl. 2: 11, 12.
1966 *Cymatorhynchia quadriplicata* (Zieten); Almeras 1966: 70-97, 73, pl. F-G, pl. 2: 5-8.
1966 *Cymatorhynchia quadriplicata* var. *vergissonensis*; Almeras 1966: 79, pl. D-E, pl. 2: 1-4.
1966 *Cymatorhynchia quadriplicata* var. *lata*; Almeras 1966: 88, pl. I, pl. 4: 6-8.
1966 *Cymatorhynchia quadriplicata* var. *depressa*; Almeras 1966: 95.
1966 *Cymatorhynchia quadriplicata* var. *trilobata*; Almeras 1966: 96.
1969 *Cymatorhynchia quadriplicata* (Zieten); Pevny 1969: 143, pl. 28: 1a-c.
1980 *Cymatorhynchia quadriplicata* (Zieten); Fischer 1980: 140, pl. 58: 1, 2.
1981 *Cymatorhynchia quadriplicata* (Zieten); Parnes 1981: 21, pl. 2: 5-13.
1987 *Cymatorhynchia quadriplicata* (Zieten); Shi 1987: 55, pl. 3: 12.
1993 *Cymatorhynchia quadriplicata* (Zieten, 1830); Shi and Grant 1993: 72-73, pl. 7: 17-19, pl. 8: 8-13, pl. 9: 1-7, 11-17, pl. 12: 1, 2, 6, 7.
1996 *Cymatorhynchia quadriplicata* (Zieten, 1830); Almeras and Elmi 1996: pl. 1: 21.

Material.—42 articulated specimens: AMNH FI-72369 (sectioned), AMNH FI-72370, bed S-7, AMNH FI-72371, AMNH FI-72372, AMNH FI-72373, AMNH FI-72374, AMNH FI-72375, bed S-10, Mughanniyya Formation, Tel el Dhahab section; AMNH FI-72376, bed 21a, AMNH FI-72377, bed 21b, AMNH FI-72378-79, bed 25, Hamam Formation, Arda section GSI M 2975, GSI M 742, GSI M 4510, GSI M 4522, GSI M 4539, GSI M 4553, GSI M 4556, GSI M 4565 (collected from the Tel el Dhahab section but beds not identified), Jordan.

Description.—Shells medium to large (Table 2), subpentagonal to subquadrate in outline, moderately dorsibiconvex, maximum width near midlength. Anterior commissure widely uniplicate. Ventral beak short, suberect to erect, deltidial plates disjunct to conjunct; costae subangular to sharp numbering about 16-17; growth lines barely visible near anterior commissure.

Ventral valve moderately to strongly convex posteriorly but less convex anteriorly; a broad, shallow sulcus originates just past one-third of valve length bearing 3-4 strong subangular costae, 5-6 costae on ventral flank; foramen round, small to medium sized, ventral interarea low.

Dorsal valve more globose posteriorly, flattened toward anterior commissure; moderately convex in lateral view; a broad, low fold originates just posterior to midlength bearing 4 costae with 6 costae on dorsal flank.

Interior. Very short, low dorsal median septum, strong dental plates parallel to subparallel anteriorly in one sectioned specimen, teeth short, somewhat bulbous; septalium not preserved, outer hinge plates subhorizontal arched slightly ventrally, sockets deep, asymmetrical, crura raduliform.

TABLE 2: Measurements (in mm) of *Cymatorhynchia quadriplicata* (Zieten, 1830).

Specimen	Length	Dorsal valve length	Width	Thickness	Fold Width
AMNH FI-72372	27.3	21.6	25.9	15.3	12.5
AMNH FI-72373	20.0	19.0	23.5	14.7	9.0
AMNH FI-72374	26.6	21.6	29.0	20.8	13.3
AMNH FI-72375	19.1	17.5	21.6	11.0	8.4
GSI M 742	23.6	21.5	24.3	19.9	10.9
GSI M 2975	27.6	23.0	29.5	21.9	14.6
GSI M 4510	26.4	23.8	30.0	20.9	12.1
GSI M 4539	23.3	21.2	24.3	15.5	10.9
GSI M 4553	24.1	21.9	26.1	15.3	12.6
GSI M 4556	21.0	19.1	21.5	14.4	9.1
GSI M 4565	15.7	14.3	15.7	10.4	7.6

Discussion.—The shells differ from *Cymatorhynchia? singularis* (USNM 303631) of Cooper (1989) in possessing less costae, a more inflated dorsal valve and less tumid ventral posterior. Buckman's (1918: pl. 17: 15a, c) *C. cymatophorina* is wider and has finer and more numerous costae (about 26 on the dorsal valve). *Cymatorhynchia quadriplicata* var. *inflata* Alméras (USNM 319165) has a more subcircular outline, more inflated dorsal valve, finer and denser ribs as well as a less developed fold and sulcus as noted by Shi and Grant (1993). They also noted that the Tethyan form varies considerably in morphology and was split into seven variants by Alméras (1966). Five of the variants (except for *C. quadriplicata* var. *densecosta* and *C. quadriplicata* var. *inflata*) have the same basic characteristics namely, less tumid dorsal than ventral valve, slightly depressed shell, subpentagonal outline, and few but strong ribs. Shi and Grant therefore regard them as a single variable species: *C. quadriplicata* (Zieten, 1830). They suggest combining the varieties *C. quadriplicata denecosta* and *C. quadriplicata* var. *inflata* into a single species: *Cymatorhynchia denecosta* Alméras, 1966. Alméras and Elmi (1996) describe a specimen of *C. quadriplicata* var. *inflata* from Toulon, France, that is larger (USNM 541691: L, 30.3; W, 29.4 [est.]; T, 24.8), much more gibbous and has finer and more numerous costae (about 26). *C. quadriplicata* var. *inflata* from Württemberg, Germany, has similar dimensions (USNM 541692: L, 30.1; W, 33.0; T, 24.4) but the costae are less numerous (20) and sturdier.

Stratigraphic and geographic range.—Bathonian to Callovian; England, France, Jordan, Ethiopia.

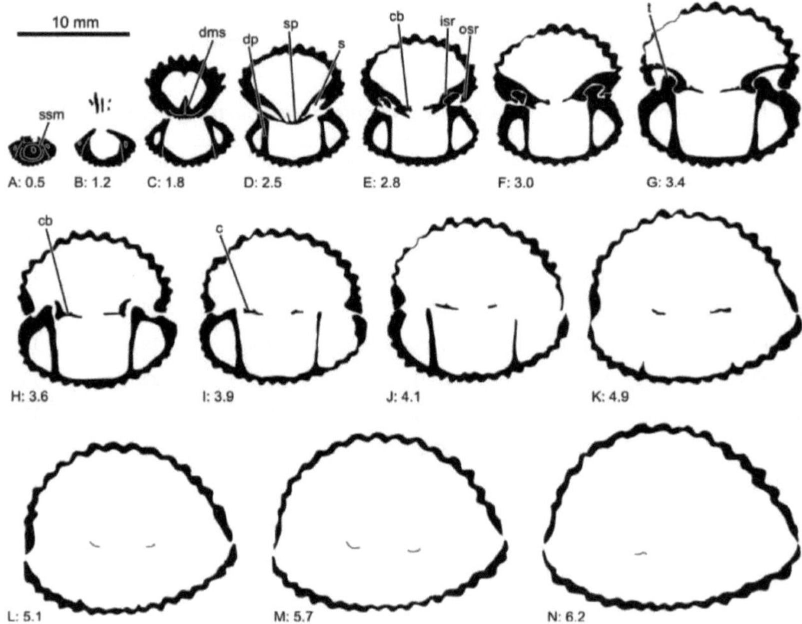

FIGURE 4: Serial sections of the rhynchonellide brachiopod *Cymatorhynchia quadriplicata* (Zieten, 1830). AMNH FI-72369. Bed 21a, Hamam Formation, Arda, Jordan. Numbers describe sectional distance from apex. Abbreviations: c, crus; cb, crural base; dms, dorsal median septum; dp, dental plate; isr, inner socket ridge; osr, outer socket ridge; s, socket; sp, septalial plate; ssm, secondary shell material; t, tooth.

Genus *Daghanirhynchia* Muir-Wood, 1935

Type species: *Daghanirhynchia daghaniensis* Muir-Wood, 1935; Callovian, Somalia.

Daghanirhynchia susanae sp. nov.
Figs. 3C, 5.

Etymology: Latinization of Susan after Susan Feldman.

Holotype: Articulated specimen AMNH FI-72380.

Type locality: Wadi Zarqa, Arda Section, Jordan.

Type horizon: Bed S-7, Mughanniyya Formation, Callovian (upper Middle Jurassic).

Material.—45 articulated shells: AMNH FI-72380, AMNH FI-72381, AMNH FI-72386, AMNH FI-72388, AMNH FI-72393, AMNH FI-723410 (sectioned) bed S-7, AMNH FI-72394, bed S-10, AMNH FI-72383, AMNH FI-72387, AMNH FI-72391, AMNH FI-72392, Mughanniyya Formation, Arda section; AMNH FI-72382, AMNH FI-72384, bed 21a, AMNH FI-72389, bed 21b, AMNH FI-72385, Hamam Formation, Arda section; GSI M 731, GSI M 732, GSI M 741, GSI M 743, GSI M 4506, GSI M 4552, GSI M 4558 (collected from the Tel el Dhahab section but beds not identified), Jordan.

Diagnosis.—Coarsely plicated *Daghanirhynchia* with long, narrow fold bearing 3 costae.

Description.—The shells are medium size (Table 3), subpentagonal in outline, dorsibiconvex, maximum width at about midlength; strongly uniplicate anterior commissure, lateral commissure straight, beak suberect to erect, costae strong and coarse and narrow, numbering 12-16 but commonly 12-13, linguiform extension moderate. Anterior margin ends abruptly, stubby.

Ventral valve gently convex to almost flat in lateral view, concave in anterior profile. Sulcus originates just anterior to umbonal region, occupied by 2–3 costae. Foramen small, deltidial plates disjunct.

Dorsal valve moderately convex in lateral profile, fold originates just anterior to umbonal region, very narrow, highly elevated above flanks, occupied by 3 costae.

Dental plates parallel in posterior region of shell but diverge anteriorly from ventral floor. Teeth short, bulbous without crenulations; dorsal median septum short, reduced to low ridge, and finally disappearing at about one-third valve length. Hinge plates horizontal, very broad, and separated; septalium not preserved; crura raduliform. Secondary thickening in umbonal chambers.

TABLE 3: Measurements (in mm) of *Daghanirhynchia susanae* sp. nov.

Specimen	Length	Dorsal valve length	Width	Thickness	Fold width
Holotype AMNH FI-72380	19.0	17.9	19.5	15.4	6.9
AMNH FI-72381	21.5	19.6	22.8	16.9	6.5
AMNH FI-72382	22.0	19.7	22.7	16.5	6.5
AMNH FI-72383	21.9	18.8	22.4	12.1	7.9
AMNH FI-72384	20.1	17.3	19.9	15.9	6.7
AMNH FI-72385	21.0	20.0	21.6	16.5	8.2
AMNH FI-72386	21.5	19.4	23.9	15.4	7.4
AMNH FI-72387	20.7	18.1	20.8	14.6	7.1
AMNH FI-72388	24.1	21.6	25.3	20.6	8.1
GSI M 743	18.3	16.3	20.4	15.1	6.4

Discussion.—*Daghanirhynchia susanae* sp. nov. differs from the type species *D. daghaniensis* Muir-Wood, 1935 in having coarser costae and a longer narrower fold. *D. macfadyeni* Muir-Wood, 1935 differs in that its fold is never raised above the level of the lateral slopes and bears a low, median fold. *D. elongata* Muir-Wood, 1935 is generally longer with finer costae and a shorter sulcus. The umbo in *D. elongata* extends further posteriorly than in *D. susanae. Daghanirhynchia macfadyeni* Muir-Wood, 1935 is characterised by a more convex ventral valve, a shallower shorter sulcus, and 4 costae on the fold.

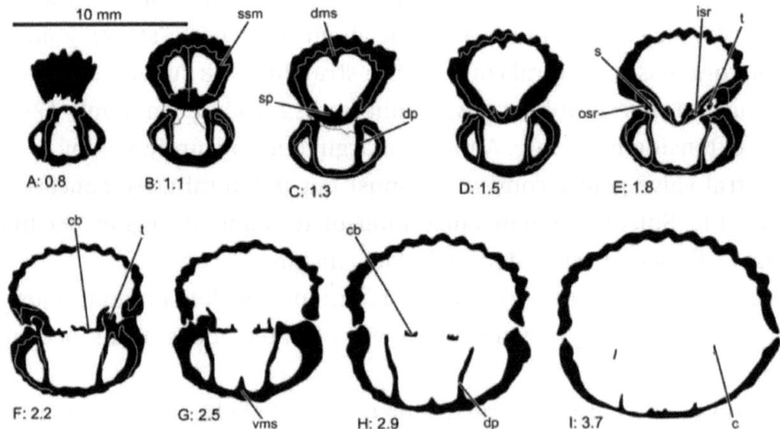

FIGURE 5: Serial sections of rhynchonellide brachiopod *Daghanirhynchia susanae* sp. nov. AMNH FI-72410. Bed 21b, Hamam Formation, Arda, Jordan. Numbers describe sectional distance from apex. Abbreviations: c, crus; cb, crural base; dms, dorsal median septum; dp, dental plate; isr, inner socket ridge; osr, outer socket ridge; s, socket; sp, septalial plate; ssm, secondary shell material; t, tooth; vms, ventral median septum.

The costae of *D. macfadyeni* are finer than in *D. susanae* but may reach the same thickness in a few specimens. *D. platiloba* Muir-Wood, 1935 differs from the new species in having a less convex dorsal valve, a longer ventral umbo, and a shallower, shorter sulcus. *D. kabeitensis* Muir-Wood, 1935 is wider than *D. susanae* and has a longer ventral umbo. The size of costae and sulcus is the same as in *D. susanae, D. farquharsoni* Muir-Wood, 1935 is equibiconvex, whereas the specimens of *D. susanae* are clearly dorsibiconvex. The sulcus in *D. farquharsoni* is shallower than in *D. susanae* and sulcus and fold are rectangular in cross section, whereas in *D. susanae* the fold and sulcus are broadly rounded in cross section.

Stratigraphic and geographic range.—Bathonian to Callovian; Jordan.

Daghanirhynchia angulocostata Cooper, 1989
Fig. 3D.

1989 *Daghanirhynchia angulocostata* sp. nov.; Cooper 1989: 26-28, pl. 6: 1-19, pl. 7: 44-53, pl. 11: 16-21, figs. 13, 14.

Material.—Two articulated specimens: AMNH FI-72395, AMNH FI-72396, bed S-7, Mughanniyya Formation (Callovian, upper Middle Jurassic), Tel el Dhahab section, Jordan.

Description.—Shells large (Table 4), outline widely triangular with maximum width anterior to midlength. Anterior commissure strongly uniplicate; costae coarse and subangular in cross section, 14-15 in number; ventral beak narrow, suberect; ventral interarea low. Foramen medium to large, longitudinally oval to suboval. Deltidial plates disjunct.

Ventral valve almost flat in lateral view, umbonal region moderately swollen. Sulcus originates just posterior to midlength and deepens anteriorly with 3-4 costae. Linguiform extension moderate with narrow, slightly convex flanks adjacent to sulcus with 6-7 costae.

Dorsal valve moderately convex in lateral profile, moderately to gently domed in anterior view; fold originates just posterior to midvalve but in some shells at midvalve, elevating slightly with 4 costae, 6 costae on dorsal flank. The internal structures of the shells are unknown due to a lack of sufficient material for serial sectioning.

TABLE 4: Measurements (in mm) of *Daghanirhynchia angulocostata* Cooper, 1989.

Specimen	Length	Dorsal valve length	Width	Thickness	Fold width
AMNH FI-72395	25.3	22.3	25.9	16.2	11.5
AMNH FI-72396	20.2	17.8	23.5	12.7	7.5

Discussion.—The Jordanian specimens compare favourably with Cooper's (1989) shells in that they are large and widely triangular with maximum width anterior to midlength. Muir-Wood's (1935) description of the genus includes the presence of fine radial striae as a feature of the genus. Cooper (1989) did not observe these features on his shells from Saudi Arabia and found none on the specimens in the USNM collections from the type locality of *D. daghaniensis* Muir-Wood, 1935 (USNM 75665, 75666). The Jordanian shells are also lacking this feature as well as the thickened deltidial plates observed by Cooper (1989) in his Saudi Arabian shells.

Stratigraphic and geographic range.—Bathonian to Callovian, Saudi Arabia, Jordan.

Daghanirhynchia? *triangulata* Cooper, 1989
Fig. 3E.

1989 *Daghanirhynchia*? *triangulata* sp. nov.; Cooper 1989: 28-29, pl. 7: 6-21, pl. 11: 11-15.

Material.—Two articulated specimens: AMNH FI-72397, bed S-7, Mughanniyya Formation (Callovian, upper Middle Jurassic), Tel el Dhahab section; AMNH 72398, bed 21a, Hamam Formation, Arda section, Jordan.

Description.—There are two shells in the collection (Table 5) tentatively assigned to *Daghanirhynchia*? *triangulata* Cooper, 1989. The specimens are triangular with maximum width considerably past midlength; slightly dorsibiconvex, strongly uniplicate anterior commissure, lateral commissure straight, beak suberect, costae strong, coarse and narrow, numbering 16-17.

Ventral valve barely convex in lateral view, almost flat, concave in anterior profile. Sulcus originates just past umbonal region, occupied by three costae. Linguiform extension long and fairly narrow; ventral interarea low; 6-7 costae on flanks.

Dorsal valve slightly convex in lateral view, strongly domed in anterior view, sides becoming steeper as lateral margins approached. Fold begins at about midlength, occupied by 4 costae, slightly less than half the valve width; 6 costae on flanks.

The internal structures of the shells are unknown due to a lack of sufficient material for serial sectioning.

TABLE 5: Measurements (in mm) of *Daghanirhynchia*? *triangulata* Cooper, 1989.

Specimen	Length	Dorsal valve length	Width	Thickness	Fold Width
AMNH FI-72397	26.5	23.1	23.5	21.3	9.7
AMNH FI-72398	22.7	20.2	22.4	13.6	7.6

Discussion.—The shell closely resembles Cooper's (1989) *Daghanirhynchia*? *triangulata* in its triangular outline and costation and is tentatively placed in that taxon until additional material can be studied.

Stratigraphic and geographic range.—Callovian, Jordan.

Genus *Pycnoria* Cooper, 1989

Type species: *Pycnoria magna* Cooper, 1989; Callovian, Saudi Arabia.

Pycnoria magna Cooper, 1989
Figs. 3F, 6.

1989 *Pycnoria magna* sp. nov.; Cooper 1989: 53-54, pl. 12: 11-36, pl. 18: 26-36, figs. 29, 30.

1991 *Pycnoria magna* Cooper; Feldman et al. 1991: 15, figs. 11A-C.

Material.—Six articulated specimens: AMNH FI-72402, AMNH FI-72404, AMNH FI-72403 (sectioned), bed S-7, AMNH FI-72400, bed S-20, Mughanniyya Formation (Callovian, upper Middle Jurassic), Tel el Dhahab section, AMNH FI-72401, bed 25, Mughanniyya Formation (Callovian, upper Middle Jurassic), Arda section; AMNH FI-7299, bed 21, Hamam Formation (Bathonian, Middle Jurassic), Arda section, Jordan.

Description.—The shells are medium to large (Table 6) with rounded sides and an obtuse apical angle. Maximum width is at or slightly anterior to mid-valve; outline subpentagonal, dorsal valve deeper and more convex than ventral valve. Lateral commissure straight, anterior commissure uniplicate. Beak short, erect to incurved; ventral interarea low, foramen hypothyrid, deltidial plates small, conjunct. Costae coarse and angular in cross-section numbering about 13-17. Some shells (AMNH FI-72399, AMNH FI-72400, AMNH FI-72401) have a relatively long linguiform extension.

Ventral valve flat in lateral profile but slightly convex in umbonal region; in anterior profile valve is shallowly concave. Low sulcus originates about one-third of distance from beak, deepening anteriorly; sulcus generally bears 3 costae and widens narrowly in some specimens but widely in others. Flanks narrow, steep bearing 4 costae.

Dorsal valve evenly convex in lateral profile, strongly to moderately domed in anterior view, with a well-defined arcuate median fold somewhat to strongly elevated above steep flanks. Fold narrow, originating at about midlength bearing 3-4 costae, usually 4. Flanks bear 5 costae.

Dental plates thick, long, and slightly divergent (less than 5°) from valve floor, shell relatively thin, median septum low, very weak. Strong, swollen slightly subquadrate hinge teeth fit into large sockets, crura raduliform, septalium

TABLE 6: Measurements (in mm) of *Pycnoria magna* Cooper, 1989; * = sectioned.

Specimen	Length	Dorsal valve length	Width	Thickness	Fold width
AMNH FI-72399	20.9	17.3	22.2	16.1	10.4
AMNH FI-72400	20.9	17.9	20.0	17.4	8.3
AMNH FI-72401	21.9	19.6	22.0	18.9	10.5
AMNH FI-72401	15.4	13.5	14.8	13.5	5.5
AMNH FI-72402	18.2	16.4	16.0	16.0	6.9
AMNH FI-72403*	21.8	20.8	21.6	15.2	9.5

appears to be narrow and short; crural bases weak. One tooth is somewhat corrugated. The general transverse outlines of Cooper's (1989) shells are similar to the Jordanian shells, the height of which is similar as is the presence of the median septum.

Discussion.—These specimens of *Pycnoria magna* compare favourably with the specimens described by Feldman et al. (1991) in that they also have 13 strong, angular, deeply incised costae with a well-defined arcuate median dorsal fold and a comparatively narrow sulcus. The Jordanian shells display more variability in that some have a more highly developed median dorsal fold and a slightly wider and more broadly arcuate anterior commissure, similar to that of *Strongyloria circularis* Cooper, 1989. Cooper's (1989) species of *Pycnoria*

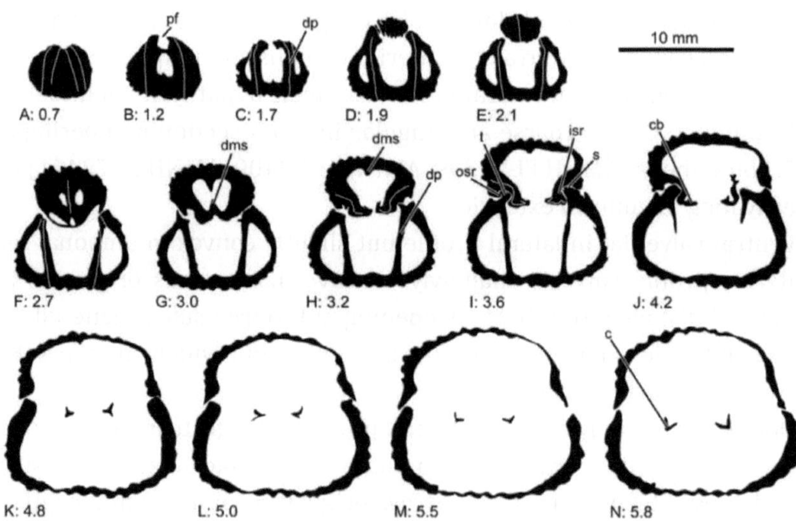

FIGURE 6: Serial sections of rhynchonellide brachiopod *Pycnoria magna* Cooper, 1989. AMNH FI-72403. Bed 25, Hamam Formation, Arda, Jordan. Numbers describe sectional distance from apex. Abbreviations: c, crus; cb, crural base; dms, dorsal median septum; dp, dental plate; isr, inner socket ridge; osr, outer socket ridge; pf, pedicle foramen; s, socket; t, tooth.

magna (USNM 303631; 243375) differs in that they have very fine concentric lines crossing the costae, a feature that may well have been removed here by exfoliation. *Pycnoria* differs from common rhynchonellides such as *Somalirhynchia* and *Daghanirhynchia* in its more solid and compact form as noted by Cooper (1989). We are in agreement with him that it differs from *Globirhynchia triangulata* in its stronger costation, narrower fold and deeper sulcus. *Burmirhynchia* is generally smaller with fewer costae (see Buckman 1918).

Stratigraphic and geographic range.—Bathonian to Callovian; Saudi Arabia, Jordan.

Subfamily Gibbirhynchiinae Manceñido, Owen, Dongli, and Dagys, 2002

Genus *Amydroptychus* Cooper, 1989

Type species: *Amydroptychus formosus* Cooper, 1989; Bajocian of Saudi Arabia.

Amydroptychus markowitzi sp. nov.

Fig. 3G.

Etymology: Named in memory of Alfred Markowitz (1926–2000), scientist, Renaissance man, and healer.

Holotype: AMNH FI-72405.

Type locality: Arda section, Wadi Zarqa, Jordan.

Type horizon: Bed 27, Mughanniyya Formation (Callovian, upper Middle Jurassic).

Material.—Two articulated specimens: AMNH FI-72405 (holotype), AMNH FI-72406, bed 27, Mughanniyya Formation (Callovian, upper Middle Jurassic); Arda section, Jordan.

Diagnosis.—*Amydroptychus* with subcircular outline, equibiconvex, medium to large foramen, and rectimarginate anterior commissure with no indication of fold or sulcus.

Description.—Medium-sized shells (Table 7), outline sub-circular and equibiconvex in longitudinal section without any indication of fold or sulcus. Ventral valve covered by 20 simple costae, dorsal valve by 21 costae. Furrows and costae are angular to rounded in cross section. Equibiconvex, apical angle acute, deltidial plates obscured, foramen medium, tubular. Anterior commissure rectimarginate; lateral commissure straight. Ventral beak erect in adults but suberect in juveniles. Twenty-one angular costae on adults and 25 on juveniles. Intercalations and bifurcations observed on juveniles but rare on adult specimens.

TABLE 7: Measurements (in mm) of *Amydropthychus markowitzi* sp. nov., *rectimarginate anterior commissure.

Specimen	Length	Dorsal valve length	Width	Thickness	Fold Width
Holotype AMNH FI-72405	22.7	20.4	22.2	13.7	*
AMNH FI-72406	14.8	13.2	15.0	8.4	*

The ventral valve is convex in lateral profile with the apex of the convexity occurring at one-third the distance from the beak; in anterior profile the valve is more sharply domed with the flanks sloping somewhat steeply until the commissure is reached.

The dorsal valve is convex in lateral view bulging more posteriorly and sloping gently toward the anterior commissure. In the juvenile specimens the bulge is less developed.

The internal structures of the shells are unknown due to a lack of sufficient material for serial sectioning.

Discussion.—The gentle folding that Cooper (1989) noted as being a conspicuous feature of *Amydroptychus formosus* is absent here. These specimens resemble *Strongyloria circularis* Cooper, 1989 but are less globose, lack an arcuately uniplicate anterior commissure, and have a narrower beak. In addition, these shells are more triangular than circular in outline as are *Strongyloria circularis* and *S. subelliptica* Cooper, 1989. Hegab (2005) described *Amydroptychus galaensis* from the Jurassic of the western side of the Gulf of Suez, Egypt, that differs from the Jordanian shells in their slightly sulcate anterior commissure (2005: 740, figs. 3, 6) and smaller number of costae (16).

Stratigraphic and geographic range.—Callovian; Jordan.

Genus *Schizoria* Cooper, 1989

Type species: *Schizoria elongata* Cooper, 1989; Bajocian of Saudi Arabia.

Schizoria elongata Cooper, 1989
Fig. 3H.

1989 *Schizoria elongata* sp. nov.; Cooper 1989: 55-56, fig. 31: 14, figs. 17-27.

Material.—Three articulated specimens: AMNH FI-72407, AMNH FI-72408, AMNH FI-72411, bed S-7, Mughanniyya Formation (Callovian, upper Middle Jurassic); Tel el Dhahab section, Jordan.

Description.—Shells small (Table 8), elongate but slightly oval, maximum width just anterior to midvalve. Dorsal valve deeper and asymmetrically more convex than ventral valve with the convexity more pronounced posteriorly;

TABLE 8: Measurements (in mm) of *Schizoria elongata* Cooper, 1989.*, rectimarginate anterior commissure.

Specimen	Length	Dorsal valve length	Width	Thickness	Fold width
AMNH FI-72407	20.5	17.3	14.0 est.	12.5	*
AMNH FI-72408	15.5	13.7	12.4	10.1	*

apical angle acute; anterior commissure on larger shell barely uniplicate but on smaller shell not preserved well enough to determine morphology; lateral commissure straight. Beak narrow, long; foramen medium, mesothyrid, tubular; deltidial plates disjunct, ventral interarea low. Costae moderate and angular in cross section but subangular and rounded on smaller shell due to weathering and exfoliation; bifurcated and intercalated mostly on umbos ranging from 24–25 in number.

Ventral valve moderately convex in lateral profile and highly domed in anterior view. Umbonal region swollen, more so in large shell. Fold with 5 costae; 6 to 7 costae on flank of ventral valve.

Dorsal valve swollen posteriorly but less so in smaller specimen; highly domed in anterior view on the larger shell. Sulcus with 4 costae; 5 to 7 costae on flank.

The internal structures of our shells are unknown due to a lack of sufficient material for serial sectioning.

Discussion.—The shells closely resemble Cooper's (1989) *Schizoria elongata* in many respects but differ in that the pedicle foramen here is mesothyrid and is slightly larger than in his specimens. *Schizoria intermedia* Cooper, 1989 has more widely spaced costae on the flanks, fold, and sulcus with no bifurcations and rare intercalations; it also has a strongly uniplicate anterior commissure. *Sphenorhynchia plicatella* differs from *Schizoria elongata* in its less elongate-oval general outline, more evenly biconvex valves, shorter and more massive umbo, and less numerous costellae.

Stratigraphic and geographic range.—Bajocian to Callovian; Saudi Arabia, Jordan.

Schizoria cf. *intermedia* Cooper, 1989
Fig. 3I.

Material.—One articulated specimen, AMNH FI-72409, bed 21b, Hamam Formation (Bathonian, middle Middle Jurassic); Arda section, Jordan.

Description.—Shell small (L = 21.6; W = 20.0; T = 15.4), elongate oval in longitudinal direction, maximum width anterior of midvalve, dorsal valve more globose than ventral valve; sides rounded, anterior commissure strongly uniplicate; ventral beak and foramen damaged; lateral commissure straight; about 17 angular to subangular costae with no bifurcations or intercalations.

Ventral valve slightly convex in lateral view, gently domed and almost flat on top, umbonal region barely swollen; sulcus broad and shallow, originating at midvalve, bearing 4 costae.

Dorsal valve moderately convex in lateral profile, and evenly but narrowly domed in anterior view; fold beginning almost at midvalve, bearing 5 costae. The internal structures are unknown due to a lack of sufficient material for serial sectioning.

Discussion.—The shell resembles *Schizoria intermedia* Cooper, 1989 but a definite taxonomic assignment cannot be made at this time due to inadequate material for serial sectioning and a damaged beak on the single specimen in the collection.

Stratigraphic and geographic range.—Bathonian, Jordan.

PALEOBIOGEOGRAPHY

Limestone beds dominate the Middle Jurassic Mughanniyya Formation of northwest Jordan with minor admixture of marl and claystone. Most of the Jurassic rocks of Jordan were deposited in shallow water, near shore environments where food supply and light were not likely to have been limiting factors (Wolff 1973) which explains the high and diverse faunal content. The environment of deposition appears to have been neritic. We hope to provide data that will enable a more precise stratigraphic correlation between the strata in northwestern Jordan and southern Israel (e.g., Hamakhtesh Hagadol or Kurnub Anticline). Lewy (1983) noted that the lithostratigraphic correlation of the strata in the Zarqa River area with those in Israel lacks the necessary biostratigraphic data for the recognition of the equivalents to the Upper Callovian Zohar and Kidod formations in Jordan. During the Callovian there was a continuous platform sedimentation regime and a regional sea level high during Middle and Upper Callovian times that inundated most of the Arabian Platform with carbonates. In Sinai (Gebel El-Maghara) these limestone rocks built up the Aroussiah Formation. In the southern Negev the sequence appears to contain more marls (Hamakhtesh Hagadol) and is known as the Zohar and Matmor formations (Hirsch et al. 1998; Hirsch 2005). In Jordan, the Mughanniyya Formation represents only the lowermost part of the Zohar Formation, as the youngest portion of the interval was eroded in infra-Cretaceous times (see Goldberg and Friedman 1974). It is of utmost interest as it represents a lower interval that is not exposed in the Jurassic of southern Israel (Hamakhtesh Hagadol). These rocks yielded a fauna that is related to Ethiopian-Somali taxa (brachiopods, bivalves [e.g., *Eligmus* Eudes–Deslongchampes, 1856], gastropods and foraminiferans [e.g., *Kurnubia* lineage]). The northeastern Ethiopian Province was part of a broad shelf surrounding the Arabo-Nubian craton in the Middle Jurassic (Lewy 1983). Shelf and marginal marine flats with mostly carbonates and evaporites dominated the broad platform of northern Gondwana, whereas southern Tethyan Seaway deep water basinal limestone, radiolarian chert, and ophiolites (Goldberg and Friedman 1974) characterise obstructed terranes to the north along the Cyprus Arc and Assyrian Suture (Krashennikov et al. 2005).

A high degree of endemism characterises the rhynchonellide brachiopod faunas of the Jurassic Ethiopian Province. In the Sinai Peninsula (Gebel el-Maghara), southern Israel (Hamahktesh Hagadol), and in Saudi Arabia (Dhruma, Tuwaiq Mountain, and Hanifa formations) *Somalirhynchia* is relatively widespread but absent in correlative rocks in Jordan. In Hamakhtesh Hagadol *Somalirhynchia* and *Burmirhynchia* occur in the Late Callovian Matmor Formation that are missing in Wadi Zarqa, Jordan. However, we have identified *Eurysites*, *Cymatorhynchia*, *Daghanirhynchia*, *Pycnoria*, *Schizoria*, and *Amydroptyhcus* from the early–middle Callovian Mughanniya Formation in Jordan. Cooper (1989) noted the high degree of endemism in a Saudi Arabian brachiopod fauna in which he described 13 new rhynchonellide genera (Dhruma, Tuwaiq Mountain, and Hanifa formations). Kier (1972) noted an extremely high degree of endemism in his study of Jurassic echinoids of Saudi Arabia; 24 out of 27 taxa were new. Muir-Wood's (1925) Jordan Valley fauna and Weir's (1925) Somalia fauna were not included here due to the lack of reliable identification since interior sections of these taxa are not available.

ACKNOWLEDGMENTS

We are grateful to Chaim Benjamini (Ben–Gurion University of the Negev, Israel) for helpful suggestions that greatly improved the paper. We thank Jann Thompson and Daniel Levin (National Museum of Natural History/Smithsonian Institution, Washington DC, USA) and Lee Davies (Natural History Museum, London, UK) for providing access to the brachiopod collection. Many thanks are due Juliane Eberhardt and Erika Scheller-Wagner (Forschungsinstitut and Naturmuseum Senckenberg, Frankfurt am Main, Germany) for technical help. We thank Fernando Alvarez (Universidad de Oviedo, Oviedo, Spain) and Attila Voros (Hungarian Natural History Museum, Budapest, Hungary) for their constructive and helpful reviews of the manuscript. The visit of MS-G to the American Museum of Natural History, New York was supported by a Lerner Grey Award.

REFERENCES

Ahmad, F. 2002. Middle Jurassic brachiopods associations from Wadi Shaban, northwest Jordan. *Basic Sciences and Engineering* 11: 745-762.

Alméras, Y. 1966. Les Rhynchonellides du Bajocien moyen de Ronzevaux pres Davaye (Saône-et-Loire): Genres *Cymatorhynchia* S. Buckman, *Lacunaerhynchia* nov. et *Septulirhynchia* nov. *Travaux du Laboratoire de Géologie de la Faculte des Sciences de l'Universite de Lyon* (new series) 13: 31-119.

------. 1987. Les brachiopodes du Lias-Dogger: paléontologie et biostratigraphie. In R. Enay (ed.), *Le Jurassique d'Arabie Saoudite Central*, 161-219. *Géobios, Mémoire spécial* 9: 1-314.

Alméras, Y. and S. Elmi. 1996. Le genre *Cymatorhynchia* (Brachiopoda, Rhynchonellacea) dans le Bajocien-Bathonien de la bordure vivarocévenole (Bassin du Sud-Est, France). *Beringeria* 18: 201-245.

Arcelin, F. and P. Roche. 1936. Les Brachiopodes Bajociens du Monsard. *Travaux du Laboratoire Géologique de la Faculté des Sciences de l'Université de Lyon 25, Memoir* 30: 1-107.

Arkell, W. J. 1952. Jurassic ammonites from Jebel Tuwayq, Central Arabia. *Philosophical Transactions of the Royal Society of London, Series B* 236: 241-313.

------. 1956. *Jurassic Geology of the World*. Edinburgh: Oliver and Boyd.

Basha, S. H. 1980. Ostracoda from the Jurassic System of Jordan including a stratigraphical outline. *Revista Española de Micropaleontología* 12: 231-254.

Basha, S. H. and M. S. Aqrabawi. 1994. Bajocian ammonites from Zerqa Wadi, Jordan. *Revista Española de Paleontología* 9: 117-123.

Blake, G. S. 1935. *The Stratigraphy of Palestine and Its Building Stones*. Printing and Stationary Office, Jerusalem.

------. 1940. Geology, soils and minerals, and hydrogeological correlation. In M. G. Ionides (ed.), *Report on the Water Resources of Transjordan and Their Development*, 43-127. Crown Agent for the Colonies, London.

Buckman, S. S. 1918. The Brachiopoda of the Namyau Beds, Northern Shan States, Burma. *Memoirs of the Geological Survey of India, Palaeontologica Indica, New Series* 3: 1-299.

Cariou, E. J. P., L. Bassoullet, L. Grossowicz, and F. Hirsch. 1997. Le Callovo-Oxfordien du Sud Levant: données biostratigraphiques nouvelle (ammonites, foraminifères) endémesme et corréations stratigraphiques. In Société géologique de France (ed.), *Réunion Spécialisée APF-SGF "De la Biostratigraphie à la Paléobiogéographie," Lyon,* (21-28 novembre 1997): 21.

Coates, J., J. P. Gottesman, M. Jacobs, and E. Rosenberg. 1963. Gas discoveres in the western Dead Sea region. *Proceedings of the Sixth World Petroleum Congress, Frankfurt am Main*, 21-36. Hamburg: Hanseatische Druckanstalt.

Cooper, G. A. 1989. Jurassic brachiopods of Saudi Arabia. *Smithsonian Contributions to Paleobiology* 65: 1-213.

Cox, L. R. 1925. A Bajocian-Bathonian outcrop in the Jordan Valley and its molluscan remains. *Annals and Magazine of Natural History,* Serie 9, Fascicle 25: 169-181.

Davidson, T. 1851-1852. The Oolitic and Liassic Brachiopoda. In T. Davidson (ed.), *A Monograph of the British Fossil Brachiopoda,* Volume 1, Part 3, 1-30, Appendix to Volume 1. London: Palaeontographical Society.

Feldman, H. R. 1986. Cladistic analysis of a Jurassic rhynchonellid brachiopod genus from the Middle Eastern Ethiopian Province. *Abstracts of the Geological Society of America Northeastern Section Meeting, Kiamesha Lake, New York* 18: 16.

------. 1987. A new species of the Jurassic (Callovian) brachiopod *Septirhynchia* from northern Sinai. *Journal of Paleontology* 61: 1156-1172.

Feldman, H. R. and C. E. Brett. 1998. Epi- and endobiontic organisms on Late Jurassic crinoids columns from the Negev Desert, Israel: implications for co-evolution. *Lethaia* 31: 57-71.

Feldman, H. R. and E. F. Owen. 1988. *Goliathyris lewyi*, new species (Brachiopoda, Terebratellacea) from the Jurassic of Gebel El-Minshera, northern Sinai. *American Museum Novitates* 2908: 1-12.

------. 1993. Vicariance biogeography of the Middle Eastern Ethiopian Province: implications for the Paleozoic. *Abstracts of the Geological Society of America Annual Meeting, Boston, Massachusetts* 25: 55.

Feldman, H. R., F. Hirsch, E. F. and Owen. 1982. A comparison of Jurassic and Devonian brachiopod communities: Trophic structure, diversity, substrate relations and niche replacement. *Journal of Paleontology* 56 (Supplement 2): 9-10.

Feldman, H. R., E. F. Owen., and F. Hirsch. 1991. Brachiopods from the Jurassic of Gebel El-Maghara, northern Sinai. *American Museum Novitates* 3006: 1-28.

------. 2001. Brachiopods from the Jurassic (Callovian) of Hamakhtesh Hagadol (Kurnub Anticline), southern Israel. *Palaeontology* 44: 637-658.

Fischer, J. C. 1980. *Fossiles de France et des Regions Limitrophes.* Paris: Masson.

Gill, G. A. and H. Tintant. 1975. Les ammonites Calloviennes du sud d'Israel. Stratigraphie et relations paleogeographiques. *Société Géologique de France, Comptes Rendus des Scéances* 4: 103-106.

Gill, G. A., J. Thierry, and H. Tintant. 1985. Ammonites Calloviennes du sud d'Israel: systematique, biostratigraphie et paleobiogeographie. *Geobios* 6: 705-751.

Goldberg, M. 1963. *Reference Section of Jurassic Sequence Exposed in Hamakhtesh Hagadol (Kurnub Anticline). Detailed Binocular Sample Description, Including Field Observations. Geological Survey of Israel,* Oil Division, Unpublished Internal Report.

Goldberg, M. and G. M. Friedman, G. M. 1974. Paleoenvironments and paleogeographic evolution of the Jurassic System in southern Israel. *Geological Survey of Israel Bulletin* 61: 1-44.

Hegab, A. A. 1988. Analysis of growth patterns in *Eudesia* (Brachiopoda) and its potential use for the identification of different species of the genus. *Bulletin of the Faculty of Science, Assiut University* 17: 25-36.

------. 1989. New occurrence of Rhynchonellida (Brachiopoda) from the Middle Jurassic of Gebel El-Maghara, northern Sinai. *Journal of African Earth Sciences* 9: 445-453.

------. 1991. The occurrence of genus *Flabellothyris* (Brachiopoda) from the Jurassic of northern Sinai. *Bulletin of the Faculty of Science, Assiut University* 20: 39-49.

------. 1992. Terebratulida (Brachiopoda) from the Jurassic of Gebel El-Maghara, northern Sinai. *Proceedings of the 8th Symposium of Phanerozoic Developments in Egypt* 8: 33-42.

------. 1993. *Eudesia* (Brachiopoda) community from the Bathonian of Gebel El-Maghara (northern Sinai): their morphologic adaptation, ontogenetic variation and paleoecology. *Palaeontographica, Abteilung A* 229: 1-14.

------. 2005. *Amydroptychus galaensis* n. sp. (Brachiopoda). A Contribution to the brachiopods of the Middle Jurassic of the Gulf of Suez. *The Fourth International Conference on the Geology of Africa* 2: 737-743.

Hirsch, F. 2005. The Jurassic of Israel. In J. K. Hall, V. A. Krasheninnikov, F. Hirsch, C. Benjamini, and A. Flexer (eds.), *Geological Framework of the Levant,* 361-392. Jerusalem: Historical Productions-Hall.

Hirsch, F. and R. Roded. 1997. The Jurassic stratigraphic nomenclature in Hamakhtesh Hagadol, northern Negev. *Geological Survey of Israel, Current Research* 10: 1014.

Hirsch, F., J.-P. Bassoullet, É Cariou, B. Conway, H. R. Feldman, L. Grossowicz, A. Honigstein, E. F. Owen, and A. Rosenfeld, 1998. The Jurassic of the southern Levant. Biostratigraphy, palaeogeography and cyclic events. In S. Crasquin-Soleau and É. Barrier (eds.), Peri-Tethys Memoir 4: Epicratonic Basins of Peri-Tethyan Platforms. *Mémoires du Muséum national d'Histoire Naturelle* 179: 213-235.

Khalil, B. and S. S. Muneizel, 1992. Lithostratigraphy of the Jurassic outcrops of north Jordan (Azab Group). *Geological Bulletin* 21: 1-50. [Internal report for the Natural Resources Authority Amman/Jordan].

Kier, P. M. 1972. Tertiary and Mesozoic echinoids of Saudi Arabia. *Smithsonian Contributions to Paleobiology* 10: 1-242.

Kitchin, F. L. 1912. Palaeontological work: England and Wales: Summation of programme for 1911. *Memoir of the Geological Survey of Great Britain and Museum of Practical Geology*: 59-60.

Krasheninnikov, F., J. K. Hall, F. Hirsch, N. Kazmin, N. Bragin, D. Klaeschen, S. Kopf, N. Silantyev, N. Vidal, R. von Huene, and S. Zverev. 2005. The Jurassic of Israel. In J. K. Hall, V. A. Krasheninnikov, F. Hirsch, and C. Benjamini (eds.), *Geological Framework of the Levant*, 700-732. Jerusalem: Historical Productions-Hall.

Lewy, Z. 1983. Upper Callovian ammonites and Middle Jurassic geological history of the Middle East. *Geological Survey of Israel Bulletin* 76: 1-56.

Manceñido, M. O., E. F. Owen, D. Sun, and A. S. Dagys. 2002. Hemithiridoidea. In R. L. Kaesler (ed.), *Treatise on Invertebrate Palaeontology. Part H, Brachiopoda, Revised*, 1326-1369. Boulder: Geological Society of America and University of Kansas Press.

Muir-Wood, H. M. 1925. Jurassic Brachiopoda from the Jordan Valley. *Annals and Magazine of Natural History* 9 (15): 181-192.

------. 1935. *The Mesozoic Palaeontology of British Somaliland, Part II of the Geology and Palaeontology of British Somaliland. Jurassic Brachiopoda*, 75-147, pls 8-13. London: Government of the Somaliland Protectorate.

Muneizel, S. S., and B. Khalil. 1993. *The Geology of Al-Salt Map Sheet No. 3154-III*. NRA. Amman: Internal report for the Natural Resources Authority.

Parnes, A. 1981. Stratigraphy of the Mahmal Formation (Middle and Upper Bajocian) in Makhtesh Ramon, southern Israel. *Geological Survey of Israel Bulletin* 74: 1-55.

Pevny, J. 1969. Middle Jurassic brachiopods in the Klippen Belt of the Central Vah Valley. *Geologicke Prace, Zpravy* 50: 133-160.

Picard, L., and F. Hirsch. 1987. *The Jurassic Stratigraphy in Israel and the Adjacent Countries*. Jerusalem: The Israel Academy of Sciences and Humanities.

Quenstedt, F. A. 1851-1852. *Handbuch der Petrefaktenkunde*. 1851: 1-528; 1852: 529-792. Tübingen: Laupp.

Quenstedt, F. A. 1858. *Der Jura*. Tübingen: Laupp.

------. 1868-1871. Brachiopoden. In F. A. Quenstedt (ed.), *Petrefaktenkunde Deutschlands*. Tübingen: Feus's.

Roche, P. 1939. Aalenian et Bajocien du Maconnais et de quelques Regions Voisines. *Travaux du Laboratoire et Geologie de la Faculte des Sciences de Lyon*, fascicle 35, série 29: 1-351.

Rollier, L. 1918. Synopsis des spirobranches (Brachiopodes) Jurassique celto-souabes, Part II: Rhynchonellides. *Mémoires de la Société Paléontologique Suisse* 42: 71-184.

Rousselle, L. 1965. Rhynchonellidae, Terebratulidae et Zeilleridae du Dogger Marocain (Moyen-Atlas Septentrional, Hauts Plateaux, Haut Atlas). *Service des Mines et de la Carte Géologique du Maroc, Notes et Mémoires* 187: 1-168.

Rybakov, M. and A. Segev. 2005. Depth to crystalline basement in the Middle East with emphasis on Israel and Jordan. In J. K. Hall, V. A. Krasheninnikov, F. Hirsch, C. Benjamini, and A. Flexer (eds.), *Geological Framework of the Levant,* 543-552. Jerusalem: Historical Productions-Hall.

Shi, X. 1987. The Middle Jurassic brachiopods from the Nyalam area, South Tibet [in Chinese with English summary]. *Contribution to Geology of the Qinghai-Xizang (Tibet) Plateau* 18: 44-69.

Shi, X. and R. E. Grant. 1993. Jurassic rhynchonellids: Internal structures and taxonomic revisions. *Smithsonian Contributions to Paleobiology* 73: 1-190.

Swarieh, A. and M. Barjous. 1993. *The Geology of Suwaylih Map Sheet No. 3154–II.* NRA. Amman: Internal report for the Natural Resources Authority.

Weir, J. 1929. Jurassic fossils from Jubaland, East Africa, collected by V.G. Glenday, and the Jurassic Geology of Somaliland. *Monographs of the Geological Department of the Hunterian Museum, Glasgow University* 3: 1-63.

Wilson, M.A., H. R. Feldman, J. C. Bowen, and Y. Avni. 2008. A new equatorial, very shallow marine sclerozoan fauna from the Middle Jurassic (late Callovian) of southern Israel. *Palaeogeography, Palaeoclimatology, Palaeoecology* 263: 24-29.

Wilson, M. A., H. R. Feldman, and E. B. Krivicich. 2010. Bioerosion in an equatorial Middle Jurassic coral-sponge reef community (Callovian, Matmor Formation, southern Israel). *Palaeogeography, Palaeoclimatology, Palaeoecology* 289: 93-101.

Wolf, W. 1973. The estuary as a habitat. An analysis of data on the soft-bottom macrofauna of the estuarine area of the rivers Rhine, Meuse, and Scheldt. *Zoologische Verhandelingen* 126: 1-242.

Zieten, C. H. von 1830-1833. *Die Versteinerungen Württembergs.* Stuttgart: Verlag des Expedition des Werks Unsere Zeit.

Chapter Thirteen

Taxonomy and Paleobiogeography of Late Bathonian Brachiopods from Gebel Engabashi, Northern Sinai

ABSTRACT

A brachiopod fauna of late Bathonian age recovered from the Kehailia Formation from Gebel Engabashi in northern Sinai consists of six species (two rhynchonellids and four terebratulids) referred to six genera, of which one genus and two species are new: *Globirhynchia sphaerica* (Cooper, 1989) new combination, *Daghanirhynchia angulocostata* Cooper, 1989, *Ectyphoria sinaiensis* new species, *Cooperithyris circularis* new genus and species, and new material: *Avonothyris* species A, and *Ptyctothyris* species A. The brachiopods described herein comprise a fauna located at the northern part of the Indo-African Faunal Realm within the Jurassic Ethiopian Province. They extend the geographic distribution of those taxa that show great affinity with the Jurassic brachiopod fauna of Saudi Arabia described by Cooper (1989). Differentiation of the endemic faunas that is so characteristic of many of these Ethiopian Province faunas is becoming more well-defined.

INTRODUCTION

The invertebrate faunas of Sinai have been studied by numerous workers over the past century. Douvillé (1916) dealt with the brachiopod, molluscan and echinoderm faunas, Cossman (in Douvillé, 1925) and Hirsch (1979) worked on the bivalves and gastropods, whereas Arkell (1952) and Parnes (1974) dealt with the ammonites. Brachiopods were studied more specifically by Farag (1957, 1959) and Farag and Gatinaud (1960a, 1960b). Farag compiled a faunal list taken from the studies of Douvillé (1916) and Arkell (1952); this is expanded upon here. In more recent studies Feldman (1987), Feldman and Owen (1988, 1993), Feldman et al. (1991, 2009), Hegab (1988, 1989, 1991a, 1991b, 1991c, 1992, 1993, 1995, 2003, 2004), and Hegab and Aly (2004) reviewed the brachiopods of northern Sinai, revised the taxonomy and discussed their paleobiogeography.

Stratigraphic, structural and mapping studies were completed by Range (1920), Moon and Sadek (1921), Hoppe (1922), Farag (1959), Al Far (1966), and Goldberg et al. (1971). Hirsch et al. (1994, 1998) presented biostratigraphic and tectonic data on northern Sinai taken, in part, from the northern Sinai Hallal borehole as well as outcrop analysis. The present paper dealing with brachiopods of the Kehailia Formation continues a long-term effort on the part of paleontologists to complete a taxonomic revision of the brachiopod faunas of the Jurassic Ethiopian Faunal Province (Feldman et al., 2001). Completion of this project will allow us to further gain insight into the paleoecology of the various marine communities (e.g., Feldman and Brett, 1998; Wilson et al., 2008) and construct a biostratigraphic framework for the Jurassic of the Middle East.

GEOLOGIC SETTING, BIOSTRATIGRAPHY AND PALEOECOLOGY

The outcrop area lies within Gebel El-Maghara, a Pliensbachian-Oxfordian sedimentary sequence approximately 2000 m thick. Paleobiogeographically this is at the junction of the Indo-African and Tethyan faunal realms in the Sinai Peninsula (Feldman et al., 1991). The rocks generally form two anticlines, the Shushet el Maghara and Homayir that are asymmetric with steep to overturned flanks, bordered by longitudinal thrust faults that lie parallel to the NE-SW trend of the fault axes (Picard and Hirsch, 1987). The stratigraphy used in this paper is based on field observations by Hegab and Feldman in addition to the work of Al Far (1966), Goldberg et al. (1971) and Picard and Hirsch (1987). In the Bathonian sequence the Kehailia-Arroussiah formations consist of shales and oolitic carbonates in the lower part with occasional sandstone interbeds. Above this lie carbonates and some shales but little sandstone, above which are shales with some thin oolitic-oncolitic limestone beds with thin sandstone interbeds. Goldberg et al. (1971; subunit 67) recognized stromatoporoids indicating proximity to the Callovian-Oxfordian facies that typifies this group. For additional detailed stratigraphic information, including columnar sections of Gebel El-Maghara and Gebel Engabashi, see Feldman et al. (1991) and Hirsch et al. (1998).

Gebel Engabashi is located within the breached anticline of Gebel El-Maghara, northern Sinai, and is characterized by numerous outcrops of Bathonian and Callovian age (see Fig. 1 for location of Gebel Engabashi; GPS coordinates: N 30°41'25"; E 33°25'07"). During the early Bathonian there was a substantial regression in the area, evidenced by paralic coals in the Safa Formation. Hegab (1991c) discussed the paleoenvironmental conditions prevailing during the deposition of the Safa Formation which he subdivided into two phases. The lower two thirds of a thick alternating shale and sandstone sequence are unfossiliferous and are characterized by primary sedimentary structures such as cross-stratification and plant remains of continental, probably fluviatile origin.

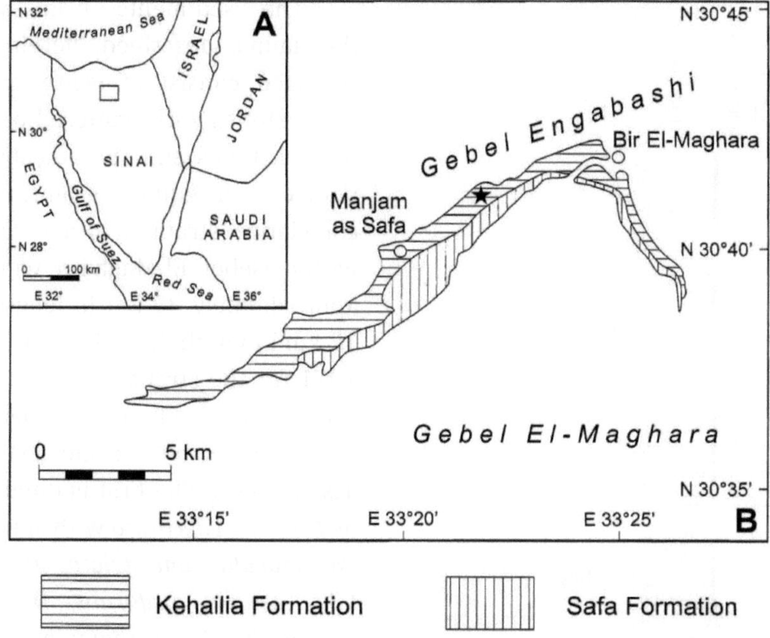

FIGURE 1: A, geographic location of study area is indicated by rectangle; B, distribution of the Bathonian sediments of the Safa and Kehailia formations of breached anticline of Gebel El-Maghara, northern Sinai, modified from Al Far (1966). The collecting locality is indicated by star.

These sediments, usually ferruginous, are characteristically brown or red due to a regressive phase resulting in rivers and streams depositing terrigenous, iron oxide-rich sediments. The upper third of the Safa Formation is characterized by shallow marine facies, consisting of marly and oolitic limestones and calcareous shales that signify the beginning of a transgressive phase. The shore line of this transgression is marked by coralline limestone with coral heads. The faunal assemblage typical of these sediments (brachiopods, mollusks, corals, and rare echinoderms) supports a shallow marine depositional environment (Hegab, 1991c).

The middle and late Bathonian strata are suggestive of a period of renewed transgression as indicated by shaly-calcareous shelf-platform sediments of the Kehailia Formation (Picard and Hirsch, 1987; Hirsch and Picard, 1988; Fig. 2). Mesozoic biostratigraphy is largely based on ammonites but few ammonites were found in these sediments so brachiopods may be the key to biostratigraphic interpretation in the region. The late Bathonian-early Callovian transition remains unclear and the alleged lowstand followed by an overall transgression (Haq et al., 1988), probably occurs much lower in the sequence (Hirsch et al., 1998).

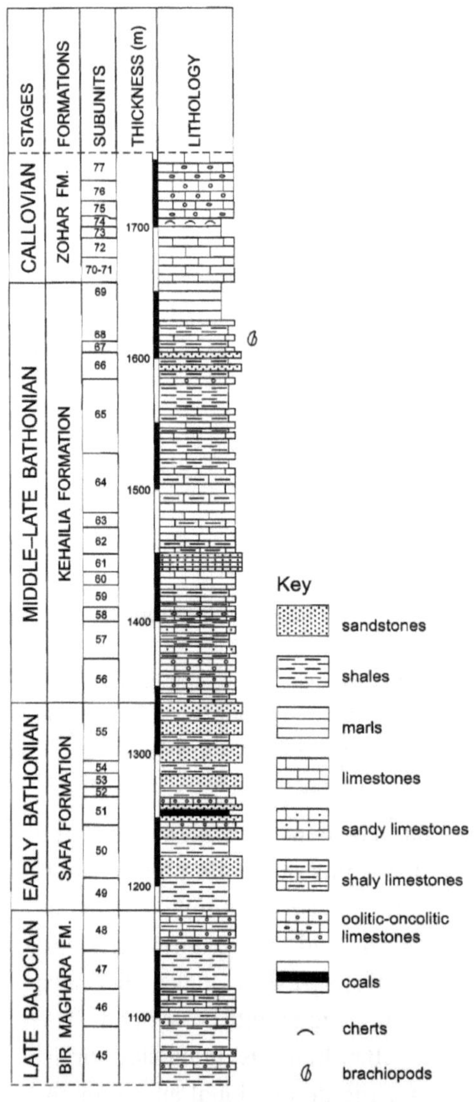

FIGURE 2: Stratigraphic composite column of breached anticline of Gebel El-Maghara, northern Sinai, showing lithology and brachiopod-yielding beds. Modified from Picard and Hirsch (1987), Feldman et al. (1991), and Hirsch et al. (1998).

The sediments from which the fauna described herein was collected consist of ochre colored wackestones with scattered patches of packstones and sporadically dispersed ferruginous grains. From correlative strata in other regions within Gebel El-Maghara which is approximately 30 km in length and 12 km in width, (see Goldberg et al., 1971 for the stratigraphy), in a calcareous and sandy series of beds, we recognize two bivalve and gastropod assemblages. The first is dominated by *Nuculana decorata* with abundant *Paleonucula tenuistriata, P. variabilis, Astarte pisiformis, Discohelix elegantula, Procerithium bouchardi, Exelissa solitudinis, Dicroloma armata,* and *D. tumida* (Hirsch, 1979). The second assemblage is dominated by *Eligmus asiaticus-Africogryphaea costellata* and is common in marly and calcareous strata with *Bucardiomya lirata,* a burrowing suspension feeder (Hirsch, 1979). This interval contains the ammonite *Bullatimorphites bullatus,* which is a Tethyan early Callovian marker, the genus *Bullatimorphites* appearing in the late Bathonian. In addition to the brachiopods, faunal constituents include: minute benthic foraminifers (lagenids and trocholinids), bivalves, gastropods and microgastropods. A foraminifer species, found in the lower part of the overlying limestones of the Gebel Arroussiah Formation, was first identified as *Meyendorffina bathonica* (in Goldberg, et al., 1971). This well-known marker of the late Bathonian in western Europe, renamed as *Orbitammina bathonica* (in Picard and Hirsch, 1987) and *Satorina apuliensis* (in Hirsch and Picard, 1988), is now considered to be a flat, new species, of *Kilianina* (in Hirsch et al., 1998) that is indicative of an early Callovian age and characterizes the southern Tethyan Realm. As no further ammonites were found preserved with the fauna, a late Bathonian age is supported by the presence of two rhynchonellid species, *Globirhynchia sphaerica* Cooper, 1989 and *Daghanirhynchia angulocostata* Cooper, 1989, also known from the Bathonian of Saudi Arabia (Cooper, 1989) as well as the late Bathonian-Callovian genus *Ectyphoria* Cooper, 1989. We feel that based on the micro- and megafauna as well as the sediments, it can be inferred that sedimentary environment was mid-neritic.

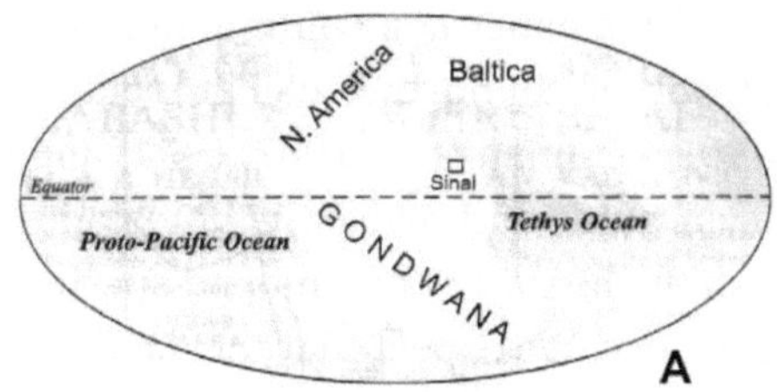

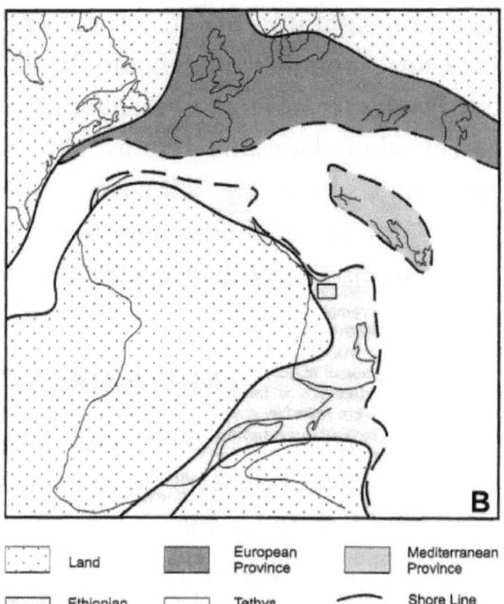

FIGURE 3: A, global paleographic map for the Middle Jurassic modified from Scotese (2004); B, paleographic map of the Ethiopian Province (Bathonian) modified from Dewey et al. (1973), Hallam (1975) and Vörös (1984). Study area is indicated by rectangle.

PALEOBIOGEOGRAPHY

The Ethiopian Province is defined geographically by a Gondwanian shelf crescent from east to north Africa including Somalia, Kenya, Tanzania, Saudi Arabia, the Levant, Egypt, and the Maghreb (the western region of north Africa including Algeria, Tunisia, Libya, Morocco, Mauritania and the western Sahara) (Hirsch et al., 1998; Fig. 3). The Ethiopian Province has been recognized from the Early Jurassic until the mid and possibly end of the Cretaceous Period by the presence of endemic taxa at the species, genus and family level. These endemics (e.g., *Somalirhynchia africana, Daghanirhynchia daghaniensis, Somalithyris bihendulensis, Striithyris somaliensis, Bihenithyris barringtoni, B. weiri)* have been noted, especially in the brachiopods, by Weir (1925) and Muir-Wood (1935). Arkell (1952, 1956) recognized endemic faunas in the ammonoid Cephalopoda and Kitchin (1912) in the trigoniacean and crassatellacean bivalves. A clearer picture of the endemism so characteristic of many of these faunas is becoming more well-defined.

This paper is a continuation of a long-term project undertaken by several workers in northern Sinai (Feldman 1986, 1987; Feldman and Owen 1988, 1993; Feldman et al., 1982, 1991) and Israel (Feldman et al., 2001), to taxonomically study and revise the brachiopod faunas. These revisions will help establish the history of brachiopod species and their evolution within the Ethiopian Province. We expect that the data compiled as a result of these studies will enable us to interpret the biogeographic history of the Ethiopian Province as well as gain insight into the structure and paleoecology of its marine communities (see, for example, Feldman and Brett, 1998). We are presently studying a zeillerid fauna from Gebel El-Maghara, northern Sinai and southern Israel, as well as rhynchonellids and terebratulids from Jordan. Upon completion of these taxonomic studies we will have completed our project of revising the brachiopods of the Ethiopian Province that will allow us to define faunal- and province-realm boundaries with greater accuracy.

MATERIALS AND METHODS

The material studied was collected in 1999 by Hegab from Gebel Engabashi, a section with the breached anticline of Gebel El-Maghara in northern Sinai. The collection consists of 28 brachiopod specimens (seven rhynchonellids and 21 terebratulids). The shells are relatively well preserved; all specimens are articulated. The specimens chosen for transverse serial sectioning were calcined and then embedded in plaster. As a result of burning, the internal elements become white and distinct from the sediment and secondary mineralization. Specimens were serially sectioned at a distance of 0.1 or 0.2 mm using a grinding apparatus (Cutrock, London). The sections were drawn using a binocular microscope (Carl Zeiss, Jena) on which was mounted a camera lucida. For every rhynchonellid specimen sectioned we drew approximately 30 sections and for the terebratulids about 70 sections from which the most important are presented. Serial sections are drawn with the ventral valve oriented uppermost (see Motchurova-Dekova et al., 2008). The photographed specimens were coated with ammonium chloride. One of the authors (HRF) had the opportunity to study Cooper's (1989) collections of Saudi Arabian specimens housed in the United States National Museum, Smithsonian Institution (USNM) and Feldman's (1991) collection from northern Sinai, housed in the American Museum of Natural History (AMNH) and the Natural History Museum (NHM) in London and compared them with the present collection. The studied brachiopod specimens are housed in the collection at the Department of Paleontology, the Faculty of Mining and Geology, the University of Belgrade, Serbia (RGF VR).

For the classification of the Rhynchonellida we refer to the Treatise on Invertebrate Paleontology, Vol. 4 (Savage et al., 2002), whereas for the Terebratulida we follow the taxonomy in Vol. 5 (Lee et al., 2006). For the supra-ordinal classification of the Brachiopoda, the classification given by Williams et al. (1996) is herein followed.

SYSTEMATIC PALEONTOLOGY

Phylum BRACHIOPODA Duméril, 1806
Subphylum RHYNCHONELLIFORMEA Williams, Carlson, Brunton, Holmer, and Popov, 1996
Class RHYNCHONELLATA Williams, Carlson, Brunton, Holmer, and Popov, 1996
Order RHYNCHONELLIDA Kuhn, 1949
Superfamily HEMITHIRIDOIDEA Rzhonsnitskaia, 1956
Family CYCLOTHIRIDIDAE Makridin, 1955
Subfamily CYCLOTHYRJDINAE Makridin, 1955
Genus GLOBIRHYNCHIA Buckman, 1918

Type species.—*Rhynchonella subobsoleta* Davidson, 1852, p. 91, pl. 17, fig. 14.

GLOBIRHYNCHIA SPHAERICA (Cooper, 1989) new combination
Figures 4, 5

Gibbirhynchia sphaerica COOPER, 1989, p. 37, pl. 10, figs. 18-30.

Material.—Five specimens, RGF VR 74/1 to RGF VR 74/5.

Description.—External characters: medium sized (see Appendix), roundly triangular to oval in outline, strongly dorsibiconvex, subspherical, typically length greater than width, rarely equal, or width greater then length, maximum width and thickness at midlength, subquadrate in anterior profile, lateral and anterior margins rounded; shell thickened markedly along margins; squama and glotta well expressed; planarea extensive, flat; beak very small, pointed, erect to slightly incurved, foramen hypothyrid, beak ridges indistinct; interarea poorly developed; anterior commissure uniplicate, high, with parallel sides; fold slightly developed, beginning at posterior third, becoming more elevated anteriorly; shallow ventral median sulcus extends from the umbonal region, widening anteriorly; shell surface covered with coarse subangular costae, on each valve numbering 11 to 12 in juvenile and 17 to 18 in adult, four costae on fold and three in sulcus; costae originating in beak region and dorsal umbo, becoming much stronger anteriorly. Internal characters: pedicle collar present; deltidial plates disjunct, relatively thick; dental plates very slightly ventrally divergent, disappearing before full development of hinge teeth; lateral cavities narrow and short; teeth massive, globular, with strong lateral denticula; crural bases crescentic, concave laterally, project above and below the level of the hinge plates, clearly separated from it; hinge plates slender and relatively narrow, ventrally convex and dorsally inclined; euseptoidum reduced to a low ridge; crura canaliform, curving ventrally, with widened and thickened distal ends (similar in appearance to a diabolo), anteriorly splitting into two approximately parallel, non-discrete plates that represent an expression of the gutter-like shape of the crura, terminating at a distance of 0.22 dorsal valve length from dorsal umbo.

Discussion.—Based on external morphology, Cooper (1989), without investigation of interior characters, attributed his new species described herein to *Gibbirhynchia.* Present serial sections through a specimen of *G. sphaerica* show clearly dorsoventrally widened distal ends of crura which are accompanied by distal splitting of the crura into two approximately parallel, non-discrete plates as noted above. This feature among the Mesozoic rhynchonellids has

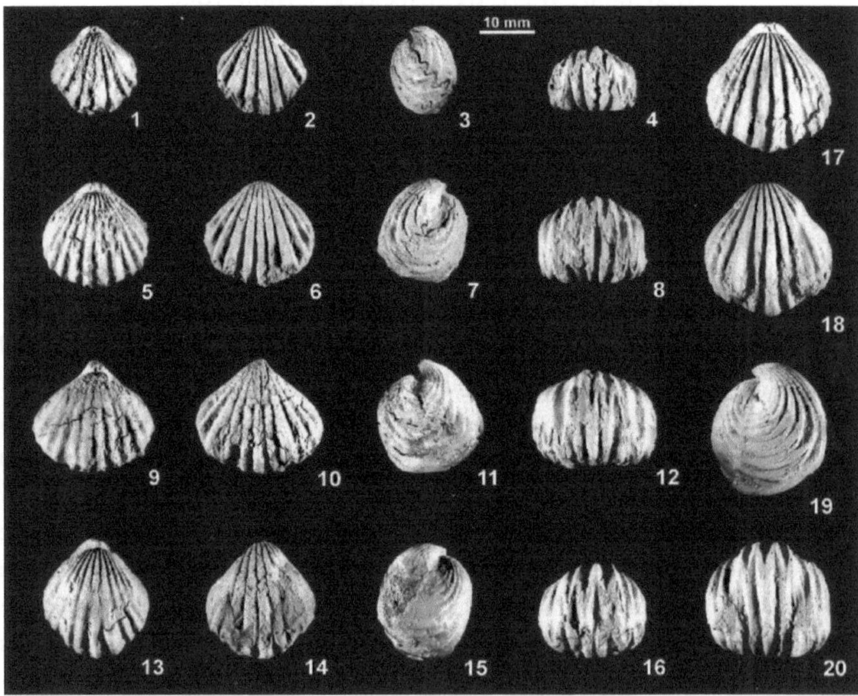

FIGURE 4: *Globirhynchia sphaerica* (Cooper, 1989) new combination, late Bathonian, Gebel Engabashi, northern Sinai. 1-4, RGF VR 74/1, dorsal, ventral, lateral, and anterior views; 5-8, RGF VR 74/2, dorsal, ventral, lateral, and anterior views; 9-12, RGF VR 74/3, dorsal, ventral, lateral, and anterior views; 13-16, RGF VR 74/4, dorsal, ventral, lateral, and anterior views of sectioned specimen; 17-20, RGF VR 74/5, dorsal, ventral, lateral, and anterior views.

been reported so far only in a few genera of the superfamily Hemithiridoidea: in the Early Jurassic *Squamirhynchia* Buckman, 1918, the Middle Jurassic *Globirhynchia* Buckman, 1918, and *Moquellina* Ching, Sun, and Ye in Ching et al., 1979, the Middle Jurassic to Early Cretaceous *Septaliphoria* Leidhold, 1921, the Early to Late Cretaceous *Cyclothyris* M'Coy, 1844, and the Late Cretaceous

FIGURE 5: Transverse serial sections of *Globirhynchia sphaerica* (Cooper, 1989) new combination, late Bathonian, Gebel Engabashi, northern Sinai. Original dimensions of the specimen (in mm): L=23.4; W=20.6; T=18.1. Numbers refer to distance in mm from the ventral apex; RGF VR 74/4.

Almerarhynchia Calzada Badia, 1974 (see Shi and Grant, 1993; Radulovic et al., 2007). Cooper (1989) based his assignment of his species to *Globirhynchia* on the type of crura. *Globirhynchia* has canaliform crura whereas the crura in *Gibbirhynchia* are raduliform. On the other hand, *Gibbirhynchia* is Early Jurassic, whereas *Globirhynchia* is Middle Jurassic in age.

Cooper (1989) erected eight species of the genus *Gibbirhynchia* from the Jurassic of Saudi Arabia. He stated that although conforming to the characters outlined for the genus in exterior details, these specimens differ to some extent from the genus in interior details and among themselves. Thus, generic attribution of those species to *Gibbirhynchia* makes Cooper's species identifications questionable. Manceñido et al. (in Savage et al., 2002, p. 1354) stated in the diagnosis of *Gibbirhynchia* that "Middle Jurassic (Bajocian-Bathonian) records of Arabia as in Cooper (1989) are probably referable to *Nastosia* Cooper, 1989." Externally, upper Bajocian *Nastosia* is very close to *Globirhynchia* but internally it is quite different; it possesses a thin septalium, long and nearly parallel, closely positioned dental plates, long and high septum, and raduliform crura.

As stated by Cooper (1989) the present species is very close to early Bathonian *Gibbirhynchia costata* Cooper (1989, p. 35, pl. 9, figs. 17-23). *Globirhynchia sphaerica* differs in having a much more spherical shape, coarser and crowded costae, and a wider beak. The specimens from northern Sinai figured herein are larger than those of Saudi Arabia figured originally by Cooper (1989) and they match the external characters described by that author.

Occurrence.—Bathonian (middle Dhruma Formation, *Micromphalites* Zone) of Saudi Arabia and late Bathonian of Gebel Engabashi within the breached anticline of Gebel El-Maghara, northern Sinai (N 30°41'25"; E 33°25'07").

Family TETRARHYNCHIIDAE Ager, 1965
Subfamily TETRARHYNCHIINAE Ager, 1965
Genus DAGHANIRHYNCHIA Muir-Wood, 1935

Type species.—*Daghanirhynchia daghaniensis* MUIR-WOOD, 1935, p. 82, pl. 8, fig. 5.

DAGHANIRHYNCHIA ANGULOCOSTATA Cooper, 1989
Figures 6, 7

Daghanirhynchia angulocostata COOPER, 1989, p. 26, pl. 6, figs. 1-19; pl. 7, figs. 44-53; pl. 11, figs. 16- 21; text-figs. 313, 14; Hegab and Aly, 2004, p. 192, pl. 4, figs. 1-4.

Material.—Two specimens, RGF VR 74/6 and 74/7.

Description.—External characters: medium sized (see Appendix), transversely oval in outline, dorsibiconvex, maximum convexity and thickening slightly posterior to midlength; deltidial plates disjunct; beak delicate, sharply pointed, slightly incurved to erect, foramen small, hypothyrid; beak ridges short, poorly developed; planarea present; squama and glotta developed; interarea small, concave; fold originating slightly posteriorly to midlength, elevating and widening anteriorly; sulcus originating slightly posterior to midlength, broad and relatively deep; anterior commissure trapezoidal; each valve ornamented with 19-21 coarse, subangular costae, five to six on fold and four to five in sulcus. Internal characters: dental plates subparallel, very short, lateral cavities small and narrow; teeth strong, globular, with lateral denticula; hinge plates narrow and slender, at beginning horizontal, anteriorly becoming slightly dorsally inclined; euseptoidum reduced to a small ridge; crural bases small, crescent-like, concave laterally, project not only in the dorsal valve, but also project above the level of the hinge plates toward the ventral valve; crura raduliform, curved toward ventral valve, anteriorly widened, straight and divergent, finish at 0.18 of dorsal valve length.

Discussion.—These shells are quite similar to those described by Cooper (1989) in general characters except that the Sinai specimens are not as widely triangular as Cooper's but this could be due to variation. They do compare favorably with some of his other illustrations (pl. 7, figs. 44-53). This species differs from *Daghanirhynchia daghaniensis* (Muir-Wood, 1935, pl. 8, fig. 5a-5c) from the Callovian of Somalia in having a much wider shell. It differs from *Daghanirhynchia magharaensis* Hegab, 2004 and *D. sadeki* Hegab, 2004 from Bathonian of northern Sinai, in having a transversely oval outline and in a larger number of costae.

FIGURE 6: *Daghanirhynchia angulocostata* Cooper, 1989, late Bathonian, Gebel Engabashi, northern Sinai. 1-4, RGF VR 74/6, dorsal, ventral, lateral, and anterior views; 5-8, RGF VR 74/7, dorsal, ventral, lateral, and anterior views of sectioned specimen.

Occurrence.—Bathonian (Dhruma Formation, *Thambites, Tulites, Micromphalites,* and *Dhrumaites* zones) of Saudi Arabia, Egypt (the Gulf of Suez) and late Bathonian of Gebel Engabashi within the breached anticline of Gebel El-Maghara, northern Sinai (N 30°41'25"; E 33°25'07").

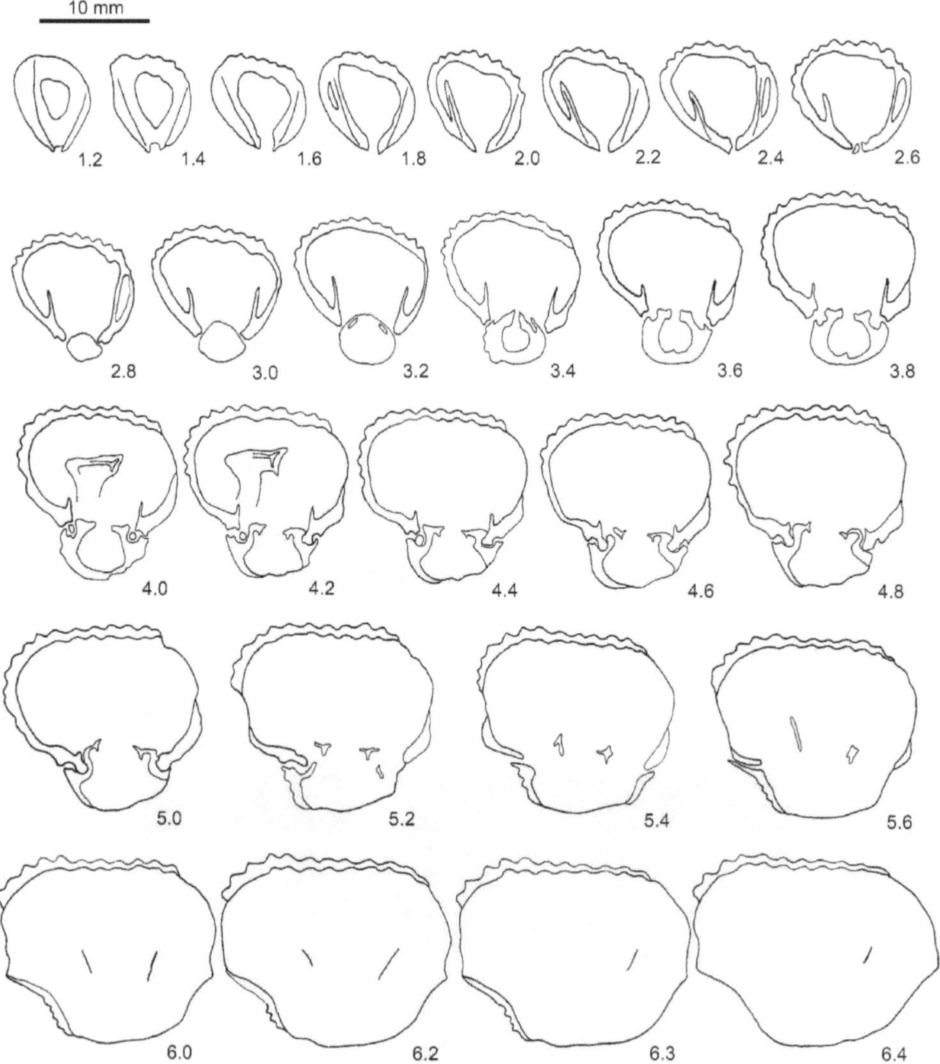

FIGURE 7: Transverse serial sections of *Daghanirhynchia angulocostata* Cooper, 1989, late Bathonian, Gebel Engabashi, northern Sinai. Original dimensions of the specimen (in mm): L=23.9; W=24.8; T=14.6. Numbers refer to distance in mm from the ventral apex; RGF VR 74/7.

Order TEREBRATULIDA Waagen, 1883
Suborder TEREBRATULIDINA Waagen, 1883
Superfamily LOBOIDOTHYRIDOIDEA Makridin, 1964
Family LOBOIDOTHYRIDIDAE Makridin, 1964
Subfamily LOBOIDOTHYRIDINAE Makridin, 1964
Genus ECTYPHORIA Cooper, 1989

Type species.—Ectyphoria inflata COOPER, 1989, p. 80, pl. 20, figs. 14-19, 27-29.

ECTYPHORIA SINAIENSIS new species
Figures 8, 9

Diagnosis.—Medium sized *Ectyphoria,* no dorsal umbonal sulcus, margins thickened, slight folding evident, pedicle collar developed, cardinal process bilobed, dorsal umbonal cavity present, hinge plates subhorizontal, attached

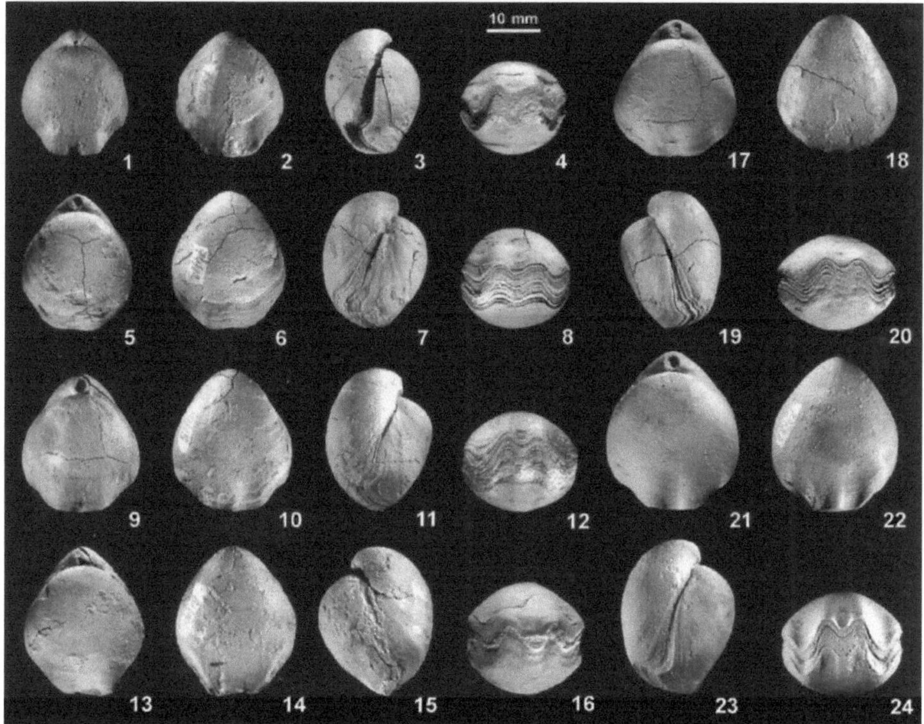

FIGURE 8: *Ectyphoria sinaiensis* new species, late Bathonian, Gebel Engabashi, northern Sinai. 1-4, paratype, RGF VR 74/10, dorsal, ventral, lateral, and anterior views; 5-8 paratype, RGF VR 74/13, dorsal, ventral, lateral, and anterior views; 9-12, paratype, RGF VR 74/14, dorsal, ventral, lateral and anterior views; 13-16, paratype, RGF VR 74/15, dorsal, ventral, lateral, and anterior views of sectioned specimen; 17-20, paratype, RGF VR 74/18, dorsal, ventral, lateral, and anterior views; 21-24, holotype, RGF VR 74/22, dorsal, ventral, lateral, and anterior views.

dorsally to small crural bases, euseptoidum reduced, transverse band highly arched, loop long 0.44 of dorsal valve length, terminal points long.

Description.—External characters: shells medium sized (see Appendix), elongate-oval to rounded-pentagonal in outline, always longer than wide,

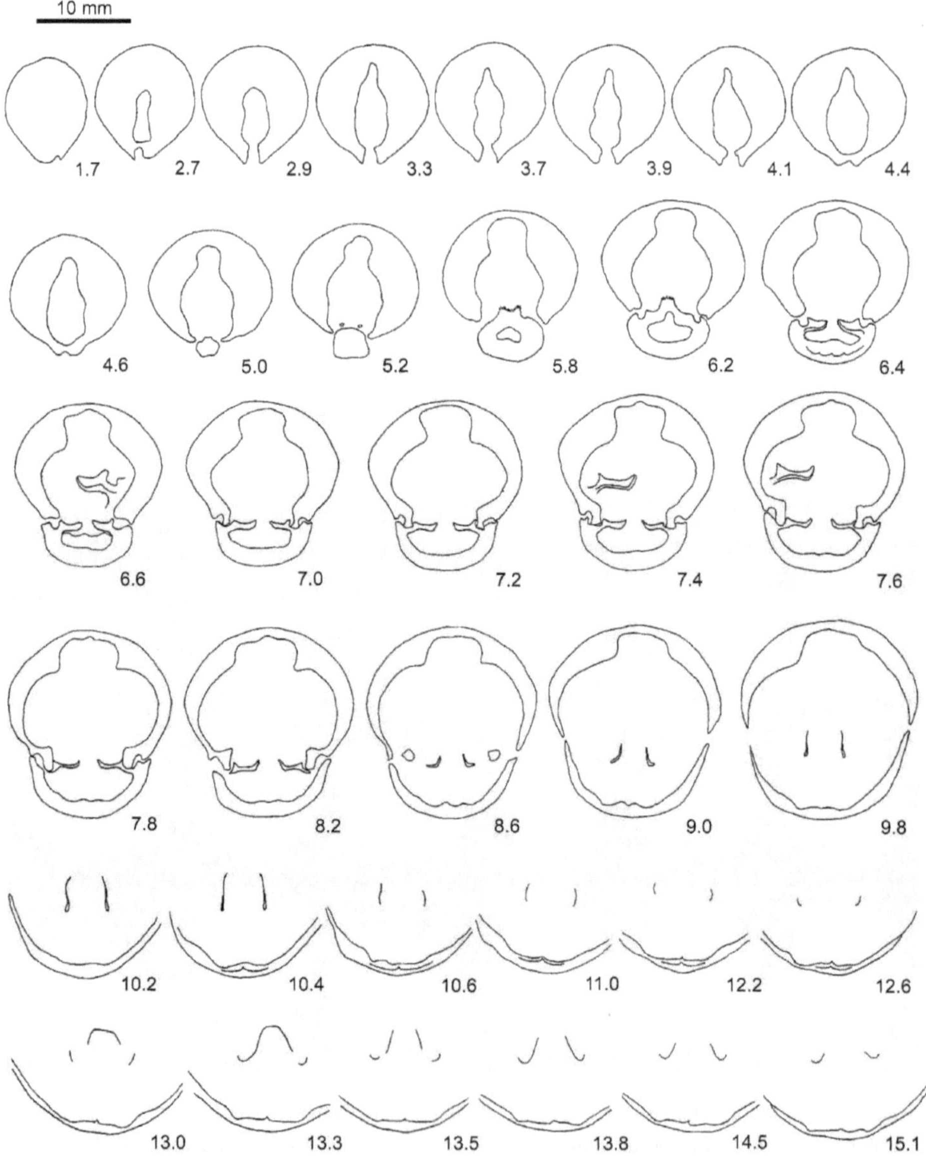

FIGURE 9: Transverse serial sections of *Ectyphoria sinaiensis* new species, late Bathonian, Gebel Engabashi, northern Sinai. Original dimensions of the specimen (in mm): L=28.4; W=22.8; T=22.0. Numbers refer to distance in mm from the ventral apex; paratype, RGF VR 74/18.

strongly and nearly equally biconvex, maximum width slightly anterior to mid length, maximum thickness at midlength; no dorsal umbonal sulcus observed; beak carinate, erect, in close contact with dorsal umbo overlapping and obscuring symphytium, foramen moderate in size, labiate, permesothyrid, beak ridges indistinct; fold and sulcus slightly developed on anterior quarter of shell; lateral commissure straight, anterior commissure slightly bisulcate; numerous growth lines concentrated toward anterior and lateral margins; marginal thickening develops in the adult stage. Internal characters: slender pedicle collar noted in ventral valve; symphytium narrow, relatively thick; cardinal process bilobed; teeth quadrate, nearly vertically inserted in relatively small sockets, with well-developed denticula; dorsal umbonal cavity developed. Hinge plates nearly horizontal; low crural bases; outer socket ridges much stronger than inner ones; crural processes high, subparallel, ventrally tapering; euseptoidum reduced to a low ridge; loop long 0.44 of dorsal valve length, transverse band wide, highly arched with long terminal points.

Etymology.—After the Sinai Peninsula from where the shells were collected.

Types.—Holotype RGF VR 74/22, paratypes RGF VR 74/8 to RGF VR 74/21; a total of 15 specimens available. *Occurrence.*—Late Bathonian of Gebel Engabashi within the breached anticline of Gebel El-Maghara, northern Sinai (N 30°41'25"; E 33°25'07").

Discussion.—Referring to the external characters, such as elongate-oval to rounded-pentagonal outline, the strongly inflated valves, the shape of beak and biplicate anterior commissure, the new species is very similar to the type species, *Ectyphoria inflata* Cooper, 1989 (p. 80, pl. 20, figs. 14-19, 27-29) from the late Bathonian (Dhruma Formation, *Dhrumaites* Zone) of Saudi Arabia. The main external difference between the new species and *E. inflata* is the presence of a dorsal umbonal sulcus in the latter. The internal characters presented in Figure 9 are comparable with those for the genus given in Cooper's serial sections (1989, fig. 36) in which the transverse band and terminal points are missing.

Genus COOPERITHYRIS new genus

Type species.—*Cooperithyris circularis* new species.

Included species.—*Sphaeroidothyris weiri* Hegab, 1992.

Etymology.—Name in honor of the late G. Arthur Cooper.

Diagnosis.—As for species.

Discussion.—Based on external morphology, particularly circular outline, short beak, and folding of the shell which is never well developed, the new genus, can be compared only with *Somalithyris* Muir-Wood, 1935, from the Oxfordian-Kimmeridgian of Somalia and Saudi Arabia. It differs externally in having a more circular outline and much less developed folding which is confined toward the anterior commissure. The internal structures of these two genera are different, with the new genus having a less developed cardinal process, smaller crural bases, subquadrate teeth with a sharper angle of insertion and longer loop. It is interesting to note that in *Cooperithyris* and *Somalithyris* (see Muir-Wood, 1936, fig. 10), and also in some Jurassic genera such are *Sphaeroidothyris* Buckman, 1918 (see Muir-Wood, 1936, fig. 10), *Ptyctothyris* Buckman, 1918 (see Muir-Wood, 1935, fig. 20), *Avonothyris* Buckman, 1918 (see Muir-Wood, 1965, fig. 637, 2a-n), *Epithyris* Phillips, 1841 (see Muir-Wood, 1965, fig. 641, 2a-j), *Tubithyris* Buckman, 1918 (see Radulovic, 1993, p. 156, fig. 4), and *Holcotchithyris* Hegab, 1995 (fig. 3) the euseptoidum is bounded by two parallel ridges. This character should not be a criterion for determining the genus or even the species.

Occurrence.—Late Bathonian of Gebel Engabashi within the breached anticline of Gebel El-Maghara, northern Sinai (N 30°41'25"; E 33°25'07").

COOPERITHYRIS CIRCULARIS new species
Figures 10, 11

Diagnosis.—Medium size, circular to elongate oval, moderately equibiconvex, sulciplicate; beak short, small, erect, foramen large, slightly labiate, permesothyrid; folding confined near anterior commissure; symphytium partly visible; cardinal process bilobed; crural bases small; hinge plates thin and wide, dorsally inclined; loop 0.48 of dorsal valve length, transverse band broad, highly arched, terminal points moderately long.

Description.—External characters: medium sized (see Appendix), circular in outline, slightly longer than wide, maximum width and thickness at mid length or slightly anterior; beak short, small, erect, in close contact with dorsal umbo, partially concealing symphytium, foramen permesothyrid, relatively large, circular, slightly labiate, pinhole size in juveniles, beak ridges indistinct; lateral commissure nearly straight, anterior commissure gently sulciplicate; folding of the shell weak, confined to near anterior commissure; surface marked by well-developed growth lines. Internal characters: low pedicle collar present; symphytium thick; cardinal process bilobed; teeth quadrate, with well-developed lateral denticula; hinge plates thin and wide, dorsally inclined; crural bases very small, project toward ventral valve, crural processes are high and slender, slightly ventrally convergent; euseptoidum strong, paralleled by two ridges; loop thin, extending 0.48 of dorsal valve length, transverse band broad and high arched, terminal points moderately long.

Etymology.—Latin *circularis,* circular, refers to the outline of the shell.

Types.—Holotype RGF VR 74/26; paratypes RGF VR 74/23 to RGF VR 74/25; a total of 4 specimens available.

Occurrence.—Late Bathonian of Gebel Engabashi within the breached anticline of Gebel El-Maghara, northern Sinai (N 30°41'25"; E 33°25'07").

Discussion.—Based on the circular outline, the shape of the beak, and sulciplicate anterior commissure the new species can only be compared with some representatives of the genus *Somalithyris* described by Cooper (1989) from the Jurassic of Saudi Arabia, namely, *Somalithyris subcircularis* Cooper, 1989 (p. 89, pl. 29, figs. 8-14) and *Somalithyris rotundata* Cooper 1989 (p. 89, pl. 29, figs. 5-17) from the Kimmeridgian (Hanifa Formation). The new species differs in having more inflated valves, a more circular outline and less pronounced folding. Another form from Saudi Arabia which is similar in general aspect to

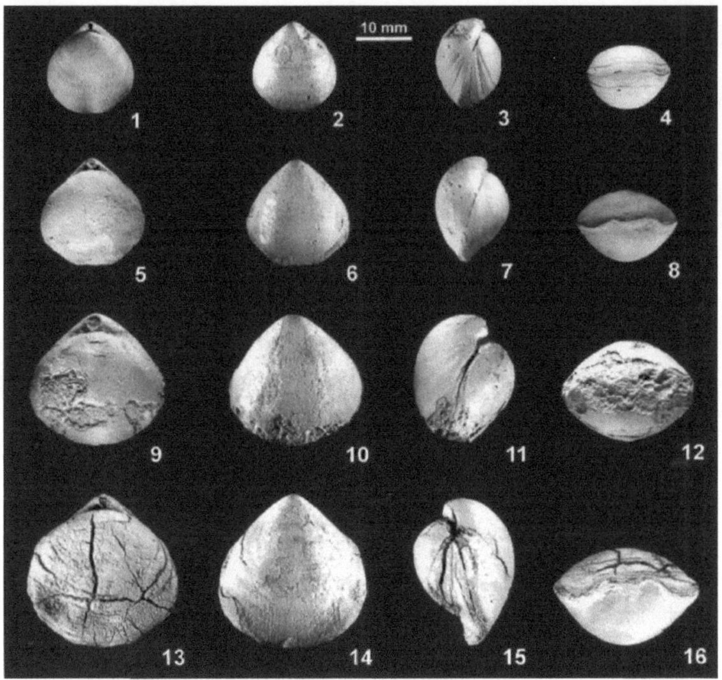

FIGURE 10: *Cooperithyris circularis* new genus and species. 1-4, paratype, RGF VR 74/23, dorsal, ventral, lateral, and anterior views; 5-8, paratype, RGF VR 74/24, dorsal, ventral, lateral, and anterior views; 9-12, paratype, RGF VR 74/25, dorsal, ventral, lateral, and anterior views of sectioned specimen; 13-16, holotype, RGF VR 74/26, dorsal, ventral, lateral, and anterior views.

the present species is the late Bajocian-early Bathonian *Sphaeroidothyris arabica* Cooper, 1989 (p. 90, pl. 29, figs. 26-34). It is distinguished by its smaller size, less inflated valves and rectimarginate anterior commissure.

From Gebel El-Maghara in northern Sinai Hegab (1992) introduced two late Callovian species: *Sphaeroidothyris weiri* (p. 40, pl. 1, figs. 12-19) and *Sphaeroidothyris magharaensis* (p. 41, pl. 1, figs. 20-22). *Sphaeroidothyris magharaensis* is considered to be a juvenile form of S. *wieri* and therefore a synonym of the latter. The interior of these fits well with those of described species.

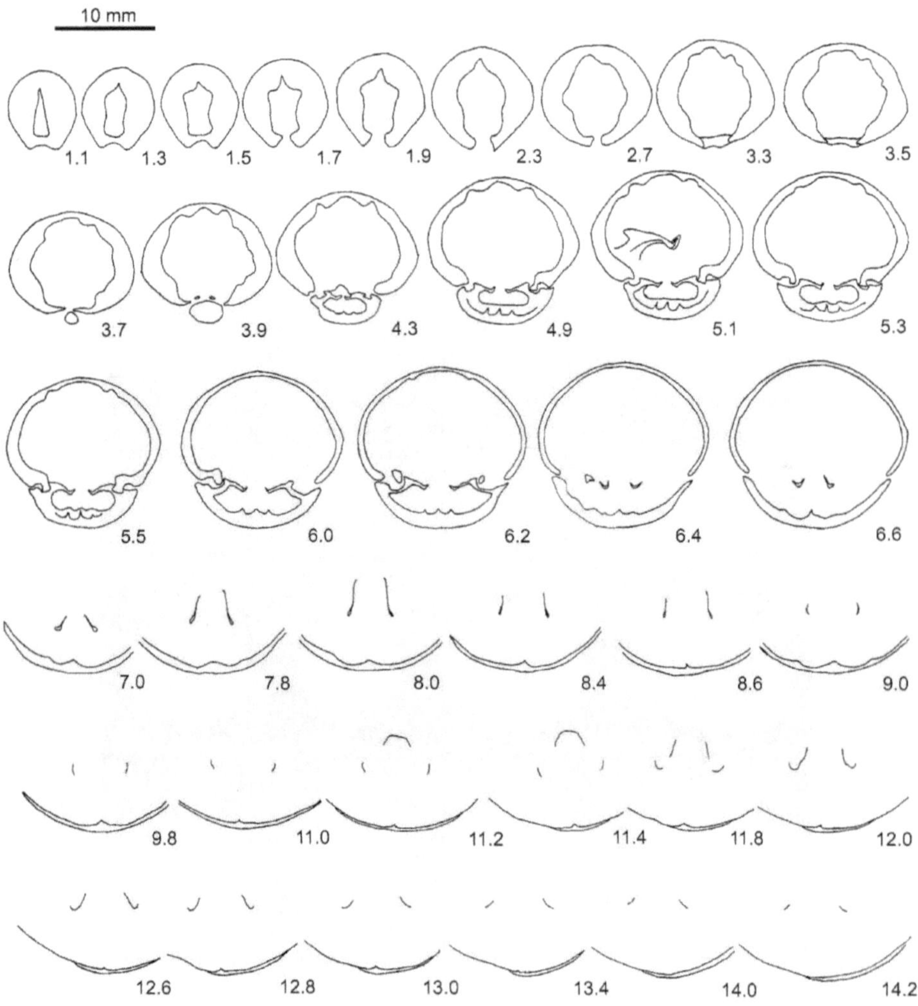

FIGURE 11: Transverse serial sections of *Cooperithyris circularis* new genus and species, late Bathonian, Gebel Engabashi, northern Sinai. Original dimensions of the specimen (in mm): L=28/4; W=22.8; T=22.0. Numbers refer to distance in mm from the ventral apex; paratype, RGF VR 74/25.

Genus AVONOTHYRIS Buckman, 1918

Type species.—Avonothyris plicatina BUCKMAN, 1918, p. 102, pl. 21, fig. 9.

AVONOTHYRIS species A
Figure 12

Avonothyris? species COOPER, 1989, p. 75, pl. 23, figs. 1-3.

Material.—One specimen with damaged beak, RGF VR 74/27.

Description.—External characters: large size (see Appendix), elongate oval in outline, equibiconvex, lateral margins slightly rounded, maximum width and thickness at mid length; beak massive, wide, erect, beak ridges short and rounded; lateral conunissure gently ventrally arched in the posterior two-thirds and sharply curved ventrally at anterior, anterior commissure strongly bisulcate; a shallow sulcus originates about midlength on dorsal valve, deepens anteriorly, bordered by rounded and parallel carinae. Ventral valve with median flattened and wide fold originating one-third the distance from the beak, separated by shorter and narrower sulci; fine growth lines, much more numerous and crowded anteriorly, and ornament fine radiating capillae (about two to three capillae per mm near the anterior margin). Internal characters: due to the paucity of material no serial sections could be prepared and internal morphology was not studied.

Discussion.—The described species show external similarities, such as outline, inflation of the valves, and especially the large size, with some species of *Avonothyris,* namely, *A. corpulenta* Buckman, 1918 and *A. variabilis* Feldman et al., 1991. In contrast to the specimen figured here, *A. corpulenta,* from the Bathonian of southern England, has a narrower beak, more widely triangular shell and slightly wider fold with convergent carinae. *A. variabilis* from the late Callovian of northern Sinai has a more subpentagonal outline and convergent and rounded carinae on the dorsal fold.

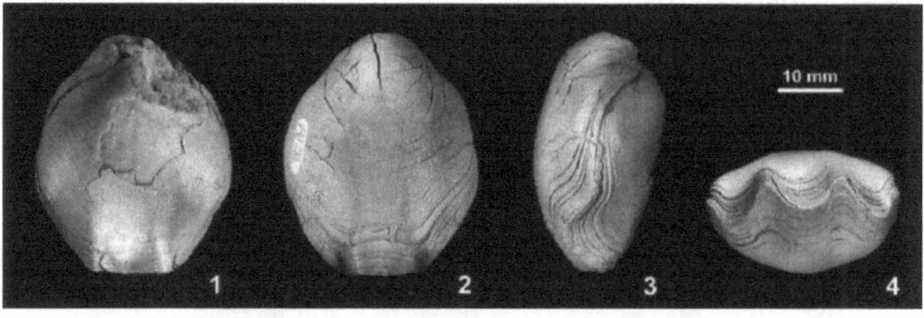

FIGURE 12: *Avonothyris* species A, late Bathonian, Gebel Engabashi, northern Sinai, 1-4, RGF VR 74/27, dorsal, ventral, lateral, and anterior views.

Externally the species is similar in dimensions, outline and the shape of folding to *Avonothyris?* species in Cooper, 1989, from the Bajocian-early Callovian (the precise age not placed) of Saudi Arabia. It differs only in having parallel carinae on the dorsal fold, a median ventral elevation, and a lateral commissure sharply curved ventrally and anteriorly. In all the described species the internal characters are unknown.

Occurrence.—Late Bathonian of Gebel Engabashi within the breached anticline of Gebel El-Maghara, northern Sinai (N 30°41'25"; E 33°25'07").

Genus PTYCTOTHYRIS Buckman, 19 18

Type species.—Terebratula stephani Davidson, 1877, p. 75, pl. I, fig. 3.

PTYCTOTHYRIS species A
Figure 13

Material.—One specimen, RGF VR 74/28.

Description.—External characters: medium sized (see Appendix), elongate subpentagonal outline, longer than wide, maximum width in anterior third of shell, maximum thickness at rnidlength, slightly ventribiconvex; beak strong, erect, in close contact with dorsal umbo totally obscuring symphytium, mesothyrid foramen, circular, relatively large, slightly labiate, beak ridges poorly developed, rounded; fold strongly elevated anteriorly, originating at midlength, bounded by two subangular carinae separated by a shallow and broad sulcus; dorsal sulcus wide, shallow, developed in anterior half, formed by two rounded carinae, with widely rounded, median carina; anterior commissure strongly biplicate; shell surface marked by numerous, well developed growth lines. Internal characters: since there is only one specimen in the collection the interior of this species is unknown.

Discussion.—The specimen studied herein is very similar in the shape of its beak, strongly biplicate anterior commissure and folding of the shell to *Ptyctothyris quillyensis* (Bayle, 1878, pl. 7, fig. 9; refigured in Almeras and Fauré, 2008, pl. II, fig. 7), from the early Bathonian *(Zigzag* Zone) of France (type

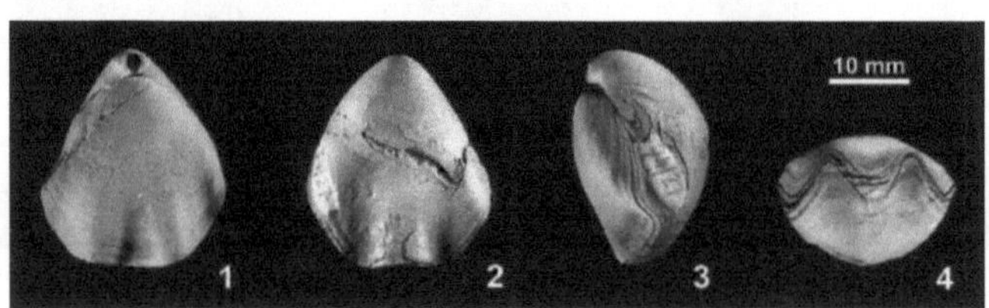

FIGURE 13: *Ptyctothyris* species A, late Bathonian, Gebel Engabashi, northern Sinai, 1-4, RGF VR

locality Quilly, Calvados), northern Sinai (Moghara Massif, Gebel Aroussieh), Somalia and Kenya. It differs from Bayle's species only in having maximum width in the anterior third (W/L=0.85 instead of 0.65 in the holotype of *P. quillyensis).* Another two representatives of the genus from northern Sinai (Gebel El-Maghara) are the late Bajocian *P. sinaiensis,* Feldman et al., 1991 and the late Callovian *P.? daghaniensis* Muir-Wood, 1935, the latter also known from Somalia. In contrast to the new species *P. sinaiensis* much weaker folding and a narrower shell. *P.? daghaniensis* has a more rounded subpentagonal outline with maximum width slightly anterior to midlength as well as weaker folding of the shell. All the species share the characteristic shape of the beak.

Occurrence.—Late Bathonian of Gebel Engabashi within the breached anticline of Gebel El-Maghara, northern Sinai (N 30°41'25"; E 33°25'07").

ACKNOWLEDGMENTS

This research was supported by a grant from the Ministry of Education and Science of the Republic of Serbia, Project No. 176015 (grant to VR and BR). Funding was also provided by grants from the National Geographic Society (to HRF). We wish to thank B. Huber (USNM) for editorial guidance. M. R. Sandy (the University of Dayton) and an anonymous reviewer provided critical remarks that helped to improve the manuscript. Thanks also go to D. Ruban (Rostov State University) who provided the authors with some paleogeographic maps.

REFERENCES

Ager, D. V. A. 1965. Mesozoic and Cenozoic Rhynchonellacea, H597-H632. In R. C. Moore (ed.), *Treatise on Invertebrate Paleontology, Pt. H. Brachiopoda 2*. Lawrence, KS: Geological Society of America and University of Kansas Press.

Al Far, D. M. 1966. *Geology and coal deposits of Gebal El Maghara* (N. Sinai). Geological Survey of Egypt, Paper 37.

Alméras, Y. and Ph. Fauré. 2008. Les Brachiopods du Jurassic moyen sur la marge de La Téthys occidentale (Maroc, Algérie occidentale). *Revue de Paléobiologie*, 27: 575-857.

Arkell, W. J. 1952. Jurassic ammonites from Jebel Tuwayq, Central Arabia. Philosophical Transactions of the Royal Society of London, Series B 236: 241-313.

------. 1956. *Jurassic geology of the world*. Edinburgh: Oliver and Boyd.

Bayle, C. E. 1878. Atlas: Fossiles principaux des terrains. *Explication de la Carte Géologique de la France, Mémoire*, 4(1), 158 pls.

Buckman, S. S. 1918. The Brachiopoda of the Namyau Beds, Northern Shan States. Burma. *Memoirs of the Geological Survey of India, Palaeontologia Indica* (new series), 3.

Calzada Badia, S. 1974. *Almerarhynchia* n. gen. *virgiliana* n. sp. del Maastrichtiense de Figols, Prepirineo catalán. *Acta Geologica Hispanica* 9: 92-97.

Ching, Yu-Gan, Ye Song-Ling, Xu Han-Kui, and Sun Dong-Li. 1979. Brachiopoda. In *Paleontological Atlas of Northwest China* (Fascicle Qinghai). Pt. l. Nanjing Institute of Geology and Palaeontology. Academia Sinica and Institute of Geological Science, Qinghai Province. Beijing: Geological Publishing House. [In Chinese]

Cooper, G. A. 1989. Jurassic brachiopods of Saudi Arabia. *Smithsonian Contributions to Paleobiology* 65.

Davidson, T. 1852. *British Fossil Brachiopoda, Oolitic and Liassic species.* Vol. 1, pt. 3, no. 2, 65-100. London: Palaeontological Society Monograph.

Davidson, T. 1877. On the species of Brachiopoda that occur in the Inferior Oolite at Bradford Abbas and its vicinity. *Dorset Natural History and Antiquarian Field Club, Proceedings.* 1: 73-88.

Dewey, J. F., W. C. Pitman, W. B. F. Ryan, and J. Bonnin. 1973. Plate tectonics and the evolution of the Alpine system. *Geological Society of America Bulletin* 84: 3137-3180.

Douvillé, H. 1916. Les terrains secondaires dans le Massif de Moghara à l'Ést de l'Isthme Suez. *Académie des Sciences de Paris, Mémoire* 54.

------. 1925. Le Calloviell dans le massif de Moghara: Avec description des fossils par M. Cossmann. *Bulletin de la Société géologique de France*, sér. 4, 25: 305-328.

Duméril, A. M. C. 1806. *Zoologie analytique ou méthode naturelle de classification des animaux.* Paris: Allais.

Farag, I. A. M. 1957. On the occurrence of Lias in Egypt. *Egyptian Journal of Geology* 1: 49-63.

------. 1959. Contribution to the study of the Jurassic formations in the Maghara massif (northern Sinai, Egypt). *Journal of Geology of the United Arab Republic* (1961 for 1959) 3: 175-199.

Farag, I. A. M., and W. Gatinaud. 1960a. Un nouveau genre de Terebratulides dans le Bathonien d'Egypte. *Journal of Geology of the United Arab Republic* 4: 77-79.

------. 1960b. Six espéces nouvelle du genre *Rhynchonella* dans le roches jurassiques d'Egypte. *Journal of Geology of the United Arab Republic* 4: 81-87.

Feldman, H. R. 1986. Cladistic analysis of a Jurassic rhynchonellid brachiopod genus from the Middle Eastern Ethiopian Province. *Geological Society of*

America, Northeastern Section Meeting, Abstracts, Kiamesha Lake, New York, 18: 16.

------. 1987. A new species of the Callovian (Jurassic) brachiopod *Septirhynchia* from northern Sinai. *Journal of Paleontology* 61: 1156-1172.

Feldman, H. R., and C. E. Brett. 1998. Epi- and endobioutic organisms on Late Jurassic crinoid columns from the Negev Desert, Israel: Implications for coevolution. *Lethaia* 31: 57-71.

Feldman, H. R., and E. F. Owen. 1988. *Goliathyris lewyi*, new species (Brachiopoda, Terebratellacea), from the Jurassic of Gebel El-Minshera, northern Sinai. *American Museum Novitates* 2908: 1-12.

Feldman, H. R., and E. F. Owen. 1993. Vicariance biogeography of the Middle Eastern Ethiopian Province: Implications for the Paleozoic. *Geological Society of America Annual Meeting, Boston, Massachusetts*, 25: 55.

Feldman, H. R., F. Hirsch, and E. F Owen. 1982. A composition of Jurassic and Devonian brachiopod communities: Trophic structure, diversity, relations and niche replacement. *Journal of Paleontology* 56: 9-10.

Feldman, H. R., E. F. Owen, and F. Hirsch. 1991. Brachiopods from the Jurassic of Gebel El-Maghara, northern Sinai. *American Museum Novitates* 3006: 1-28.

------. 2001. Brachiopods from the Jurassic (Callovian) of Hamakhtesh Hagadol (Kurnub Anticline), southern Israel. *Palaeontology* 44: 637-658.

Feldman, H. R., M. Schemm-Gregory, F. Ahman, and M. A. Wilson. 2009. New implications on biogeography and taxonomy of the zeillerid terebratulids *Eudesia* and *Sphriganaria* (Brachiopoda, Middle Jurassic). Abstracts, 79th Annual Meeting of the German Paleontological Society, Bonn, Terra Nostra, Schriften der Alfred Wegener Stiftung, 3: 33.

Goldberg, M., A. Barzel, P. Cook, and Y. Mimran. 1971. Preliminary columnar section of Gebel Maghara. Report No. MM/1/71, Abstracts, Israel Geological Society Meeting.

Hallam, A. 1975. *Jurassic environments*. Cambridge: Cambridge University Press.

Haq, B., J. Hardenbol, and P. Vail. 1988. Mesozoic and Cenozoic chronostratigraphy and cycles of sea-level chauge. *SEPM Special Publication* 42: 71-108.

Hegab, A. A. 1988. Analysis of growth patterns in *Eudesia* (Brachiopoda) and its potential use for the identification of different species of the genus. *Bulletin of the Faculty of Science, Assiut University* 17: 25-36.

------. 1989. New occurrence of Rhynchonellida (Brachiopoda) from the Middle Jurassic of Gebel El-Maghara, northern Sinai. *Journal of African Earth Sciences* 9: 445-453.

------. 1991a. New genus *Praeudesia* (Brachiopoda) from the Jurassic outcrops of Gebel El-Maghara, northern Sinai, Egypt. *Bulletin of the Faculty of Science, Assiut Uuiversity* 20: 1-17.

------. 1991b. The occurrence of genus *Flabellothyris* (Brachiopoda) from the Jurassic of northern Sinai. *Bulletin of the Faculty of Science, Assiut University* 20: 39-49.

------. 1991c. Biostratigraphic zonation of Bathonian-Callovian rocks from Gebel El-Maghara, Northern Sinai, Egypt. *Journal of African Earth Sciences* 13: 183-192.

------. 1992. Terebratulida (Brachiopoda) from the Jurassic of Gebel El-Maghara, northern Sinai. *Proceedings of the 8th Symposium of Phanerozoic Developments in Egypt* 8: 33-42.

------. 1993. *Eudesia* (Brachiopoda) community from the Bathonian of Gebel El-Maghara (northern Sinai): Their morphologic adaptation, ontogenetic variation and paleoecology. *Palaeontographica* 229: 1-14.

------. 1995. *Holcotchithyris* new genus, Terebratulida (Brachiopoda) from Jurassic rocks of Northern Sinai. *Journal of Basic and Applied Sciences* 1-3: 259-270.

------. 2003. Brachiopoda from the Middle Jurassic at Bir El-Maghara, northern Sinai, Egypt. *The Third International Conference on the Geology of Africa, Assiut* 2: 427-442.

------. 2004. Rhynchonellida (Brachiopods) from the Middle Jurassic of northern Sinai, Egypt. *Abstracts, 7th International Conference on the Geology of the Arab World, Cairo University*, 255-263.

Hegab, A. A., and M. F. Aly. 2004. Contribution to the Middle Jurassic Brachiopoda (Rhynchonellida) from Gulf of Suez, Egypt. *Egyptian Journal of Paleontology* 4: 183-198.

Hirsch, F. 1979. Jurassic bivalves and gastropods from northern Sinai and southern Israel. *Israel Journal of Earth Sciences* 28: 128-163.

Hirsch, F. and L. Picard. 1988. The Jurassic facies in the Levant. *Journal of Petroleum Geology* 2: 277-308.

Hirsch, F., B. Conway, H. R. Feldman, E. F. Owen, and A. Rosenfeld. 1994. The Jurassic of the Levant: Biostratigraphy, palaeogeography and global events. *Abstracts, Fourth International Congress of Jurassic Stratigraphy and Geology, Mendoza, Argentina* 4: 21.

Hirsch, F., J.-P. Bassoulett, É. Cariou, B. Conway, H. R. Feldman, L. Grossowicz, A. Honigstein, E. F. Owen, and A. Rosenfeld. 1998. The Jurassic of the southern Levant. Biostratigraphy, palaeogeography and cyclic events. In S. Carasquin-Soleau and É. Barrier (eds.), Peri-Tethys memoir 4: epicratonic basins of Peri-Tethyan platforms. *Mémoires du Muséum National d'Histoire Naturelle* 179: 213-235.

Hoppe, W. von. 1922. Jura und Kreide der Sinai-balbinsel. *Zeitschrift der Deutschen Geologischen Gesellschaft Palaestina Yereins* 45: 61-79; 97-219.

Kitchin, F. L. 1912. Palaeontological work: England and Wales: Summation of programme for 1911. *Memoir of the Geological Survey of Great Britain and Museum of Practical Geology*, 59-60.

Kuhn, O. 1949. *Lehrbuch der Palaozoologie*. Stuttgart: E. Schweizerbart.

Lee, D. E., D. I. Mackinnon, T. N. Smirnova, P. G. Baker, Y.-G. Jin and D.-L. Sun. 2006. Terebratulida, 1965-2255. In R. L. Kaesler (ed.), *Treatise on Invertebrate Paleontology. Pt. H. Brachiopoda 5*. Lawrence, KS: Geological Society of America and University of Kansas Press.

Liedhold, C. 1921. Beitrag zur genaueren Kenntnis und Systematik einiger Rhynchonelliden des reichsländischen Jura. *Neues Jahrbuch für Mineralogie, Geologie und Paläontologie (Beilage Band)* 44: 343-368.

M'Coy, F. 1844. *A Synopsis of the Characters of the Carboniferous Limestone Fossils of Ireland*. Dublin: University Press.

Makridin, V. P. 1955. Nekotorye ûrskie rinhonellidy Evropejskoj časti SSSR. *Zapiski Geologičeskogo Fakulteta Har'kovskogo Universiteta* 12: 81-91. [In Russian]

Makridin, Y. P. 1964. *Brahiopody ûrskih otloženij Russkoj platformy i nekotoryh priležaŝih k nej oblastej*. Moskva: Nedra. [In Russian]

Moon, F. W. and H. Sadek. 1921. Topography and geology of northern Sinai. *Petroleum Research Bulletin, Cairo* 10.

Motchurova-Dekova, N., Y. Radulovic, and M. A. Bitner. 2008. Orientations of the valves when presenting serial sections of post-Palaeozoic brachiopods: Tradition and utility versus the revised Treatise. *Lethaia* 41: 493-495.

Muir-Wood, H. M. 1935. Jurassic Brachiopoda. In W. A. Macfadyen et al. (eds.), *The Mesozoic Palaeontology of British Somaliland. Pt. 2. The Geology and Palaeontology of British Somaliland*, 75-147 London: Government of the Somaliland Protectorate.

------. 1936. *A Monograph of the Brachiopoda of the British Great Oolite Series. Pt. I. The Brachiopoda of the Fuller's Earth*. Palaeontographical Society of London, Monograph 89.

------. 1965. Mesozoic and Cenozoic Terebratulidina, H762-H816. In R. C. Moore (ed.), *Treatise on Invertebrate Paleontology, Pt. H. Brachiopoda 2*. Lawrence, KS: Geological Society of America and University of Kansas Press.

Parnes, A. 1974. Biostratigraphic correlation of the Middle Jurassic in Makhtesh Ramon, Gebel El-Maghara and in Morocco. *Abstracts, Israel Geological Society, Annual Meeting*, 26.

Phillips, J. 1841. Figures and descriptions of the Palaeozoic fossils of Cornwall, Devon, and west Somerset. *Geological Survey of Great Britain, Memoir*. London: I. Longman and Co.

Picard, L. and F. Hirsch. 1987. *The Jurassic stratigraphy in Israel and adjacent countries.* Jerusalem: The Israel Academy of Sciences and Humanities.

Radulovic, Y. 1993. Middle Jurassic brachiopods from Novo Korito area (eastern Serbia, Carpatho-Balkanides). *Geološki anali Balkanskoga poluostrva* 57: 141-159.

Radulovic, B., N. Motchurova-Dekova, and Y. Radulovic. 2007. New Barremian rhynchonellide brachiopod from Serbia and the shell microstructure of Tetrarhynchiidae. *Acta Palaeontologica Polonica* 52: 761-782.

Range, P. 1920. Die Geologie der Isthmuswueste. *Zeitschrift der Deutschen geologische Gesellschaft* 72: 233-242.

Rzhonsnitskaia (Ržonsnitsaâ), M. A. 1956. Systematization of Rhynchonellida. XX *Congreso Geológico Internacional, Mexico, Resúmenes de los Trabajos presentados, Report* 20: 125-126.

Savage, N. M., M. O. Manceñido, E. F. Owen, S. J. Carlson, R. E. Grant, A. S. Dagys, and S. Dong-Li. 2002. Rhynchonellida. In R. L. Kaesler (ed.), *Treatise on Invertebrate Paleontology, Pt. H Brachiopoda* (revised) 4, 1027-1276. Lawrence, KS: Geological Society of America and University of Kansas Press.

Scotese, C. R. 2004. A continental flipbook. *Journal of Geology* 112: 729-741.

Shi, Xiao-Ying, and R. E. Grant. 1993. Jurassic rhynchonellids, internal structures and taxonomic revisions. *Smithsonian Contributions to Paleobiology* 73.

Vörös, A. 1984. Lower and Middle Jurassic brachiopod provinces in the Western Tethys. *Annales Universitatis Scientiarum Blidapestinensis de Rolando Eötvös nominatae, Sectio Geologica* 24: 207-233.

Waagen, W. H. 1882-1885. Salt range Fossils, Pt. 4 (2) Brachiopoda. *Memoires of the Geological Survey of India* 13: 391-546.

Weir, J. 1925. Brachiopoda, Lamellibranchiata, Gastropoda and belemnites. In The collection of fossils and rocks from Somaliland made by messrs. B. K. N. Wyllie and W. R. Smellie. Monographs of the Geological Department of the Hunterian Museum, Glasgow University, 3: 1-63.

Williams, A., S. J. Carlson, C. H. C. Brunton, L. E. Holmer, and L. Popov. 1996. A supra-ordinal classification of the Brachiopoda. *Philosophical Transactions of the Royal Society of London, B* 351: 1171-1193.

Wilson, M. A., H. R. Feldman, J. C. Bowen, and Y. Avni. 2008. A new equatorial, very shallow marine sclerozoan fauna from the Middle Jurassic (late Callovian) of southern Israel. *Palaeogeography, Palaeoclimatology, Palaeoecology* 263: 24-29.

APPENDIX A:
Original Publication and Contributing Authors

1. A Comparison of Jurassic and Devonian Brachiopod Communities: Trophic Structure, Diversity, Substrate Relations and Niche Replacement

Howard R. Feldman, Francis Hirsch, and Ellis F. Owen*

Third North American Paleontological Convention, Proceedings, Montreal, Canada, 1:169-174. Vol. 1, August 1982

American Museum of Natural History, New York, NY, Geological Survey of Israel, Jerusalem, and British Museum (Natural History), London

2. A New Species of the Jurassic (Callovian) Brachiopod *Septirhynchia* from Northern Sinai

Howard R. Feldman

Journal of Paleontology 61, no. 6 (November 1987): 1156-1172.

3. *Goliathyris lewyi,* New Species (Brachiopoda, Terebratulacea) from the Jurassic of Gebel El-Minshera, Northern Sinai

Howard R. Feldman and Ellis F. Owen*

American Museum Novitates 2908 (11 February 1988): 1-12

American Museum of Natural History, New York, NY

4. Brachiopods from the Jurassic of Gebel El-Maghara, Northern Sinai

Howard R. Feldman, Ellis F. Owen,* and Francis Hirsch

American Museum Novitates 3006 (22 May 1991): 1-19

American Museum of Natural History, New York, NY

5. Epi- and Endobiontic Organisms on Late Jurassic Crinoid Columns from the Negev Desert, Israel: Implications for Co-evolution

Howard R. Feldman and Carlton E. Brett

Lethaia (Oslo) 31 (1998): 57-71.

6. The Jurassic of the Southern Levant: Biostratigraphy, Palaeogeography and Cyclic Events

Francis Hirsch, Jean-Paul Bassoulet, Élie Cariou, Brian Conway, Howard R. Feldman, Lydia Grossowicz, Avraham Honigstein, Ellis F. Owen,* and Amnon Rosenfeld

Peri-Tethys Memoir 4: Epicratonic basins of Peri-Tethyan platforms, *Mémoires du Muséum National d'Histoire Naturelle* (Paris) 179 (1998): 213-235.

7. Brachiopods from the Jurassic (Callovian) of Hamakhtesh Hagadol (Kurnub Anticline), Southern Israel

Howard R. Feldman, Ellis F. Owen,* and Francis Hirsch

Paleontology 44, no. 4 (2001): 637-658

8. A New Species of *Coenothyris* (Brachiopoda) from the Triassic (Upper Anisian-Ladinian) of Israel

Howard R. Feldman

Journal of Paleontology 76, no. 1 (January 2002): 34-42

9. Paleoecology, Taphonomy, and Biogeography of a *Coenothyris* Community (Brachiopoda, Terebratulida) from the Triassic (Upper Anisian-Lower Ladinian) of Israel

Howard R. Feldman

American Museum Novitiates 3479 (25 July 2005): 1-19.

American Museum of Natural History, New York, NY

10. A New Equatorial, Very Shallow Marine Sclerozoan Fauna from the Middle Jurassic (Late Callovian) of Southern Israel

M. A. Wilson, H. R. Feldman, J. C. Bowen, Y. Avni

Palaeogeography, Palaeoclimatology, Palaeoecology 263 (2008): 24-29

11. Bioerosion in an Equatorial Middle Jurassic Coral-Sponge Reef Community (Callovian, Matmor Formation, southern Israel)

Mark A. Wilson, Howard R. Feldman, Elyssa Belding Krivicich

Palaeogeography, Palaeoclimatology, Palaeoecology 289 (2010): 93-101

12. Jurassic Rhynchonellide Brachiopods from the Jordan Valley

Howard R. Feldman, Mena Schemm-Gregory*, Fayez Ahmad, and Mark A. Wilson

Acta Palaeontologica Polonica 57, no. 1 (2012): 191–204.

13. Taxonomy and Paleobiogeography of Late Bathonian Brachiopods from Gebel Engabashi, Northern Sinai

Howard R. Feldman, Vladan J. Radulovic, Adel A. A. Hegab, and Barbara V. Radulovic

Journal of Paleontology 86, no. 2 (2012) 238-252

* Deceased

INDEX

Printed by Libri Plureos GmbH in Hamburg,
Germany